(physicist)

PARTICLE ASTROPHYSICS

AIP CONFERENCE PROCEEDINGS 203

PARTICLE ASTROPHYSICS

THE NASA COSMIC RAY PROGRAM FOR THE 1990s AND BEYOND

GREENBELT, MD 1989

EDITORS:

W. VERNON JONES
NASA HEADQUARTERS

FRANK J. KERR
UNIVERSITIES SPACE
RESEARCH ASSOCIATION

JONATHAN F. ORMES
NASA/GODDARD
SPACE FLIGHT CENTER

AIP

American Institute of Physics **New York**

L.C. Catalog Card No. 90-55077
ISBN 0-88318-763-9
DOE CONF 8911161

Printed in the United States of America.

CONTENTS

NOTE: Italics indicate the person who presented the paper.

Preface

The Cosmic and Heliospheric Physics discipline is indeed fortunate to have had several space missions selected in 1989: the POsitron Electron Magnet Spectrometer (POEMS) was selected in February for definition as a secondary payload for the Earth Observing System to be launched in the mid-to-late 1990s; the Solar and Magnetospheric Particle Explorer (SAMPEX) was selected in April for construction and development leading to launch on a Scout vehicle by June 1992; the Particle Astrophysics Superconducting Magnet Facility (Astromag) was selected in June with three science instruments, WiZard, LISA, and SCIN/MAGIC, for definition leading to launch in early 1999 during the assembly phase of Space Station Freedom; the Heavy Nuclei Collector (HNC) was also selected in June for definition as an attached payload for the assembly phase of Space Station Freedom; and the Advanced Composition Explorer (ACE) was selected in October as an Explorer mission leading to launch about 1997.

The optimism associated with the mission selections is tempered by the realization that space history has shown that selection for definition or flight does not always lead to an actual flight opportunity. In view of this fact, the Cosmic and Heliospheric Management Operations Working Group (which assists in tactical planning for the discipline) suggested that the *ad hoc* Cosmic Ray Program Working Group should be re-convened in order to provide input to the long-range strategic planning activities of both the National Aeronautics and Space Administration Space Physics Subcommittee and the National Academy of Sciences Astronomy and Astrophysics Survey Committee.

The November 6–8, 1989 workshop was meant to provide grassroots input to the long-range planning activities. The workshop attendees were encouraged to assess the goals and aspirations of the discipline in terms of realistic and timely opportunities, while recognizing that some of the opportunities that may have seemed to be pie-in-the-sky even a few months ago may now deserve careful consideration. For example, there has been a lot of planning activity for Lunar-based research following President Bush's statement on July 20, 1989, in honor of the 20th anniversary of the Apollo 11 landing on the moon, to the effect that the U.S. should consider a Human Exploration Initiative involving a permanent Lunar base in the first decade of the 21st century.

This volume reports on the deliberations of the workshop and the findings of the Cosmic Ray Program Working Group. The meeting was organized by the Universities Space Research Association, in cooperation with NASA Headquarters and the Goddard Space Flight Center. We wish to acknowledge the excellent work of Ms. Claudette Sharps in the organization of the meeting and the assembly of this volume.

W. Vernon Jones
Frank J. Kerr
Jonathan F. Ormes

List of Participants

James H. Adams, Jr.
Naval Research Laboratory

Steven P. Ahlen
Boston University

Daniel N. Baker
NASA/Goddard Space Flight Center

Louis M. Barbier
NASA/Goddard Space Flight Center

James J. Beatty
Boston University

David Berley
National Science Foundation

W. Robert Binns
Washington University

Adri Burger
Bartol Research Institute

David Burks
Science Applications International Corp.

Michael L. Cherry
Louisiana State University

Eric R. Christian
USRA/Goddard Space Flight Center

David Clements
Bartol Research Institute

Henry J. Crawford
University of California, Berkeley

Alan C. Cummings
California Institute of Technology

James A. Earl
University of Maryland

Robert W. Ellsworth
George Mason University

Joseph A. Esposito
Bartol Research Institute

Paul Evenson
Bartol Research Institute

Kenneth Fox
Science Applications International Corp.

Thomas K. Gaisser
Bartol Research Institute

M. Garcia-Munoz
Enrico Fermi Institute

George Gloeckler
University of Maryland

Robert L. Golden
New Mexico State University

Jordan A. Goodman
University of Maryland

T. Gregory Guzik
Louisiana State University

Richard Heinz
Indiana University

Martin H. Israel
Washington University

Jack R. Jokipii
University of Arizona

Frank C. Jones
NASA/Goddard Space Flight Center

W. Vernon Jones
NASA Headquarters

Frank J. Kerr
University of Maryland/USRA

David B. Kieda
University of Utah

Donald A. Kniffen
NASA/Goddard Space Flight Center

Keith E. Krombel
USRA/Naval Research Laboratory

William R. Kropp
University of California, Irvine

James C. Ling
NASA Headquarters

Frank B. McDonald
University of Maryland

Robert E. McGuire
NASA/Goddard Space Flight Center

Marvin Marshak
University of Minnesota

Glenn M. Mason
University of Maryland

James Matthews
University of Michigan

Richard A. Mewaldt
California Institute of Technology

Peter Meyer
Enrico Fermi Institute

Barrett Milliken
Enrico Fermi Institute

Dietrich Müller
Enrico Fermi Institute

Jonathan F. Ormes
NASA/Goddard Space Flight Center

Thomas A. Barnell
NASA/Marshall Space Flight Center

P. Buford Price
University of California, Berkeley

Reuven Ramaty
NASA/Goddard Space Flight Center

Donald V. Reames
NASA/Goddard Space Flight Center

Michael H. Salamon
University of Utah

Eun-Suk Seo
NASA/Goddard Space Flight Center

Maurice M. Shapiro
University of Maryland

S. Shawhan
University of Iowa

Rein Silberberg
Naval Research Laboratory

Manfred Simon
University of Siegen

James A. Slavin
NASA/Goddard Space Flight Center

George F. Smoot
Lawrence Berkeley Laboratory

Henry W. Sobel
University of California, Irvine

Edward C. Stone
California Institute of Technology

Robert E. Streitmatter
NASA/Goddard Space Flight Center

James Sullivan
Massachusetts Institute of Technology

Simon P. Swordy
Enrico Fermi Institute

Yoshiyuki Takahashi
University of Alabama, Huntsville

Evelyn Tuska
Bartol Research Institute

Allan J. Tylka
USRA/Naval Research Laboratory

Giuseppe Vacanti
Iowa State University

T. von Rosenvinge
NASA/Goddard Space Flight Center

C. J. Waddington
University of Minnesota

William R. Webber
University of New Hampshire

John P. Wefel
Louisiana State University

Mark E. Wiedenbeck
Enrico Fermi Institute

R. Jeffrey Wilkes
University of Washington

Akira Yamamoto
National Laboratory for High Energy Physics
Japan

Workshop Report

THE NASA COSMIC RAY PROGRAM FOR THE 1990'S AND BEYOND

Interim Report of the NASA Cosmic Ray Program Working Group

S. P. Ahlen[a], W. R. Binns[b], M. L. Cherry[c], T. K. Gaisser[d], W. V. Jones[e], J. C. Ling[e], R. A. Mewaldt[f], D. Muller[g], J. O. Ormes[h], R. Ramaty[h], E. C. Stone[f], C. J. Waddington[i], W. R. Webber[j], and M. E. Wiedenbeck[g]

a) Boston University, Boston, MA 02215
b) Washington University, St. Louis, MO 63130
c) Louisiana State University, Baton Rouge, LA 70803
d) Bartol Research Foundation, Newark, DE 19716
e) NASA Headquarters, Code ES, Washington DC 20546
f) California Institute of Technology, Pasadena, CA 91125
g) University of Chicago, Chicago, IL, 60637
h) Goddard Space Flight Center, Greenbelt, MD 20771
i) University of Minnesota, Minneapolis, MN 55455
j) University of New Hampshire, Durham, NH 03824

ABSTRACT

The interim report of the 1989 NASA Cosmic Ray Program Working Group is presented. The report summarizes the cosmic ray program for the 1990's, including the recently approved ACE, Astromag, HNC, POEMS, and SAMPEX missions, as well as other key elements of the program. New science themes and candidate missions are identified for the first part of the 21st Century, including objectives that might be addressed as part of the Human Exploration Initiative. Among the suggested new thrusts for the 21st century are: an Interstellar Probe into the nearby interstellar medium; a Lunar-Based Calorimeter to measure the cosmic ray composition near $\sim 10^{16}$ eV; high precision element and isotope spectroscopy of ultraheavy ($Z \geq 30$) elements; and new, more sensitive, studies of impulsive solar flare events.

1. INTRODUCTION

This interim report of the 1989 NASA Cosmic Ray Program Working Group is based on the results of a workshop on the NASA Cosmic Ray Program for the 1990's and Beyond which was held at the Goddard Space Flight Center on November 6-8, 1989. The workshop involved more than 50 members of the cosmic ray research community who contributed to the discussion and identification of future objectives and missions. The Cosmic Ray Program Working Group met immediately after the workshop in order to formulate the program presented in this interim report.

The program outlined here builds on that of three previous reports[1,2,3] prepared by the Cosmic Ray Program Working Group in 1982, 1985, and 1987. Responding to the evolving programmatic opportunities and constraints of the overall NASA program at the time, those reports developed a coherent program of research in cosmic rays that would address the general recommendations for the discipline outlined in the 1982 Astronomy Survey Committee (G. B. Field, chairman).

In the last year, most of the missions described in those earlier Cosmic Ray Program Working Group reports have become part of the NASA program for the next decade, including the recent selection of the Astromag Facility and the Heavy Nuclei Collector for Space Station Freedom, the Advanced Composition Explorer, the Positron/Electron Magnetic Spectrometer for the Earth Observing System, and SAMPEX, a small Explorer. Thus, by the beginning of the next century many of the scientific objectives discussed in those reports should have been accomplished and there will be opportunities for new scientific thrusts.

There are now several long range studies underway to identify new directions for the NASA program, including another decadal Astronomy and Astrophysics Survey (J. N. Bahcall, chairman) by the National Academy of Sciences/National Research Council and a Space Physics Planning Study (G. L. Siscoe, chairman) by the NASA Space Physics Advisory Subcommittee. In addition, NASA has been asked to develop a Human Exploration Initiative that would require continued monitoring of the solar and geospace environment and could include the establishment of a lunar base that would offer the opportunity for observations from the lunar surface.

Because the planning horizon for these studies is well beyond that of the previous reports of the Cosmic Ray Program Working Group, the group was reconvened in order to consider the new opportunities and outline the objectives of a longer range program. This interim report briefly reviews the program currently approved for the next decade and formulates new science thrusts and identifies candidate missions that would be key attributes of a vigorous cosmic ray program in the first decades of the next century.

Table 1 summarizes the major scientific themes that are addressed by the program described in this report.

2. MISSION REQUIREMENTS FOR THE CURRENT PROGRAM

The NASA cosmic ray program for the 1990's will be centered around the development and launch of a number of recently selected missions. These missions and other key elements of the program are described briefly below and summarized in Table 2.

Table 1 - Science Themes

The Origin and Evolution of Matter - The diverse populations of high energy particles arriving near Earth provide samples of matter from other regions of the solar system and the galaxy. Comprehensive studies of the composition of these particles will make it possible to:

- Determine and compare the composition of the solar corona, the local interstellar medium, and galactic cosmic ray sources.
- Identify nucleosynthesis and galactic evolutionary effects that distinguish solar system and galactic matter.
- Search for evidence of special sources of cosmic rays, such as supernovae and Wolf-Rayet stars.
- Determine the time scales for the nucleosynthesis, acceleration, and confinement of cosmic ray nuclei in the galaxy.
- Identify the origins of antiprotons and positrons in cosmic rays, and search for anti-nuclei and other exotic particles of galactic or extragalactic origin.

Nature of the Heliosphere and the Interstellar Medium - The particles and fields of the nearby interstellar medium are presently hidden from our view by the solar wind that encapsulates us within the bubble called the heliosphere. Exploratory *in situ* measurements within and beyond the boundaries of the heliosphere will:

- Investigate the large-scale structure and dynamics of the heliosphere as it responds to solar variations on a variety of time scales.
- Investigate "anomalous" and galactic cosmic ray acceleration at the solar wind termination shock, including possible *in situ* observations.
- Measure directly particles and fields in nearby interstellar space, and investigate their role in the energy balance of the galaxy, and in shaping the heliosphere.
- Study particle propagation in interplanetary and interstellar space.

Cosmic Acceleration Processes - Particle acceleration is ubiquitous in nature and is one of the fundamental problems in space physics. Comprehensive studies of solar, interplanetary, and galactic particles spanning many decades in energy will test shock acceleration models on a wide range of scales, and allow us to:

- Study particle acceleration and interaction processes in solar flare events with orders of magnitude increased sensitivity.
- Search for evidence of continuous cosmic ray acceleration by supernova shock waves.
- Investigate the acceleration mechanisms and sites responsible for the highest energy particles in our galaxy.

Table 2 - The Cosmic Ray Program for the 1990's

The following missions, recently selected for development and launch during the coming decade, are the focus of the NASA cosmic ray program for the 1990's.

- **ACE** - An Advanced Composition Explorer to measure the elemental and isotopic composition of H to Ni nuclei over six decades in energy/nucleon, from solar wind to galactic cosmic ray energies.

- **Astromag** - A superconducting magnetic spectrometer facility for the Space Station, including powerful instruments that will extend particle and anti-particle spectroscopy into the GeV and TeV energy ranges.

- **HNC** - A Heavy Nucleus Collector for the Space Station that will measure the abundances of the heaviest elements in the periodic table.

- **POEMS** - A POsitron Electron Magnetic Spectrometer for the Earth Observing System that will measure electrons and positrons from the Galaxy and the Sun, and also solar γ-rays and neutrons.

- **SAMPEX** - A Solar, Anomalous, and Magnetospheric Particle Explorer to be launched as part of the Small Explorer Program to study low energy particles from a polar orbit.

In addition, the continuing cosmic ray program also includes the following key elements:

- **New Cosmic Ray Instruments** to be carried on spacecraft soon to be launched for the Ulysses, CRRES, WIND, and NOAA-I missions.

- **Global Heliosphere Initiative** - Extended exploration of the heliosphere and monitoring of the Sun with instruments on the existing interplanetary network of Voyager, Pioneer, Galileo, ICE, and IMP-8 spacecraft, soon to be joined by Ulysses and others.

- A vigorous **Balloon-Flight Program** will make it possible to develop and test new instrumentation, and to initiate new investigations of cosmic ray elements, isotopes, antiprotons, and positrons.

- An **Accelerator-Based Program** for testing and calibrating new detectors and for measuring nuclear cross sections critical to astrophysics and the Human Exploration Initiative

2.1 Small Instruments Outside the Magnetosphere or in Earth Orbit

Small Instruments on other Missions - Most cosmic ray experiments that have flown in space were carried on near-Earth or interplanetary spacecraft that included several small particle and fields experiments as well as other instrumentation. New instruments of this kind presently scheduled for launch in the early 1990's include (1) a cosmic ray element and isotope spectrometer to be launched on Ulysses in 1990 and carried out of the ecliptic and over the solar poles, where it will conduct a solar latitude survey of a variety of energetic particle components; (2) a similar spectrometer to be carried into Earth-orbit on CRRES in 1990; (3) a solar flare and galactic cosmic ray isotope spectrometer to be launched on WIND in 1992 and carried to the L_1 Lagrangian point as part of the ISTP program; and (4) a solar flare isotope spectrometer to be carried on NOAA-I. These instruments, the first two of which were originally scheduled for launch in the 1980's, will play important roles in the broader objectives of these missions and take the long-awaited next steps in energetic particle spectroscopy beyond ISEE-3.

ACE - The instruments on the Advanced Composition Explorer (ACE), one of two recently selected Explorer-class missions, have greatly increased collecting power and coverage, allowing coordinated and comprehensive measurements of the elemental, isotopic, and charge-state distributions of energetic particles of solar, interplanetary, interstellar and galactic origins. Definitive measurements will be made of the abundance of essentially all long-lived isotopes from H to Zn ($1 \leq Z \leq 30$), spanning the energy range from that of the solar wind ($\lesssim 1$ keV/nucleon) to that of galactic cosmic rays (several hundred MeV/nucleon). The prime objective of ACE will be to determine and compare the elemental and isotopic composition of several distinct samples of matter, including the solar corona, the interplanetary medium, the local interstellar medium, and galactic matter. To accomplish this, ACE includes six high-resolution spectrometers, each designed to provide the optimum, charge, mass, and charge-state resolution in its particular energy range, and each having a geometry factor 10 to 100 times greater than previous or planned experiments. ACE is presently expected to be launched in ~1997 to an orbit about the L_1 Lagrangian point.

SAMPEX - The Solar, Anomalous, and Magnetospheric Explorer (SAMPEX) is one of four recently selected missions for the new Small Explorer (SMEX) program. Scheduled for launch on a Scout rocket in 1992, SAMPEX will measure energetic electron and ion composition from ~0.4 to ~200 MeV/nucleon in a near-polar orbit. Key objectives of SAMPEX include determination of the charge state of the "anomalous" cosmic rays (predicted to be a singly-ionized sample of the interstellar medium), and measurement of relativistic electrons precipitating from the magnetosphere which undergo interactions in the middle atmosphere that affect ozone depletion.

2.2 Large Detectors in Earth Orbit

Astromag is a superconducting magnetic spectrometer facility for particle astrophysics that has been selected as a Space Station Freedom attached payload. Earlier, the Cosmic Ray Program Working Group recommended the development and flight of a large, high-intensity magnet for the mass spectroscopy of high energy cosmic rays. To this end, the Astromag project was extensively studied and then named as a Space Station Facility. Three first-generation experiments have now been selected for the first use of Astromag: Wizard, LISA and SCIN/MAGIC.

Wizard utilizes precision tracking chambers, time-of-flight scintillators, transition radiation detectors, and state-of-the-art calorimetry to search for primordial antimatter and to determine the spectra of antiprotons, positrons, and light nuclei ($1 \leq Z \leq 8$) to high energies. Measurements with unprecedented precision will be made up to energies of the order of a TeV. These measurements will bear on a wide range of fundamental issues, including (1) the implications of cosmic ray antiproton and positron fluxes and possible antinuclei for Grand Unified Theories and super-symmetry theories in cosmology, (2) other possible origins for the enhanced abundances of antiprotons and positrons in cosmic rays; and (3) the nature of the acceleration and propagation mechanisms of galactic cosmic rays.

LISA, a Large Isotope Spectrometer for Astromag, will make the first direct measurements of cosmic ray isotopes at energies above 1 GeV/nucleon. Particle mass will be determined with velocity measurements provided by a series of Cerenkov counters and rigidity measurements provided by scintillating optical fiber trajectory detectors. Isotope measurements of nuclei from Be to Ni will include studies of iron-group isotopes which can determine the time delay between nucleosynthesis and cosmic ray acceleration, and studies of radioactive nuclei such as ^{10}Be, ^{14}C, ^{26}Al and ^{37}Cl over a range of relativistic time-dilation factors to probe the density and distribution of interstellar material. In addition, LISA will measure cosmic ray element abundances with high precision to nearly 1 TeV/nucleon and perform a sensitive search for heavy cosmic ray antimatter.

SCIN/MAGIC will study the Spectra, Composition, and Interactions of Nuclei (SCIN) at very high energies using Magnet Interaction Chambers (MAGIC) that include passive track-recording devices such as x-ray film, plastic track detectors, and nuclear emulsions. The abundances of protons, helium, and heavier nuclei will be measured up to energies approaching 10^{15} eV, and electron spectra will be measured to almost 10^{13} eV. This instrument will also study interactions of nuclei in a unique regime where the collision energy-densities may be high enough to produce a phase change in the nuclear matter to a "quark-gluon" plasma.

Following the completion of the experiments selected for the initial use of Astromag, there are a number of second generation experiments requiring continued use of this facility. These include higher energy isotope experiments, more sensitive antimatter searches and use of the magnetic field for a new approach to high-energy gamma-ray spectrometry. The facility and its large-volume high-intensity magnetic field may also be used for other investigations such as plasma physics experiments.

HNC, the Heavy Nucleus Collector selected for deployment on Space Station Freedom, is a large-area (16 m^2) array of glass track detectors which will determine the elemental composition of heavy ($Z \gtrsim 50$) cosmic ray particles with high precision. With an exposure of several years, the HNC will measure the ratio of uranium to thorium with sufficient accuracy to answer the crucial question of whether cosmic rays are freshly synthesized material recently ejected from supernovae, or simply old interstellar material accelerated indirectly by passing supernova shock waves. HNC will also search for transuranium nuclei such as Pu, Np, and Cm that would be expected in fresh nucleosynthesis products.

POEMS, the POsitron Electron Magnet Spectrometer, will make the first space-based investigation of the positron spectrum in the energy range below 1 GeV, thereby complementing investigations on Astromag at higher energies. These investigations should resolve the question of whether cosmic ray positrons are accelerated in the sources as "primaries" or produced as "secondaries" by interactions of heavier cosmic rays during their transport through the galaxy. POEMS will also measure solar gamma-rays, neutrons, and electrons, and it will investigate cosmic ray modulation by tracking the effects of charge-sign-dependent processes such as drifts in the interplanetary magnetic field. POEMS is currently scheduled to fly on the first Eos polar-orbiting platform.

2.3 The Global Heliosphere Initiative

During the 1990s an impressive array of spacecraft instrumented for cosmic ray investigations and related studies of heliospheric and solar phenomena will be in place. Pioneers 10 & 11 and Voyagers 1 & 2 will be exploring the regions beyond 50 AU from the Sun, with Pioneer 10 heading down the heliospheric tail and the others heading towards the solar apex, returning data from the mid-latitude regions above and below the ecliptic plane. In the inner heliosphere, the IMP-8 and ICE spacecraft will be monitoring the solar wind, cosmic ray, and solar particle intensity near 1 AU, close to the solar input that drives the dynamics of the heliosphere. These veteran spacecraft will soon be joined by others, including Ulysses, as it embarks on a solar latitude survey of the spectra and composition of a variety of particle species.

This network of spacecraft represents a uniquely powerful configuration for studying the large scale structure and dynamical processes in the heliosphere and for locating and characterizing the heliospheric boundary. The opportunities which this network provides for simultaneous measurements at a variety of heliospheric radii and latitudes will not be duplicated in the foreseeable future. The instruments at 1 AU provide critical baseline measurements for the spacecraft at large radii and high latitudes, making it possible to distinguish spatial and temporal variations in the observations.

During the 1980s, measurements from this array of spacecraft revolutionized our understanding of the heliosphere and its interactions with energetic particles. Equally significant returns can be expected from continued tracking and analysis of data from these spacecraft in the future. These missions should continue to have strong support in NASA's program for the 1990s and beyond.

2.4 Missions of Opportunity

Currently under discussion are joint US-Soviet flights of two cosmic ray experiments on the Soviet space station MIR that could take place within the next few years.

CRN, the Cosmic Ray Nuclei detector, was flown several years ago on Space Shuttle Challenger, where it successfully demonstrated its capabilities for determining the charge and energy spectra of very high energy cosmic ray nuclei, albeit with limited statistics due to an exposure time of only a few days. CRN should be considered for a second, much longer deployment in space. A one year deployment would allow CRN's transition radiation detectors to realize their potential for determining the high energy cosmic ray composition of nuclei from boron to iron up to energies of ~10 TeV/nucleon (corresponding to an energy of nearly 10^{15} eV for Fe nuclei).

Also under discussion for flight on MIR is a small version (1 to 3 m^2) of the glass detectors to be used in the Heavy Nucleus Detector (HNC) selected for flight on Freedom. For a two year exposure in MIR's 51.6 degree inclination orbit the yield of $Z>60$ ultraheavy nuclei for a 2 m^2 detector would be ~15% of that expected for a 5 year exposure of the full HNC detector in the 28.5 degree orbit planned for Freedom.

3. MISSION REQUIREMENTS FOR THE LONG-RANGE PROGRAM

The long-range program described below and summarized in Table 3 includes new science thrusts and candidate missions for the first part of the 21st century.

3.1 Cosmic Ray Studies with an Interstellar Probe

The particles and fields of the local interstellar medium are excluded from the heliosphere by the solar wind and its embedded magnetic field. In our present view the solar wind flows radially outward to a termination shock, surrounded at somewhat greater distances by a heliopause, the boundary between the solar wind plasma and the interstellar gas. Theoretical models, supported by observations from Pioneer, Voyager, and other spacecraft, show that the interplanetary magnetic field, by means of processes known collectively as "solar modulation", shields the inner heliosphere from any direct knowledge of the composition, spectrum, and energy density of cosmic rays in interstellar space below several hundred MeV/nucleon, as well as significantly modifying the spectra of higher

Table 3 - New Thrusts for the 21st Century

- An **Interstellar Probe** is required to extend direct measurements of cosmic rays and related phenomena beyond the heliosphere and to assess their role in the energy balance and dynamics of the local interstellar medium and the galaxy.

- A **Lunar-based Calorimeter** would determine the composition and investigate the origin of cosmic rays up to $\sim 10^{16}$ eV, where a "knee" in the energy spectrum suggests the onset of effects due to acceleration limitations, galactic escape, or new cosmic ray sources.

- **Ultra-Heavy Nuclei** - Two new missions would study the synthesis of cosmic ray nuclei in the upper 2/3 of the periodic table, and measure nucleosynthesis and acceleration time scales with radioactive "clocks" that include U, Th and heavier nuclei.

  - A large area electronic spectrometer in Earth orbit or on a lunar base that would measure ultra-heavy elements from $Z \simeq 30$ to $Z \simeq 100$.

  - An Explorer mission to measure the isotopes of ultra-heavy nuclei in solar flares and in galactic cosmic rays.

- **Solar Flare Studies at 0.4 AU** - A solar particle spectrometer on the Mercury Orbiter mission would provide greatly increased sensitivity for studying solar flare acceleration and injection from a vantage point close to the Sun.

Other Initiatives

- As a complement to the energetic particle measurements from the Advanced Composition Explorer (ACE) and other missions, a Small Explorer mission to measure the spectra of γ-rays and neutrons emitted in solar flares would provide valuable additional information.

- To insure continued progress in defining the large-scale structure and dynamics of the heliosphere it is important that appropriate particle and fields instrumentation be included on all future interplanetary spacecraft.

energy particles up to about 10 GeV/nucleon. To observe interstellar cosmic rays one must get outside the heliopause. We propose a dedicated Interstellar Probe mission that would cross the termination shock and heliopause and make a significant penetration into interstellar space, thereby providing invaluable *in situ* measurements of the particles and fields in nearby regions of the Galaxy.

Present estimates place the solar wind termination shock at 40 to 100 AU, with the heliopause approximately a factor of two more distant. One or more of the Pioneer or Voyager spacecraft may well cross the termination shock within their lifetimes, but it is much less likely that the heliopause is within reach of Voyager's potential range of ~130 AU, to be reached in ~2015. Although these spacecraft may locate the termination shock and heliopause and return unique exploratory data, detailed studies will require modern instrumentation designed to comprehensively observe the interstellar medium. Recent studies by JPL show that an Interstellar Probe could be accelerated to a velocity of 10 to 20 AU/yr, thereby overtaking the Voyagers within a few years, reaching a distance of 200 AU within 10 to 20 years, and then possibly continuing out to 500-1000 AU in an extended mission.

An Interstellar Probe to the nearby interstellar medium could make a wide range of *in situ* measurements, including: the structure of the distant heliosphere; the acceleration of the "anomalous" cosmic rays and other species at the solar wind termination shock; the unmodulated interstellar spectra and composition of low-energy cosmic ray nuclei and electrons; the composition of the local interstellar gas; the interstellar magnetic field and other parameters required to understand cosmic ray transport; and the cosmic ray contribution to the heating, pressure, and energy balance of the local interstellar medium. In addition, measurements could be made of the ~1 MeV positrons from ^{56}Co decay in supernovae that are likely responsible for the galactic 0.5 MeV γ-ray flux, and a search could be made for cosmic ray anisotropies due to recent supernovae or nearby sources. In view of the fundamental importance of such measurements to a wide range of studies that include cosmology, stellar and galactic evolution, nucleosynthesis, and space plasma physics, it is important that NASA give serious consideration to the development of this truly exploratory mission.

3.2 Very High Energy Cosmic Rays

The spectrum and composition of cosmic rays at energies above 10^{15} eV/nucleus is an area of intense interest. There is a distinct change in slope, or "knee" in the all-particle energy spectrum near 10^{16} eV, suggesting that at higher energies nuclei may have a fundamentally different origin and composition than at lower energies. Because of the low fluxes, however, direct measurements of nuclei with 10^{15} to 10^{17} eV are very difficult. Current balloon- and space-based experiments are too small to provide good statistics in this region; presently the only information comes from indirect Earth-based measurements of extensive air showers.

There are suggestions of very interesting astrophysics and particle physics at these high energies. It is presently thought that acceleration of particles to $\gtrsim 10^{16}$ eV would require larger scale shocks than are presently thought to be possible in the galactic disk, and cosmic rays may also escape more efficiently from the galaxy at these energies. In addition, balloon observations have given indications of a nuclear matter phase change in the interactions of very high energy cosmic ray nuclei, and ground-based measurements have suggested that the ratio of hadronic to electromagnetic interactions of photons may be rapidly increasing at these energies.

A Lunar-Based Calorimeter - In order to measure the primary spectrum and composition at 10^{16} eV with reasonable statistical precision, a collecting power of ~100 m^2sr and an exposure time of ~5 years will be required. The corresponding weight of such a calorimeter, if designed for space flight, would be ~100 tons. Since this is probably too heavy to be flown in orbit, a lunar-based detector, fabricated primarily from lunar material, probably offers the only practical means for building the experiment. The bulk of the calorimeter material would be compressed lunar regolith; only relatively light plastic scintillators, electronics, and support structure would be carried up from Earth. A calorimeter with a collecting power of ~100 m^2sr could use several hundred tons of lunar soil as the calorimeter material, but only a few per cent of this mass would need to be brought to the Moon. Elements from protons through iron would be measured with good precision at energies above the "knee" in the all-particle spectrum.

A Space-Based Calorimeter - A large space-based calorimeter (25 m^2sr with a 3-4 year exposure) would provide a major improvement over the SCIN/MAGIC experiment on Astromag for charges Z = 1 to 5 and over CRN for heavier nuclei, and, in addition, would be an important intermediate step leading to a still larger lunar calorimeter. A 25 m^2sr calorimeter, weighing ~25 tons, could be flown as a Large Attached Payload on a second generation Space Station, or it would be a candidate for a joint US-Soviet mission or other international collaboration. Such a possibility warrants detailed study.

3.3 Ultraheavy Nuclei Studies

A Large-Area Spectrometer for Ultraheavy Elements - Measurements of the elemental composition of elements beyond the iron peak are of special interest because of the information they carry about neutron-capture nucleosynthesis processes that have forged the upper 2/3 of the periodic table, and because they include a number of radioactive species that can be used as "clocks", particularly among the "actinide" group of elements that include Th, U, and possibly transuranic nuclei. The Heavy Nucleus Collector (HNC) selected for Space Station Freedom will use passive track detectors with an area of 16 m^2 to measure UH nuclei with $Z \gtrsim 50$. A total of ~40 actinides (with a sizeable uncertainty) is expected by HNC in a four year exposure.

It appears that the next step beyond HNC would be a detector with at least an order of magnitude increase in collecting power. Such a detector could provide accurate measurements of the clocks in the actinide region, and provide a sensitive search for transuranic nuclei that would signify recent nucleosynthesis contributions to cosmic rays. This next instrument should also be capable of providing fully resolved measurements of elements with $30 \leq Z \leq 50$. A large part of the required increase in collecting power might be achieved by placing this instrument in a high-inclination orbit, or outside the magnetosphere, perhaps on a lunar base. Because passive detectors require recovery, it presently appears that a large array of electronic detector modules is the best approach for achieving the required collecting power. One such modular concept, called "C-Shell", is based on Cerenkov and time-of-flight measurements. As there are a number of possible approaches, including a lunar-based instrument or in-orbit assembly of large arrays, it is important that NASA study options for satisfying this very significant objective.

An Explorer for Solar/Galactic Ultraheavy Isotopes - The currently approved investigations of energetic heavy nuclei will explore in depth the nucleosynthesis of solar and galactic material for elements from hydrogen to nickel ($Z \leq 28$). The extension of composition studies to provide isotopic abundances of ultraheavy elements ($Z > 28$) will make it possible to investigate the neutron-capture nucleosynthesis processes responsible for production of more than 3/4 of the stable nuclides. Preliminary information about isotopes up to $Z \simeq 40$ will be provided by instruments on ACE, but the detailed study of these rare species, and those above $Z = 40$, will require new large area instrumentation exposed outside the Earth's magnetosphere for a period of several years. These objectives can be met by an Explorer-class spacecraft dedicated to the measurement of the isotopic composition of ultraheavy elements in both solar energetic particles and galactic cosmic rays. Presently, the mass resolution needed for these measurements has been realized only for low energy particles ($\lesssim$1 GeV/nucleon) which are prevented from reaching low Earth orbit, except at high latitudes, by the geomagnetic field. To achieve the necessary exposure outside the magnetosphere, these measurements should be carried out in an orbit at $\gtrsim$20 Earth radii. Alternative approaches include use of a polar orbiting platform in low Earth orbit (for which low energy particles can be collected over ~30% of the orbit) or lunar based instruments.

The large-area sensor systems needed for ultraheavy isotope measurements up to $Z=40$ and possibly higher can be realized as extensions of detectors now under development or planned for balloon experiments and for ACE. For galactic cosmic ray studies, options include measurement of dE/dx vs. total energy using ionization detectors (either solid state or gas) and measurement of Cerenkov emission vs. range (or energy). For solar flare isotopes, large time-of-flight vs. total energy sensor systems should provide the needed resolution and statistics. Continuing development of these types of detectors using high altitude balloon exposures and accelerator beam tests is essential for realizing the goals of ultraheavy isotope investigations.

3.4 Solar Energetic Particle Studies

Impulsive Solar Flare Studies from a Mercury Orbiter - The Mercury Orbiter (MeO) mission offers a unique opportunity for obtaining solar energetic particle (SEP) observations that can answer long-standing, fundamental questions about the flare process and the solar corona itself.

An MeO solar energetic particle instrument could make fundamental observations of the two major types of solar flares: (1) large solar particle events, and (2) small, impulsive events which show enrichments in ^{3}He, heavy ions, electrons, and (sometimes) gamma rays. The large solar particle events are associated with Type II and IV radio bursts caused by coronal shock waves. However, since the energetic particles are scattered many times on their path to 1 AU, observations at Earth orbit cannot separate the effects of particle injections extended in time versus particle scattering and storage near the Sun. Overall, the most convincing physical picture is that in large solar particle events the long injection time scales reflect acceleration at the Sun due to large shocks moving through the corona, accelerating particles out to tens of solar radii. While this and other possible scenarios cannot be conclusively proved or disproved with current observations, a solar particle instrument on MeO, located at 80 solar radii, would immediately resolve this fundamental question since the shocks could come near to, or even pass by, the spacecraft in the main acceleration phase.

Although the large solar particle events produce the greatest fluxes of accelerated nuclei, they occur much less frequently than small impulsive flares. Exploratory studies of impulsive flares have been carried out at 1 AU, but detailed studies of this most common type of solar particle event have been limited by their small intensity which is adversely affected by propagation to 1 AU. However, at 0.4 AU the flux risetime is extremely short, allowing the time of particle injection at the Sun to be determined to within a few minutes, providing crucial information for comparing with MeO gamma-ray and neutron observations. Observations at 1 AU, on the other hand, cannot establish particle injection times more accurately than ~15-20 minutes due to wandering of the interplanetary magnetic field. In addition, at 0.4 AU peak fluxes are several hundred times greater than those at 1 AU, due to the combined effects of magnetic field divergence and velocity dispersion. Thus, MeO observations can make unambiguous determinations of abundances and charge states from flares which cannot even be observed at Earth orbit because of dilution with other classes of particles.

The low frequency of large events makes it likely that no such event would occur during the short period when Solar Probe is closer than 0.4 AU to the Sun. On the other hand, MeO will spend years close to the Sun, ensuring that many solar particle events of both classes will be observed.

A Small Explorer To Conduct Solar Flare Gamma-Ray Observations in Conjunction with the Advanced Composition Explorer - One of the most important results of recent solar flare studies is the finding that ions and relativistic electrons are routinely accelerated in impulsive solar flares, and that such acceleration is an intrinsic property of the flare energy release process. This was first demonstrated by Solar Maximum Mission (SMM) observations, which revealed the very impulsive nature of the solar flare gamma-ray emission. Recent energetic particle observations of solar flares support and elaborate on this result: impulsive flares reveal enhanced ^{3}He, heavy ion and relativistic electron abundances, as well as highly stripped Fe ions whose charge state is much more characteristic of hot flare plasma than of the million-degree corona. Prior to the SMM it was thought that ion acceleration in solar flares is a just by-product of the main energy-release event, manifest only in large interplanetary proton events.

The simultaneous observation of accelerated particles, gamma rays, and neutrons from impulsive solar flares would have significant scientific merit. Gamma-ray line emission and accelerated particles provide the only unambiguous signatures of ion acceleration in solar flares. Among the many correlated studies, perhaps the most important will be the simultaneous determination of the elemental and isotopic abundances of the particles suffering nuclear interactions at the Sun and those which escape from the Sun. The former can be determined from high-resolution gamma-ray spectroscopy; the latter from accurate charged particle observations. Abundance determination is perhaps the most powerful technique in cosmic evolution studies, and is central to all of particle astrophysics. But only for solar flares can abundances be simultaneously determined inside a source and in the accelerated particles which escape from it. In addition, the correlated gamma-ray and particle observations can also provide unique information on particle acceleration, escape and transport, which are central issues to the study of cosmic ray origin.

The required gamma-ray observations could be carried out with an array of Ge detectors flown on a Small Explorer launched while ACE and possibly Wind are operating in the mid to late 1990's. With a mechanical cooler, this instrument could remain operational for up to a decade, thereby carrying out solar observations over the entire cycle 23.

3.5 The Explorer Program

This highly successful program has traditionally been the means by which relatively inexpensive missions dedicated to well-focussed objectives could be flown. Recently a line of Small Explorers was added in an effort to provide more rapid access to space for new, low-cost experiments. Within the Cosmic and Heliospheric Branch the Advanced Composition Explorer (ACE) and a Small Explorer (the Solar, Anomalous, and Magnetospheric Particle Explorer - SAMPEX) were recently selected for development. This report also identifies an Ultraheavy Isotope Explorer and a Small Explorer for solar gamma-rays as two additional concepts that can provide high-priority science within the Explorer framework. As another example, a dedicated Explorer to conduct coordinated measurements of solar flare nuclei, electrons, neutrons, γ-rays, and x-rays during solar cycle 24, preferably inside of $\sim$0.5 AU, would be an important next step beyond ACE, MeO, and the solar flare gamma-ray Small Explorer discussed above.

As indicated by these examples, we are confident that there will continue to be a need for Explorer-class missions within the cosmic physics discipline. These requirements could be met by the suggested establishment of an independent line of Space Physics Explorers.

4. OTHER PROGRAM ELEMENTS

4.1 Small Instruments on Spacecraft Leaving 1 AU

The Heliosphere is a dynamic structure that undergoes major changes over the 22-year solar cycle, significantly affecting the flow of cosmic rays into the inner solar system. The Global Heliosphere Initiative makes use of a number of spacecraft in deep space to study the three dimensional properties of the heliosphere at ever-increasing distance from the Sun. After 1996, however, there will be no observations in the middle heliosphere (5 to 30 AU) unless appropriate instruments are included on future spacecraft leaving 1 AU. Unfortunately, this cost-effective approach is threatened by the exclusion of even modest instruments on future missions, and by the exclusion of any cruise-phase measurements. This recent trend undermines a proven and successful strategy for achieving important objectives at minimal incremental costs to the mission. Appropriate cruise-phase science should therefore be included in the operating plans for future spacecraft leaving 1 AU.

4.2 Balloon-Borne Instruments

The balloon program remains an indispensable part of particle astrophysics. Its main functions are (i) development of new experimental techniques, (ii) test and verification of space instrumentation, and (iii) pursuit of specific science goals that are commensurate with the capabilities of the balloon vehicle. Furthermore, balloon borne investigations play an irreplaceable role in the training and education of students and young scientists: they provide hands-on experience and the opportunity for quick response to new scientific questions. Their low cost, as compared to space vehicles, permits the taking of some risks that are inherent in truly innovative approaches. The following are examples of activities for which there is no practical alternative to the use of balloons:

(1) Experimental techniques: development of advanced particle detectors, such as aerogel and ring-imaging Cerenkov counters, scintillating fiber detectors, transition radiation detectors, and high pressure gas ionization chambers. In some cases, such techniques are adapted from applications in the laboratory and at accelerators, in others they are specifically developed for particle astrophysics applications.

(2) Verification of space instrumentation: test of the performance of magnet spectrometer systems and the resolution of trajectory measuring devices and time-of-flight systems within a near-space environment.

(3) Science goals: As the load carrying capability of balloons is quite substantial (~2-3 tons), and flight durations of days or perhaps even weeks are possible, a number of significant investigations can be pursued, including measurements of isotopc abundances to ~1 GeV/nucleon, observations of electrons, positrons and antiprotons over a fairly wide energy region, and measurements of the energy spectra of the more abundant particle species to ~100 GeV/nucleon.

An adequate level of support for the balloon program and enhanced funding for state-of-the-art detector developments are essential for the future of the discipline.

4.3 Cosmic Rays and the Human Exploration Initiative

Manned exploration missions will require a better understanding of the effects of human exposure to galactic cosmic ray and solar energetic particle fluxes. This will involve questions of long-term exposure to galactic cosmic rays (for example, on a long-duration Mars mission or an extended tour of duty at a lunar base) and transient exposure to the particle fluxes from intense solar flares. Determination of realistic shielding requirements for extended missions demands accurate measurements of the temporal and spatial dependence of the observed cosmic ray fluxes, spectra, and composition, and depends on an accurate assessment of the attenuation of cosmic ray fluxes by shielding. Real-time and near-real-time measurements throughout the heliosphere will be provided by CRRES, Wind, ACE, and the network of spacecraft comprising the Global Heliosphere Initiative. Improved and updated measurements, together with an expansion of the existing program of interstellar and interplanetary propagation studies, will make possible significantly improved predictions of baseline flux levels, long term variations over the solar cycle, and short term variations due to heliospheric fluctuations.

An accurate assessment of shielding efficiency requires improved knowledge of nuclear interaction cross sections, particularly those for neutron production and those below a few hundred MeV/nucleon where present data are sparse. To accomplish these goals will require a vigorous program to measure the relevant heavy ion cross sections at the Bevalac.

4.4 Theory and Laboratory Investigations

Theory - There is a continuing need for theoretical investigations to support the scientific objectives of NASA's particle astrophysics program. Selected topical meetings that bring together theorists with different but related interests, as well as experimentalists, will be an effective addition to the ongoing Theory Program of the Cosmic and Heliospheric Branch. A timely example would be an international topical conference on acceleration and transport of energetic particles at all scales and energies including interplanetary, heliospheric, galactic and extragalactic.

Accelerator Facilities - The Bevalac heavy ion accelerator at Lawrence Berkeley Laboratories has become an essential support facility for the field. The scientific significance of the observed abundances of cosmic ray nuclei depends critically on detailed knowledge of the nuclear cross sections, which can only be measured systematically at the Bevalac. Although a modest beginning in the measurement of these cross sections has been made, much more work is needed if full advantage is to be taken of present and future observations of cosmic ray abundances. The development and calibration of particle detectors also requires exposure to Bevalac heavy ion beams. NASA programs have traditionally been allocated some 200-300 hours of running time per year. This has become inadequate to meet the need, and additional running time is urgently needed to meet present requests. The anticipated shutdown of the Bevalac in 1994-5 will represent a serious loss to the field and it is not clear how it will be replaced. The only comparable facility is the GSI(Darmstadt)-SIS facility, expected to come on line in 1990. Whether NASA programs can obtain running time on this nuclear physics facility is unknown but probably unlikely. It is therefore important that NASA work with the relevant agencies and governments to ensure that sufficient access to appropriate accelerator beams is available for instrument calibrations and cross section measurements in the future.

The Cosmic Ray Neutron Monitor Network Over the past four decades the world-wide network of neutron monitors, now consisting of 60 stations, has provided the only continuous measurements of the high rigidity (>1 GV) galactic cosmic ray flux, as well as making measurements of high energy particle fluxes in solar flare events. Neutron monitor measurements of the modulated cosmic ray spectrum now extend over nearly two complete 22-year solar magnetic cycles. Although neutron monitors respond mainly to secondaries resulting from the interaction of high energy protons in the upper atmosphere, recent measurements have been made of direct neutrons from solar flares, thus providing invaluable insights into the solar flare acceleration process. Observation of these very rare neutron events requires a combination of large geometrical factor, high proton cutoff rigidity, and good time resolution. It is critical to maintain the present neutron monitor network to support space-based studies of cosmic and heliospheric phenomena.

4.5 Infrastructure and Resources

The cosmic ray astrophysics program, like all programs of space exploration, is focused towards a small number of missions, such as ACE, Astromag, or a future Interstellar Probe. A significant fraction of the resources available at NASA must be dedicated to the support of such missions. However, each mission takes many years from inception to launch. These time scales present a serious problem for the participation of the scientific community, in particular at universities. Although a first rate program requires contributions from innovative young scientists and students, such individuals often find themselves in a critical "publish or perish" phase of their career and may turn towards other fields unless a scientific infrastructure exists that permits creative work on a time scale shorter than that of the longer term space missions. This work might include data analysis from previous space ventures, but it must also permit experimental activities that are at the cutting edge of technology.

It is therefore essential for the future of the field that opportunities and resources be available to the scientific community for the training of young scientists, for innovative detector development in the laboratory, for Guest Investigator programs on flight missions, for short-turnaround observations in sub-orbital missions such as balloons, and for quick access to space as provided by Small Explorers or payloads attached to the Space Station. And it is essential that an adequate level of basic funding be made available for such work.

Acknowledgements: We thank G. M. Mason for his contributions to this report.

REFERENCES

1) *Cosmic Ray Program for the 1980's*, Report of the Cosmic Ray Program Working Group, M. H. Israel and J. F. Ormes, Co-chairmen, August, 1982.
2) *The Particle Astrophysics Program for 1985 - 1995*, Report of the NASA Cosmic Ray Program Working Group, M. H. Israel and J. F. Ormes, Co-chairmen, December, 1985.
3) *The Particle Astrophysics Program*, Report of the NASA Cosmic Ray Program Working Group, M. H. Israel and J. F. Ormes, Co-chairmen, October 1987.

Currently Planned Space Mission Program

OVERVIEW OF THE 1989 MISSION SELECTIONS: A QUALITY THRUST IN PARTICLE ASTROPHYSICS

W. Vernon Jones
National Aeronautics and Space Administration, Code ES
Space Physics Division, Washington, DC 20546

ABSTRACT

The space missions selected over the past year in response to NASA Announcements of Opportunity include an ensemble of unprecedented particle astrophysics investigations: (1) a polar-orbiting, Scout-class Small Explorer mission; (2) a Delta-class Explorer mission in a halo orbit at the L1 Lagrangian point; (3) a secondary payload on the Earth Observing System polar platform; (4) the Astromag facility on the early Space Station Freedom with three complementary first-generation investigations; and (5) a fourth Space Station Freedom payload to investigate actinide nuclei. There is a one-to-one mapping of these missions with the top priorities of the Cosmic Ray Program Working Group. Carrying them to completion will ensure U.S. preeminence in particle astrophysics well into the first decade of the next century. Investigations made in conjunction with a potential lunar base might provide the next quantum step in the 21st century.

I. INTRODUCTION

The Cosmic Ray Program Working Group (CRPWG) was established in 1981 for the purpose of formulating a coherent program of research in particle astrophysics that would implement the recommendations contained in the 1982 National Academy of Sciences report of the Astronomy Survey Committee (the so-called Field Report) entitled Astronomy and Astrophysics for the 1980's. The scientific goals for particle astrophysics were discussed extensively in the first CRPWG report issued in 1982. Less than two years later, however, in January 1984, President Reagan announced plans for the Space Station. Since this opened up a new avenue for research, the CRPWG was re-convened with the charge to re-assess its findings in light of the new Space Station program.

The second CRPWG report, released in 1985, confirmed the scientific objectives established in the 1982 report. It also stressed the need for: (1) a Particle Astrophysics Magnet Facility for the Space Station (Astromag) to study high energy galactic cosmic rays, especially antiprotons, positrons, and nuclear isotopes; (2) an Explorer-class mission outside the magnetosphere to study the isotopes of low energy galactic cosmic rays and solar energetic particles; (3) reflight of the Cosmic Ray Nuclei (CRN) experiment for 1 - 2 years on the Space Station; and (4) exposure of the Heavy Nuclei Collector (HNC) on the LDEF-1b mission, i.e., reflight of the LDEF spacecraft.

Following the 1985 CRPWG report, NASA established the Astromag Definition Study Team for the purpose of defining the scope of potential science missions and strawman payloads for the Astromag facility. By the time this team released its interim report in August 1986, several changes had occurred that would seriously impact the particle astrophysics program. First, the HNC mission was cancelled because of the uncertainty associated with retrieval of the LDEF carrier in the aftermath of the Challenger accident. Second, the CRN experiment reflight on the shuttle was cancelled during the restructuring of the Spacelab program. Finally, the phased-assembly of the Space Station was announced, which led to concerns that Astromag might not be accommodated until well into the 21st century.

For the third time in this decade, the CRPWG was re-convened in 1987 with the objective of re-assessing the particle astrophysics program in view of the dramatically changed NASA environment in which the program must be carried out. The third CRPWG report, issued in October 1987, again reaffirmed the science objectives established in the 1982 report and the top priority missions outlined in the 1985 report. The following list, taken *verbatim* from the 1987 report, summarizes the priorities that have guided NASA Headquarters management of the particle astrophysics discipline for the past several years.

- With expanded development of balloon-flight instrumentation and balloon-flight operations it will be possible to investigate new aspects of isotopic and elemental composition, and the energy spectra of antiprotons and positrons.

- An Explorer mission is required for a comprehensive set of precision measurements of the isotopic and elemental composition of solar, interplanetary, and galactic particles over a wide range of energies.

- A magnetic spectrometer in space remains an irreplaceable goal for key studies of isotopic composition, anti-particles, electrons, and positrons.

- Instruments launched into orbit on Scout rockets and other missions of opportunity can provide low-cost, rapid, focused investigations of several cosmic ray species, such as measuring the charge state and isotopic composition of anomalous cosmic rays.

- Reflight of the Cosmic Ray Nuclei Experiment (CRNE) for a year or more would extend measurements of cosmic ray composition to important new energies.

- Study of the elemental composition of UH nuclei remains important, although demise of the Long Duration Exposure Facility (LDEF) suggests a need to reconsider approach.

- In addition, a sizeable body of data from previous missions requires continued analysis and interpretation efforts.

The 1989 mission selections in response to NASA Announcements of Opportunity are almost completely in phase with these priorities. With the exception of the CRNE reflight, the CRPWG goals will be fully realized by carrying out these selected investigations; exploratory discussions with the Soviets have been carried out on the possibility of flying CRN on the MIR-2 station. In addition to these space missions, several new payloads have been developed for the Balloon Program, whose launch operations are experiencing unprecedented success, including the development of long-duration (1 - 3 week) flights at both mid-latitudes and in Antarctica. The funding for data analysis of previous missions has been maintained at approximately their pre-1987 levels.

The CRPWG has been re-convened in 1989 to assess the current status of the discipline and to look beyond the recent mission selections to the future. The workshop held in conjunction with the CRPWG meeting has provided an opportunity for grassroots input into the strategic planning process for the U.S. particle astrophysics program in the 21st century, specifically for the current NASA Space Physics Subcommittee long-range planning activity and the 1989-90 National Academy of Sciences Astronomy and Astrophysics Survey Committee study.

The following Section II summarizes the key scientific issues for the field of particle astrophysics. Section III addresses the accomplishments of the previous space missions. Section IV outlines the potential contributions of the missions selected in 1989, which are tantamount to a quality thrust in this discipline.

II. SCIENTIFIC RATIONALE

A. Understanding the Solar System

Measurement of the composition of particles injected into the interplanetary medium by solar flares provides direct data on the current solar elemental and isotopic constituents. This provides information for understanding the history of solar system material and the contributions of solar particle radiations to other solar system reservoirs, including planetary magnetospheres, the interplanetary medium, lunar soil, and meteorites. It also adds new dimensions to studies of solar flare acceleration and propagation processes. Studies of solar energetic particle (SEP) composition, acceleration, and propagation also contribute to our understanding of the origin and propagation of cosmic rays at various sites in the interstellar medium.

Although the Sun contains more than 99% of the solar system material, we have only limited knowledge of its elemental composition and almost no direct knowledge of its isotopic composition. The isotopic abundances in the Cameron table of "solar system" abundances are based almost entirely on measurements of terrestrial material, while meteoritic measurements are the standard for solar system elemental abundances.[1]

Differences in isotopic composition of the SEP and solar wind particles point to major differences in these two classes of particles emitted by the Sun. Of the few elements studied to date, neon is of particular interest because of the variety of Ne isotope abundances that have been observed in the solar system, including solar wind, lunar, meteoritic, and planetary material. The SEP observations give a $^{22}Ne/^{20}Ne$ ratio of about 0.12, significantly different from the $^{22}Ne/^{20}Ne = 0.073$ ratio measured in the solar wind. There is evidence for similar differences in SEP and solar wind particles implanted in lunar and meteoritic samples.[2] Furthermore, observations of nitrogen isotopes implanted in lunar soil imply that the solar wind $^{15}N/^{14}N$ ratio has increased by about 15% over the life of the moon.[3] Explanations suggested for this surprising variation include: (1) continuing high-energy nuclear reactions on the sun; (2) accretion of interstellar material that alters the solar composition; and (3) a mixture of two separate components of solar system nitrogen with differing $^{15}N/^{14}N$ ratios. The observed isotopic variations in solar system material can be explained only partially by fractionation effects during acceleration and propagation of solar flares or the solar wind.

The solar atmosphere is the source region for the SEP, whose charge state is determined by the plasma condition at the acceleration site. Therefore, a detailed measurement of the charge state distribution of individual ionic species can be used, for example, to determine the electron temperature at the acceleration site. The ionic charge to mass ratio is the principal factor for organizing spectral and composition variations in large solar flares. The composition variations show a power law dependence on this ratio. By correcting for such effects, unfractionated coronal abundances for a wide range of elements can be obtained.[4]

A special class of small solar flares rich in ^{3}He has been observed to exhibit $^{3}He/^{4}He$ ratios ranging from a few percent to greater than unity. The latter imply enrichments of 10^{3} - 10^{4} over typical solar values and, therefore, represent one of the most extreme fractionation process known to occur at astrophysical sites. These ^{3}He-rich solar flares are generally small events that provide a sample of coronal material from more restricted high-temperature regions, with clues to conditions in the solar corona and to the mechanisms of particle acceleration in solar flares. It is currently thought that the enrichment may occur in two steps:[5] first the particles are heated via plasma instabilities; then the particles above a certain threshold in velocity or rigidity are further accelerated. Both steps depend strongly on the ionic charge state of the ions, so determination of the ionic charge state, especially for heavy ions, is vital for understanding these events.

In the energy range below ~50 MeV/nucleon there are at least six elements (He, C, N, O, Ne, and Ar) with anomalously high fluxes believed to represent neutral particles that drift into the heliosphere, become ionized by the solar wind or UV radiation, and then get accelerated, possibly at the solar wind termination shock.[6] This reasonable scenario leads to the expectation that the anomalous particles are singly ionized, in which case they can be readily distinguished from the fully-ionized galactic cosmic rays.[7] Therefore, isolation of the singly charged nuclei provides a direct sample of the local interstellar medium. Comparison of the isotopic composition of the singly-charged anomalous component with that of the solar system nuclei, including those accelerated in solar flares, carries information about the galactic evolution in the solar neighborhood over the time since formation of the solar nebula.

B. Astrophysics

The galactic cosmic rays are probably a sample of the interstellar medium, but their isotopic composition does not reflect that of the solar system (based on the only three elements for which isotopic composition at the cosmic-ray source is well determined, i.e., neon, magnesium, and silicon). On the other hand, the solar system itself is a sample of the interstellar medium that happened to condense about 4.6 billion years ago. Current thinking about this apparent paradox suggests that the difference may be attributed to one (or some combination) of the following three reasons:[8,9] (1) the cosmic rays sample a different part of the galaxy - perhaps some come from closer to the galactic center where there is more nucleosynthesis activity; (2) the cosmic rays are a recent (20 million year old) sample, whereas the solar system is a sample from 4.6 billion years ago, i.e., the galaxy's composition is evolving; (3) the solar system is not typical of the interstellar medium, i.e., it may have been "contaminated" by a supernova shortly before its condensation.

Since the galactic cosmic rays are fully ionized, the time scale between nucleosynthesis and acceleration can be revealed by measurements of isotopes such as ^{56}Ni, ^{57}Co, ^{55}Fe, and ^{59}Ne, which decay only by electron capture.[10] These isotopes clearly address the question of whether cosmic rays represent a sample of the interstellar medium or newly synthesized material, possibly from supernova.

Particle acceleration to extremely relativistic energies is a widespread phenomenon known to occur in the earth's bow shock, the sun, galactic objects (SS433), the galaxy (cosmic rays), and centers of active galaxies (radio jets). There are now specific models of shock acceleration for cosmic rays which predict cutoffs in the spectra of individual elements somewhere between 10 and 1000

TeV/nucleon, i.e., at energies below the "knee" in the all-particle spectrum. Neither CRNE on Spacelab-2 for a week[11] nor the accumulated JACEE balloon flights over the past decade[12,13] have penetrated the knee region, where dramatic changes in chemical composition have been postulated to explain the abrupt change in spectral slope of the overall cosmic ray spectrum.[14]

The question of how the ubiquitous cosmic rays are stored in the interstellar medium is but one example of the more general problem of their origin, interactions with the galactic interstellar plasma, and impact on galactic dynamics. One can measure the containment time-scales and infer the interaction mechanism of particles with magnetic fields and vice versa. This helps in understanding the role played by energetic particles in creating hydromagnetic waves, turbulence, etc. and, in turn, their effect on the motions of the particles. An understanding of such effects is essential to understanding the acceleration of cosmic rays. The spectral changes that take place between the source and Earth must be understood before the source spectrum can be determined.

Given the great uncertainty in any prediction of exotic particles, one could not advocate an expensive mission principally for the purpose of searching for them. On the other hand, these potentially high-payoff studies can be made as a by-product of fundamental particle astrophysics measurements in the "bread-and-butter" areas of spectra and composition.

III. COVERAGE OF PREVIOUS MISSIONS

Figure 1 shows the current state of observations for the various cosmic ray particles (identified by the abscissa) as a function of energy (ordinate). Low energy particle detectors have been included on numerous spacecraft (e.g., early Explorers, Pioneers, and Voyagers), some of which are still returning data. However, cosmic ray isotopes have been resolved only with the minimal capabilities of the International Sun-Earth Explorer, ISEE-3, and to some extent the Interplanetary Monitoring Platform, IMP-8. High energy data have been collected with numerous high-altitude balloon experiments, but the only high energy space missions have been the HEAO-C3 investigation of 1 - 10 GeV/nucleon ultra-heavy (Z > 30) elements and the Spacelab-2/CRN measurements of heavy (Z = 3 - 30) element spectra from 50 - 500 GeV/nucleon.

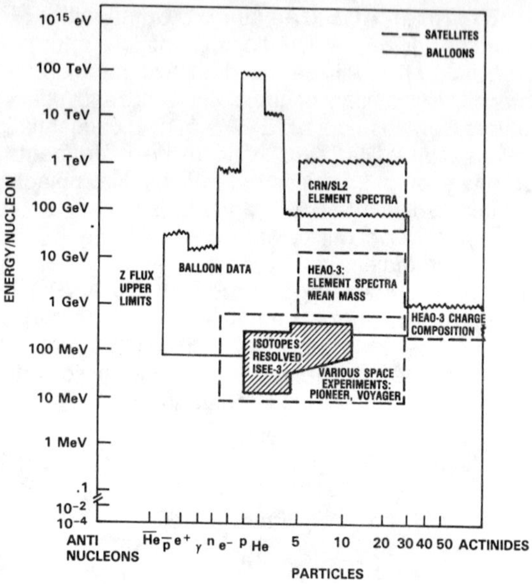

Fig. 1 Schematic of coverage of prior missions. Courtesy of J. F. Ormes.

It should go without saying that the quality of the observations is not the same in all regions, for example, the high energy balloon observations are more limited in statistics than the satellite observations. Therefore, the former should, for the most part, be considered exploratory in nature. Space exposures are required for definitive measurements of most parameters being studied in the high energy regime. The most profound point to be made with Fig. 1 is that there is amazingly little information on the isotopic composition of cosmic rays, which is actually the foundation of particle astrophysics. Both ACE and Astromag (complementary investigations that operate in different energy regimes) will excel in fulfilling this need, which until now has been limited to the small number of isotopes observed with ISEE-3.

IV. COVERAGE OF THE MISSIONS SELECTED IN 1989

A. Low Energy Missions

Figure 2 shows the particle and energy coverages of the low-energy missions selected in 1989. Their unique contributions to this quality thrust in particle astrophysics are briefly outlined below.

Small Explorer/SAMPEX (Solar, Anomalous, and Magnetospheric Particle Explorer) in its low-earth polar orbit offers a unique capability for determination of the ionic charge state of the anomalous cosmic ray component, thought to be matter from the local interstellar medium; if this is indeed the case, then SAMPEX will provide the first isotopic analysis of known local interstellar matter. It also has a parallel focus on studies of precipitating magnetospheric electrons, which are not being investigated by the other cosmic ray missions. The SAMPEX instruments selected in April 1989 are under development for a March-June 1992 launch, so they will provide isotopic analysis of solar flares during the peak activity following the current solar maximum. This will be the first mission in the logical sequence SAMPEX-WIND-ACE, which will provide measurements of both solar and galactic cosmic rays throughout the next solar cycle.

Explorer/ACE (Advanced Composition Explorer), which was selected in September 1989, is the flagship of low energy cosmic ray investigations. Its L1 point, halo-orbit will permit definitive measurements of the ionic charge state and isotopic composition of the solar wind and solar energetic particles, as well as the isotopic composition of $Z > 3$ galactic cosmic rays at energies up to about 0.5 GeV/nucleon. The sensitivity of the instruments range from 10 to 1000 times the sensitivity of prior missions with similar objectives. The ACE mission will likely be launched in the 1997 time frame, near solar minimum, and with its proposed five-year mission will provide data over about half of a solar cycle, i.e., through the 2002 solar maximum.

Earth Observing System/POEMS (POsitron Electron Magnet Spectrometer) focuses on measurements of the positron and electron components over the energy range 5 MeV to 5 GeV. It will use that critical data to trace processes occurring within the geospace environment, in solar flares, and elsewhere in the galaxy. It will also perform complementary measurements of heavier particles, e.g., protons and ^4He (elements), ^2H and ^3He (isotopes), and $Z > 3$ nuclei. The data will be used to investigate the origin (primary or secondary) of galactic positrons, to study the charge sign dependence of solar modulation over a large fraction of the solar cycle, to measure positron and neutral particle emission from solar flares, and to monitor the temporal variations of the charged particle intensities and energy spectra in the Earth Observing System (EOS) orbit. The POEMS payload was selected in February 1989 as a secondary investigation for Platform-A of the EOS to be launched in the mid-to-late 1990's.

Upcoming missions with low energy cosmic ray instruments under development prior to the 1989 selections include ISTP/WIND, Ulysses, CRRES/CRIE, and TIROS/EHIC. The previously-approved EPACT (Energetic Particle Acceleration-Composition-Transport) instrument on the ISTP/WIND spacecraft will perform measurements that overlap in time with those of both ACE and SAMPEX, so it will provide data in a complementary phase of the solar cycle The EPACT sensitivity is considerably less (factor of 10 - 100) than ACE but a factor of 3 greater (due to the duty cycle) than SAMPEX for elemental and isotopic composition of solar and galactic cosmic rays. The combination of SAMPEX, WIND, and ACE will cover nearly a full (11-year) solar cycle with minimum duplication in time. The POEMS investigation could, of course, cover an entire 11-year solar cycle if the 15 year EOS lifetime is realized. As a group, these experiments will provide the global solar phase-particle-coverage needed in low energy cosmic rays.

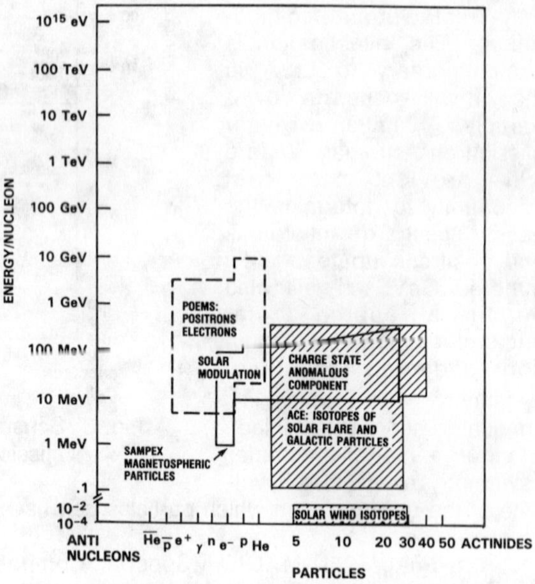

Fig. 2. Schematic coverage of the low energy missions. Courtesy of J. F. Ormes.

B. High Energy Missions

Figure 3 shows the particle and energy coverage of the high-energy missions selected as Space Station Attached Payloads in June 1989. They consist of three complementary investigations using the Astromag facility and one non-Astromag investigation. The unique contributions of these missions to this quality thrust in particle astrophysics are briefly outlined below.

Space Station Freedom/Astromag (Particle Astrophysics Magnet Facility) is the flagship of the high energy cosmic ray regime. It will be developed as a joint NASA-ASI (Italian Space Agency) facility for launch in early 1999. Astromag is designed to accommodate at least two experiments operating simultaneously, and the experiments can be changed out and/or serviced as required. The first-generation observations will provide unprecedented information about nucleosynthesis, cosmic ray origin, acceleration regions, and modes of propagation. They will also conduct highly visible searches for antimatter with unprecedented sensitivity; the detection of antimatter would, of course, revolutionize the theories of cosmology.

Astromag/LISA (Large Isotope Spectrometer for Astromag) will perform isotopic observations of $Z > 3$ galactic cosmic rays in an energy region 10 times greater than the complementary ACE, with about the same statistics as ACE. Together LISA and ACE will fulfill a long-awaited dream of the particle astrophysics community, namely, isotopic measurements of galactic cosmic rays up to iron nuclei over several decades in energy. The sensitivity of LISA for detecting antimatter is a factor of about 100 greater than previous experiments, which should be sufficient to demonstrate whether a matter-antimatter asymmetry exists in the local super-cluster. The LISA instrument fulfills the objectives of the high-Z strawman payload described in the Astromag Study Team Report. It would need to operate for at least two years on Astromag.

Astromag/WiZard (Investigation to "Measure the Cosmic Rays including Anti-protons, Positrons, Nuclei and to conduct a search for Primordial Antimatter") will be carried out by a U.S.-Italy-West German collaboration. The investigation is complementary to LISA in that it will focus on low-Z particles. In a two-year mission on Astromag, WiZard will provide the first opportunity to determine the energy spectra of antiprotons and positrons up to a few hundred GeV. It will also permit a search for primordial anti-helium at an unprecedented level of sensitivity, 10^{-8}, which is adequate for determining whether a matter-antimatter asymmetry exists in the volume of the universe from which particles can have reached us since the beginning of its expansion.

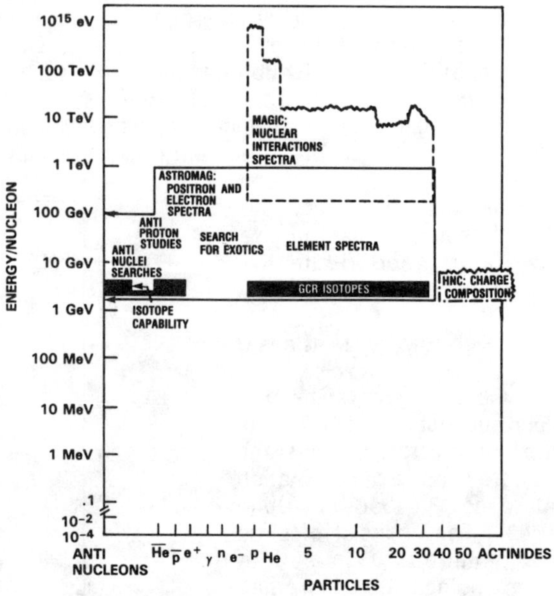

Fig. 3 Schematic coverage of the high energy missions. Courtesy of J. F. Ormes.

Astromag/SCIN-MAGIC (Spectra, Composition, and Interactions above 10 TeV - Magnet Interaction Calorimeter) is a US-Japan collaboration which requires only a three-month exposure in the Astromag facility. It utilizes the magnetic field of Astromag for a unique study of the characteristics of nuclear interactions at energies exceeding plausible thresholds for production of quark-gluon plasmas. This is one of the most fundamental problems in contemporary high energy nuclear physics with applications to understanding high density stellar objects, e.g., neutron stars and black holes. This mission will also extend the proton, helium, and heavy nuclei spectra up to 10^{15} eV, which approaches the energy region where the cosmic ray power law energy spectrum has a change in slope. The short, 90-day on-orbit requirement makes it a good companion experiment for the Astromag facility. It could be run during the initial checkout of Astromag or during change-out of major experiments.

HNC (Heavy Nuclei Collector), which had been cancelled following the Challenger accident, was re-selected in June 1989 for flight on the early Space Station Freedom. The HNC is a large-area array of passive, etchable glass track detectors capable of investigating the elemental abundances of the extremely rare $Z > 50$ nuclei in the cosmic radiation. It emphasizes the relative abundances of the actinide nuclei, Th, U, etc., which provide the only practical primary clock for determining the age of cosmic ray matter. It will also extend the search for $Z > 110$

nuclei and other exotic particles by a factor of about 100 over previous searches. A 2.5 - 4 year exposure is needed for this investigation.

V. CONCLUDING REMARKS

Collectively, the cosmic ray missions selected in 1989 will cover the charge, isotope, and energy ranges needed for a quantum leap in particle astrophysics and, indeed, accomplish some objectives of experiments initially proposed and selected for flight in the early 1970's. The state-of-the-art instruments will collect the data necessary for meeting the highest priority goals of the cosmic physics discipline, some of which have attracted the interest of high energy physicists, thereby, helping unify the efforts of scientists ranging from accelerator physicists to astrophysicists. As we look to the 1990's and beyond, our highest priority must be to bring these missions to fruition. They are the foundation for other particle astrophysics missions to be developed for launch in the first decade of the 21st century. Identification of the latter is the major goal of this workshop.

VI. ACKNOWLEDGEMENTS

Credit for the 1989 mission selections for Particle Astrophysics ultimately belongs to the collective efforts and response of the scientific community with high quality research proposals that met the goals laid down for the discipline during the years-long studies of The Cosmic Ray Program Working Group under the joint leadership of Martin A. Israel and Jonathan F. Ormes.

VII. REFERENCES

1. A. G. W. Cameron, in *Essays in Nuclear Astrophysics*, C. A. Barnes, D. D. Clayton, and D. N. Schramm, eds., Cambridge University Press (1982).
2. D. C. Black, Astrophys. J. 266, 889 (1983).
3. J. Geiss and P. Bochsler, Geochim Cosmochim. Acta 46, 529 (1982).
4. H. H. Breneman and E. C. Stone, Astrophys. J. Letters 299, L57 (1985).
5. G. M. Mason, D. V. Reames, B. Klecker, D. Hovestadt, and T. T. von Rosenvinge, Astrophys. J. 303, 849 (1986).
6. G. Gloeckler, Rev. Geophys. Space Phys. 17, 569 (1979).
7. L. A. Fisk, B. Koslovsky, and R. Ramaty, Astrophys. J. Letters 190, L35 (1974).
8. S. E. Woosley and T. A. Weaver, Ap. J. 243, 561 (1981).
9. M. Casse and J. A. Paul, Ap. J. 258, 860 (1982).
10. A. Soutoul, M. Casse, and E. Juliusson, Ap. J. 219, 753 (1989).
11. J. M. Grunsfeld, J. L'Heureux, P. Meyer, D. Muller, and S. P. Swordy, Ap. J. 327, L31 (1988).
12. W. V. Jones, Y. Takahashi, B. Wosiek, and O. Miyamura, Ann. Rev. Nucl. Part. Sci., Vol. 37, p. 71 (1987).
13. T. H. Burnett, et al. (The JACEE Collaboration), "Energy Spectra of Cosmic Rays above 1 Tev/Nucleon," Astrophys. J. Letters, in press.
14. T. Hara, et al., 18th ICRC Conference Papers (Bangalore, India, 1983, Tata Institute of Fundamental Research, Bombay), Vol. 9, p. 1983.

THE ENERGETIC PARTICLES: ACCELERATION, COMPOSITION, AND TRANSPORT (EPACT) EXPERIMENT ON THE ISTP/WIND SPACECRAFT

D. V. Reames, T. T. von Rosenvinge, R. Ramaty,
NASA/Goddard Space Flight Center

G. M. Mason, D. C. Hamilton,
Dept. of Physics and Astronomy, University of Maryland

M. A. Forman,
Astronomy Program, State University of New York

W. R. Webber
Astromomy Dept., New Mexico State University

ABSTRACT

The EPACT experiment will measure abundances, spectra and angular distributions of particles from 20 keV/amu to 500 MeV/amu. At high energies, isotopes will be resolved up through Z=26, at intermediate energies elements with 1<Z<82 will be observed and at low energies element abundances above Z=2 will be resolved for the first time.

SCIENTIFIC MOTIVATION

The last generation of energetic-particle experiments revealed a diverse family of plasmas in which particle acceleration takes place. We now distinguish populations of energetic particles accelerated in solar flares, in interplanetary (IP) traveling shocks, at the Earth's bow shock, in planetary magnetospheres, at co-rotating interaction regions and at the heliospheric boundary, as well as the galactic cosmic rays. In each of these particle populations, the abundances of elements and isotopes, their energy spectra, their time profiles and the distribution of their arrival directions provide information on the nature of the source and on the mechanisms of particle acceleration (see reviews by Meyer[1], Lin[2], Mason[3], Webber[4] and Reames[5]).

In some cases it is possible to observe particle acceleration *in situ*, and to correlate the energetic particles with the thermal plasmas from which they are drawn and with the magnetic field or shock parameters that are involved. In the case of solar flares, we can relate the properties of the directly-observed particles with those inferred in the flare site from X-ray, gamma-ray, radio and optical observations. For distant astrophysical objects where direct particle measurements are impossible, the quality of these inferences may be crucial to understanding the underlying physics. Two specific problems addressed by EPACT are given below.

a) Solar Energetic Particles

For many years, large proton events were thought to typify particle acceleration at the sun. As the sensitivity and resolution of our instruments increased, however, it became clear that the numerous smaller flare events have abundances that are quite unlike those in the large proton events. The time-intensity plot in Figure 1 shows the contrast that is often seen. A large proton increase occupies the right half of the figure, terminating near the passage of the IP shock labeled SC. During this event the intensity of O remains about an order of magnitude below that of Fe for several days so that Fe/O ~ 0.1. In the two impulsive-flare events occurring on Nov 20 and Nov 22, sudden increases in electrons are accompanied by only modest proton increases, but the Fe intensity rises dramatically and Fe/O ≥ 1.0. The electron-rich, Fe-rich, ^{3}He-rich events are distinct[6] from the large events which have abundances near those in the solar corona. We now believe that the enhancements that occur in the impulsive-flare events are the result of wave-particle interactions[7] in the flare-heated material while the more-normal abundances in the large events result from acceleration by a shock passing through ambient material in the corona or the solar wind.

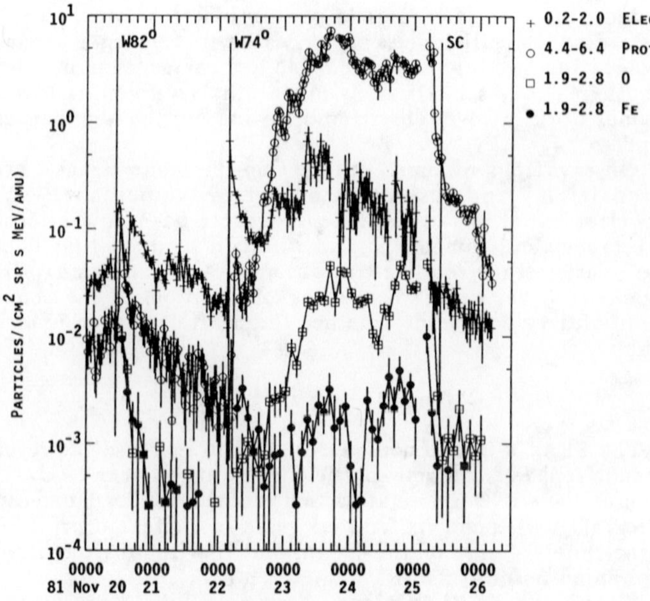

Fig. 1. Impulsive-flare and shock events.

For the impulsive-flare events, we are able to make good associations between particles and flares[8,9] and to study the relationship between particles and the radio and X-ray properties of the flares[10,11,12]. While it seems that enhanced abundances occur in nearly all impulsive flares, our present instruments only have the sensitivity to observe about 20% of the flares with electron increases. Since the ^{3}He/^{4}He ratio actually increases in smaller events, the small flares that we don't see may be more bizarre than the larger ones that we do. Does the mechanism saturate? Are all impulsive-flare events ^{3}He-rich?

The enhancement of elements increases with Z in impulsive-flare events as shown[3,2] in Figure 2. This increase with Z probably comes from a decrease in the particle gyrofrequency (~Q/A) so that different species interact with different regions of the plasma-wave spectrum in the acceleration region. What about the elements with Z>>26? Does the wave spectrum continue to increase toward lower frequencies (higher Z)? The sharp resonance process proposed by Fisk[7] may already be ruled out for the enhancement of the heavy elements.

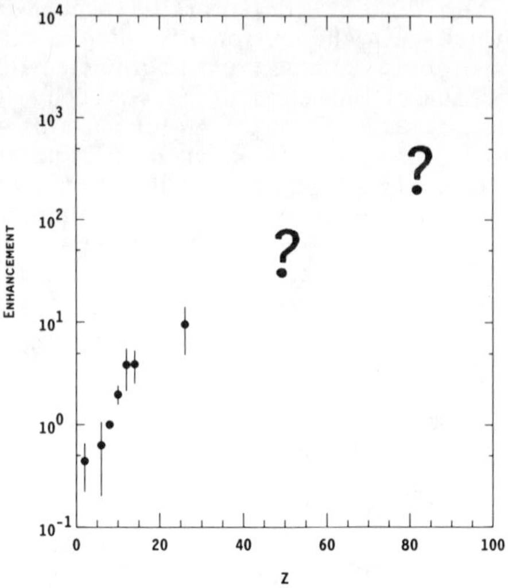

Fig. 2. Heavy-element enhancements.

b) Acceleration by Shocks

Most of energetic-particle populations that we observe have probably been accelerated by collisionless shock waves traversing a plasma. The theory of this acceleration is highly evolved[13,14] but the separation of spatial and temporal effects in the observations is often elusive. Even at low energies, most of the previous work involves only H and He and not the abundances and spectra of heavy elements.

Observations of many interesting phenomena have been hampered by lack of sensitivity and resolution of our instruments. Even the large IP shock-associated event in the right half Figure 1 has a low intensity of Fe. In the case of the anomalous component, it is difficult to study time dependences because long time averages are required to observe the particles at all. For the low-energy particles from the Earth's bow shock, observations of the abundances of the elements should clearly determine whether the particle source is terrestrial or is the solar wind.

EPACT

The EPACT Experiment is designed to increase the resolution, sensitivity and dynamic range of energetic-particle observations near 1AU. The experiment will be flown on the WIND spacecraft which is scheduled for launch in December, 1992. The spacecraft will spend its first 1.5 years in elongated orbit between 10 and 240 R_E on the sunward side of the Earth, and it will subsequently orbit the sunward L1 Lagrangian point at 240 R_E.

EPACT is a multi-telescope system involving 8 individual telescopes housed in three separate groups. The ELectron-Isotope TElescope (ELITE) system covers the highest energy range with three telescopes: the Isotope Telescope resolves isotopes of elements with Z>6 and 2 Alpha-Proton-Electron (APE) telescopes cover energies above 4 and 20 MeV/amu, respectively. The Low-Energy Matrix Telescope (LEMT) system contains 3 identical domed telescopes with large geometry factors to measure the elements $1 \le Z \le 82$ above about 2 MeV/amu. The Supra-Thermal Energetic Particle (STEP) telescope system uses 2 identical telescopes to measure particle masses down to ~20 keV/amu. All of the telescopes use Si solid-state detectors because of their high reliability. Such detectors have been in operation for up to 15 years in space without gain changes or other degradation. The detectors are not susceptible to catastrophic damage from micrometeorites and have a high radiation tolerance.

Each of the telescopes is served by onboard microprocessors that can identify particle species from the 16-bit pulse-height measurements either by table look up (at low energies) or by calculation using power-law expressions for the range-energy relation. The onboard identification proceeds at rates that are 10000 times faster than could be accomplished by telemetering the raw data to the ground at the experiment bit rate of <500 bits/sec.

All of the telescopes provide spin-sectoring information on the arrival directions of the particles. LEMT provides coarse altitude and azimuth information on particles and can accumulate selected species in alt-azimuth bins at high rate.

a) ELITE

A cross section of the IT is shown in Figure 3. The two front detectors in the IT two-dimensional position sensing accomplished by and orthogonal set of 85 metalized strips on each side of the detector. Strips are placed at a 0.5 mm pitch and are interconnected by 50-micron resistive strips that are part of the metalization pattern. Five preamplifiers are spaced evenly along each resistive chain to minimize the charge collection time without excessive power consumption.

The IT has a geometry factor of 8 cm^2 sr and covers the range $2 \le Z \le 26$. At Fe, typical mass resolution is 0.3 amu and the energy coverage is 10-360 MeV/amu.

The APE telescopes extend coverage to minimum-ionizing particles in 2 energy ranges. APE-A measures electrons of 0.2-2 MeV and protons and ^{4}He from 4.3-25 MeV/amu and APE-B covers electrons from 1-10 MeV and H & ^{4}He from 19-500 MeV/amu. Both APEs resolve He isotopes and measure elements up to Fe to provide redundancy. With geometry factors of 1.2 and 1.6 cm^2 sr, respectively, and fast response times, they are designed to operate in intense events without saturation.

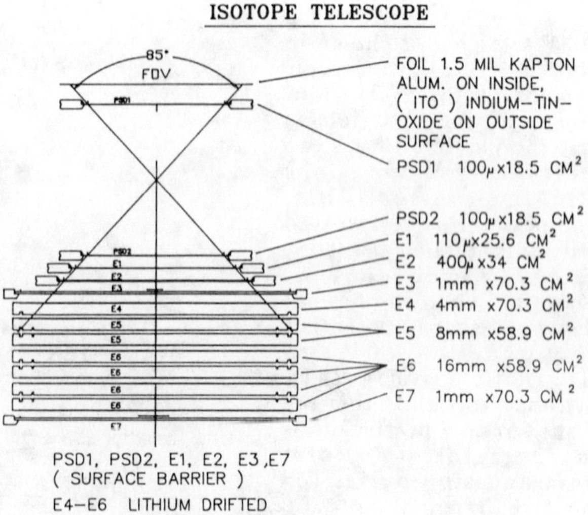

Fig. 3. The isotope telescope.

b) LEMT

One of the 3 identical LEMTs is illustrated in Figure 4. Coarse position information is provided by dividing the 38 cm^2 detector beneath the dome with a pattern of 5 orthogonal metalized strips on each side. Knowing which dome detector is traversed by a particle and its position in the second detector determines arrival direction to about ±15° and allows correction for the path length in the dome detector. Two of the 3 LEMTs are inclined above and below the ecliptic to provide coverage of 4π sr. The total geometry factor of the LEMTs is over 50 cm^2 sr, a factor of 100 times larger than that of the VLET system on ISEE-3 that has contributed to nearly all of the measurements of ^{3}He-rich events during the last 10 years.

Because of its high sensitivity, LEMT is designed to observe the elements in the 30≤Z≤90 region. Abundance peaks near Z≈50 and Z≈82 will be clearly resolved above about 3 MeV/amu and will be seen in large events even without enhancements.

Each LEMT can select events for pulse-height analysis at rates of 200 KHz and can pulse-height analyze, identify and bin particles at rates over 10 KHz.

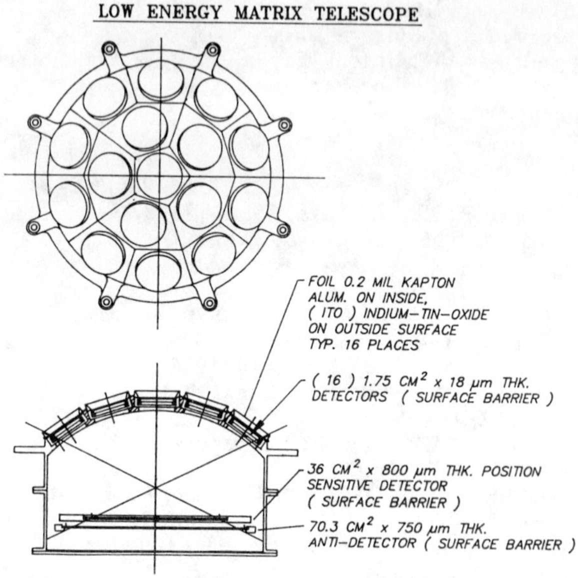

Fig. 4. The LEMT telescope.

c) STEP

The configuration of a STEP telescope is shown in Figure 5. STEP measures time-of-flight (TOF) and energy on each particle so that the particle's mass can be deduced. However, the resolution is such that, isotopes of He are resolved and individual elements are resolved at higher Z. Large-area micro-channel plates and shaped potentials are used to collect electrons from a thin entrance foil and from the front surface of the total-energy detector to form start and stop pulses for the TOF system.

Each STEP has a rectangular field of view and with a 17° x 44° opening angle and has geometry factor of 0.4 cm^2 sr.

STEP brings element and He-isotope abundance measurements to the region well below 1 MeV/amu where only H and ^{4}He have been observed previously. This region has high intensities of ions in IP shocks and solar flares.

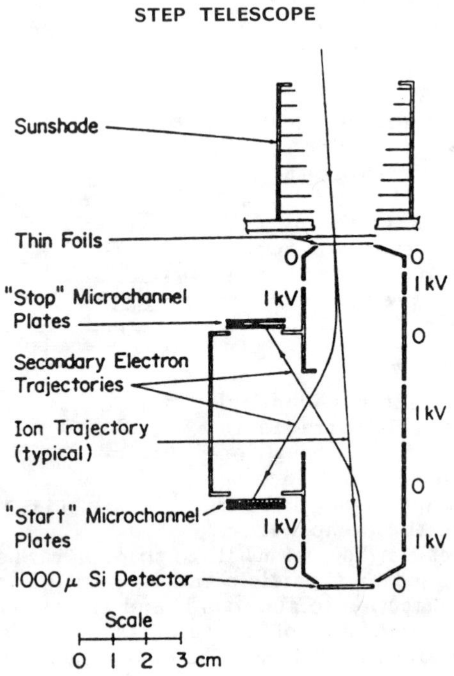

Fig. 5. The STEP telescope.

REFERENCES

1. Meyer, J.P., *Ap. J. Suppl.*, **57**, 151 (1985).
2. Lin, R.P., *Rev. Geophys.*, **25**, 676 (1987).
3. Mason, G.M., *Rev. Geophys.*, **25**, 685 (1987).
4. Webber,W.R. in *Cosmic Abundances of Matter*, AIP Conf. 183, C.J. Waddington ed. (1989), p 100.
5. Reames, D.V., *Ap. J. Suppl.* (1990, in press).
6. Reames, D.V., *Ap. J. (Letters)*, **330**, L71 (1988).
7. Fisk, L. A., *Ap. J.*, **224**, 1048 (1978).
8. Reames, D.V., von Rosenvinge, T.T., and Lin, R.P., *Ap. J.*, **282**, 716 (1985).
9. Reames,D.V. and Stone,R.G., *Ap. J.*, **308**, 902 (1986).
10. Kahler, S.W., Lin, R.P., Reames, D.V., Stone, R.G., and Liggett, M., *Solar Phys.*, **107**, 385 (1987).
11. Reames, D.V., Dennis, B.R., Stone, R.G., and Lin, R.P., *Ap. J.*, **327**, 998 (1988).
12. Reames, D.V., *Ap. J. (Letters)*, **325**, L53 (1988).
13. Lee, M. A., *J. Geophys. Res.*, **88**, 6109 (1983).
14. Lee, M.A. and Ryan, J.M., *Ap. J.*, **303**, 829 (1986).

THE COSMIC RAY AND SOLAR FLARE ISOTOPE EXPERIMENTS IN THE CRRES, NOAA-I AND "ULYSSES" SATELLITES

M. Garcia-Munoz
Laboratory for Astrophysics and Space Research,
Enrico Fermi Institute, The University of Chicago,
Chicago, Illinois 606 37

ABSTRACT

The ONR-604 instrument has been designed to measure the energy spectra and the isotopic composition at 1 AU of the elements Hydrogen to Nickel in the energy range 40-500 Mev/n and will be carried by the Combined Release and Radiation Effects Satellite (CRRES) to be launched July, 1990 in a highly eccentric orbit between a low perigee and a synchronous orbit point. It will measure galactic cosmic rays, solar energetic particles, and trapped and pseudo-trapped particles in the Earth magnetosphere. Also at 1AU, thc Energetic Heavy Ion Composition (EHIC) instrument, designed to study mainly the elemental and isotopic composition of solar energetic particles over the charge range H to Ni, in the energy range 0.5 to 200 MeV/n, will be placed in a Sun-synchronous circular polar orbit of 833 or 870 km altitude by a NOAA-I satellite scheduled to be launched late 1990 or early 1991. The University of Chicago High Energy Telescope (HET) experiment which is part of the COSPIN consortium on the "Ulysses" mission will measure the energy spectra and the isotopic composition of the elements Hydrogen to Nickel in the energy interval 20-450 MeV/n in the heliosphere, both in the ecliptic plane and at high heliographic latitudes in an orbit around the Sun that will have an aphelion near a Jupiter orbit point and a perihelion of about 1.4 AU. It will be launched October 1990.

THE ONR-604 EXPERIMENT ON THE CRRES SATELLITE

CRRES Mission Description

The CRRES (Combined Release and Radiation Effects Satellite) mission is a joint program of the National Aeronautics and Space Administration and the U. S. Air Force Space Test Program.

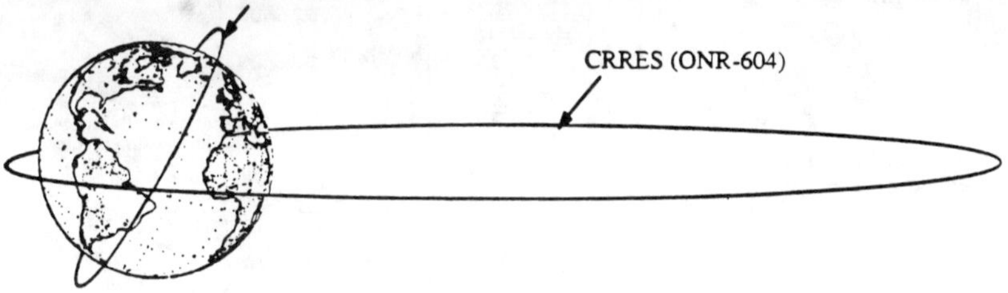

Figure 1: CRRES and NOAA-I Orbits

The scientific objective of the CRRES mission is the study of the near-Earth radiation environment and the effects of this environment on space flight hardware components[1]. This will be accomplished by placing an ensemble of experiments in a spacecraft that will move in a highly eccentric elliptic orbit with an initial perigee of 350 km. and an apogee at a geosynchronous altitude of 35,786 km. The inclination will be near 18 degrees and the period about 10 1/2 hours (Figure 1). The launch is scheduled for July, 1990 for a mission duration of at least 3 years.

The ONR-604 Experiment.

The ONR-604 experiment is a part of the CRRES program and has been developed by the University of Chicago under the joint sponsorship of the Office of Naval Research and the National Aeronautics and Space Administration.

Scientific Objectives. The main objectives of the ONR-604 experiment are:

1. Determination of the spectra and isotopic composition of the galactic cosmic rays and solar energetic particles in the outer part of the orbit where the measurements are not perturbed by the trapped radiation background. The ONR-604 can separate isotopes with energies 40 to 500 MeV/n and therefore can measure galactic cosmic ray and energetic solar isotopes during a large fraction of the CRRES orbital period.
2. The high energy processes in solar flares and, in particular, the particle acceleration mechanisms in the Sun.
3. The solar conditions leading to the explosive phase of solar flares.
4. Constraints on stellar nucleosynthesis models of nuclear species and their application to the Sun.
5. To provide a mapping of heavy nuclei intensities along the CRRES orbit which are going to be correlated with the single event upsets registered by the microelectronics experiment in CRRES.

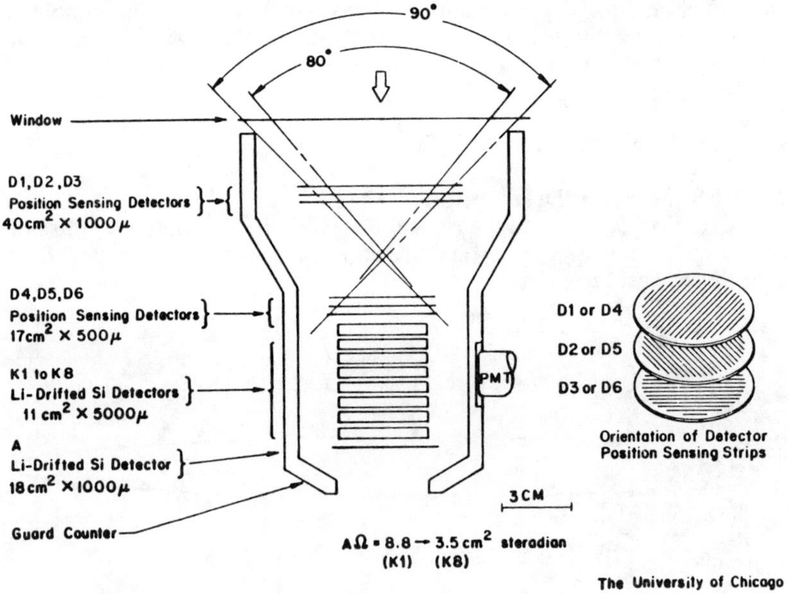

Figure 2. ONR-604 Telescope

Telescope Description. A complete description of the ONR-604 experiment is given by Simpson et al.[2]. Figure 2 shows a cross-section of the ONR-604 telescope It consists of a stack of 15 solid state detectors surrounded by a guard plastic scintillator viewed by a photomultiplier. The first 6 stack detectors are position sensing detectors, developed at the University of Chicago, and form a subsystem capable of determining the trajectory of crossing particles. The particle analysis is based in the multiple dE/dx technique and the instrument has the mass resolution to separate isotopes of the elements Hydrogen through Nickel in the general energy range 40 to 500 Mev/n.

THE EHIC EXPERIMENT ON THE NOAA-I SATELLITE.

The Energetic Heavy Ion Composition (EHIC) experiment, also under the joint sponsorship of the Office of Naval Research and the National Aeronautics and Space Administration, is a cooperation between the University of Chicago and the Herzberg Institute of Astrophysics of the National Research Council (Canada). The instrument will be carried by a NOAA-I satellite in a Sun-synchronous circular polar orbit of 833 or 870 km altitude and 99 degree inclination (see Figure 1). The launch is scheduled for late 1990 or early 1991. The expected duration of the NOAA-I mission will be between 2 and 3 years. EHIC will thus make detailed measurements both in the polar magnetic field regions connected to interplanetary space (~30-40% of the orbit) as well as in portions of the inner magnetosphere.It will then measure energetic solar flare particles in the polar regions and trapped energetic particles in the magnetosphere The instrument has been designed to measure the elemental and isotopic composition of energetic particles between H and Ni over the energy range 0.5 to 200 MeV/nucleon.

The primary scientific objective for this experiment is to obtain elemental and isotopic composition data which can be used to test models of a) solar flare ion acceleration, transport in interplanetary space, and entry into the magnetosphere. and b) nucleosynthetic processes leading to the elemental and isotopic composition of the Sun. EHIC will provide data on the fluxes and energy spectra of heavy ions, which can cause single bit upsets and damage to integrated circuits, found in the Sun synchronous orbits used by NOAA and other weather satellites.

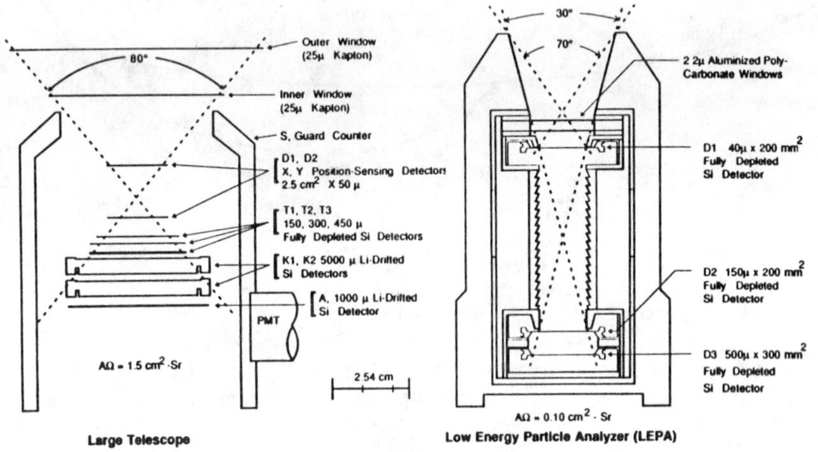

Figure 3. EHIC Telescopes

A cross section of the EHIC telescopes is shown in Figure 3. The elemental and isotopic measurements of EHIC are based on the multiple dE/dx technique. In order to achieve isotopic resolution of heavy ions, particle trajectories are measured in the Large Telescope using double sided position-sensitive detectors also developed at the University of Chicago.

THE HET EXPERIMENT IN THE "ULYSSES" SATELLITE.

"Ulysses" Mission Description.

The "Ulysses" mission is a joint program of the National Aeronautics and Space Administration and the European Space Agency. The scientific objective of the "Ulysses" mission is the study in three dimensions of the interplanetary medium and the Sun by making measurements in a large range of solar latitudes extending to about 80 degrees. These studies will include the properties of the solar wind, the interface of this wind with the solar corona, solar radio bursts and plasma waves, solar X-rays, solar energetic particles and galactic cosmic rays and interplanetary and interstellar dust and gas. Experiments designed to accomplish these objectives have been placed in a spacecraft that will use Jupiter gravity-assist to achieve a high solar latitude trajectory with aphelion near a point of the Jovian orbit and a perihelion at about 1.4 AU (Figure 4). The launch is scheduled for October 1990 and the mission will last a minimum of five years.

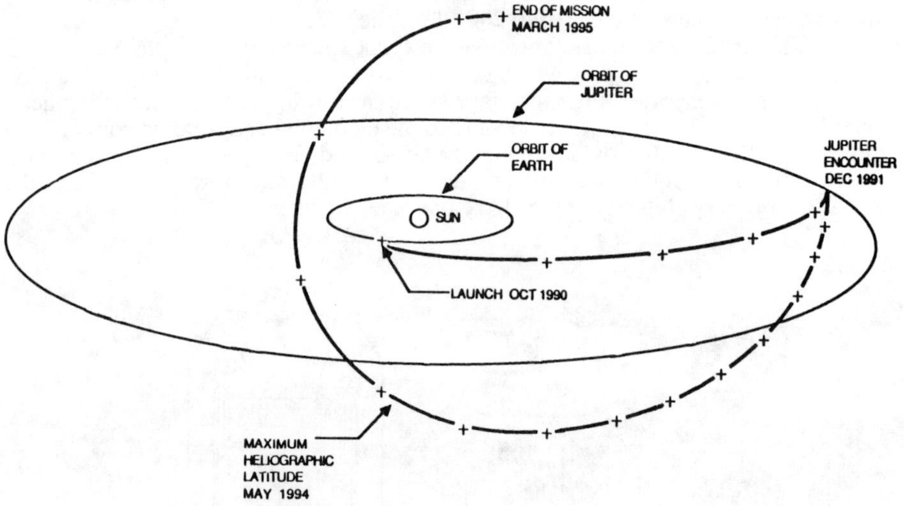

Figure 4. "Ulysses" Orbit

The "Ulysses" HET Experiment.

The University of Chicago High Energy Telescope (HET) is sponsored by the National Aeronautics and Space Administration and is part of the COSPIN consortium in "Ulysses" which is made up of six telescopes designed for the measurement of energetic charged nuclei and electrons. The COSPIN consortium consists of: University of Chicago (USA), Herzberg Institute of Astrophysics (Canada), Imperial College (England), Space Science Division of ESA (Holland) and University of Kiel (West Germany).

Scientific Objectives.

The scientific objective of the HET is to map in three dimensions the heliospheric distribution of energetic isotopes of Hydrogen to Nickel in the general energy interval 30-450 Mev/n and determine in this way the time and space characteristics of the interaction of energetic particles with the solar wind plasma and the acceleration and propagation of cosmic rays in the galaxy.

HET Description.

The HET is very similar to the ONR-604 telescope (Figure 5). The main differences is that the HET detector stack has six instead of the eight detectors in ONR-604, and that the lower set of position sensing detectors (D4, D5, D6) are 1000 micrometers thick in HET instead of 500 micrometer as in ONR-604. Like ONR-604 the HET can separate the isotopes of the elements Hydrogen to Nickel. A complete description of the HET is included in the description by Simpson et al.[3] of the SIM experiment of the COSPIN consortium.

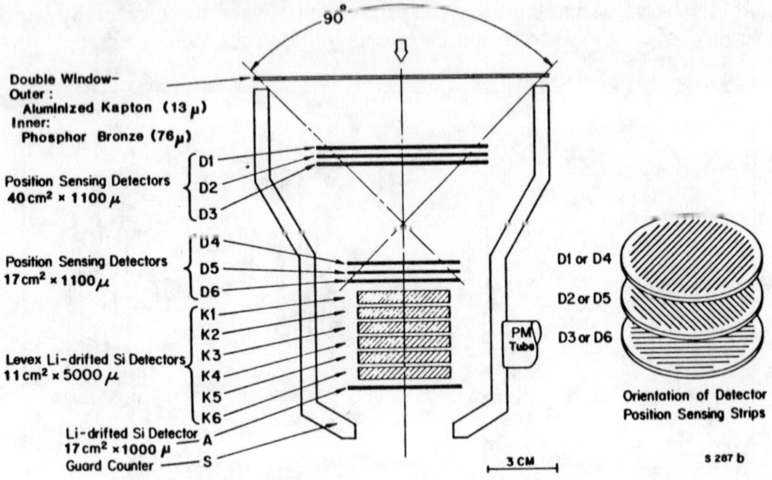

Figure 5: "Ulysses" High Energy Telescope

ACCELERATOR CALIBRATIONS OF ONR-604, EHIC AND HET.

The ONR-604, EHIC and HET experiments have been calibrated at the Bevalac accelerator of the Lawrence Berkeley Laboratory (LBL). In addition EHIC has been also calibrated at the tandem Van de Graff accelerator of the Chalk River Nuclear Laboratories.The calibration at LBL was with primary Fe and Ne beams and the secondary beams obtained by spallation of primaries.

Figure 6A shows ONR-604 histograms for the elements Ti through Mn obtained in a run at 5 degree beam incidence angle showing the good isotopic separation obtained. In order to test the efficiency of the trajectory normalization provided by the ONR-604 position sensing system, data from accelerator beam runs at two different incidence angles, 5 and 20 degrees, have been mixed and analyzed using the trajectory normalization provided by the instrument. The result is shown in Figure 6B where it can be seen that the isotopic resolution is comparable to that of Figure 6A which corresponds to a single incidence angle. Similar isotopic separation has been obtained for the HET and the EHIC telescopes

The ONR-604 and EHIC were also exposed to a 100 mCurie Ce^{144}-Pr^{144} electron radioactive source to investigate the perturbation effects in measurements carried out in the middle of very high electron fluxes.

BASELINE MEASUREMENTS AT 1 AU BY ONR-604 AND EHIC

By the time that the two Chicago experiments will be in orbit they will be the only high resolution cosmic ray telescopes available at 1 AU to support as baseline the cosmic ray measurements made by the outer heliosphere network of space probes Pioneer 10/11 and Voyager 1/2. As mentioned, the "Ulysses" spacecraft is due for launch in October 1990 on its way to Jupiter and it will carry among the COSPIN experiments, besides the HET, a Low Energy Telescope which extends downwards the HET energy range so that both of them cover essentially the same energy and particle species ranges as EHIC and ONR-604. The similarity between HET and ONR-604 will make the cross reference between them specially fruitful in the separation of temporal from spatial effects in the years 1990-1996.

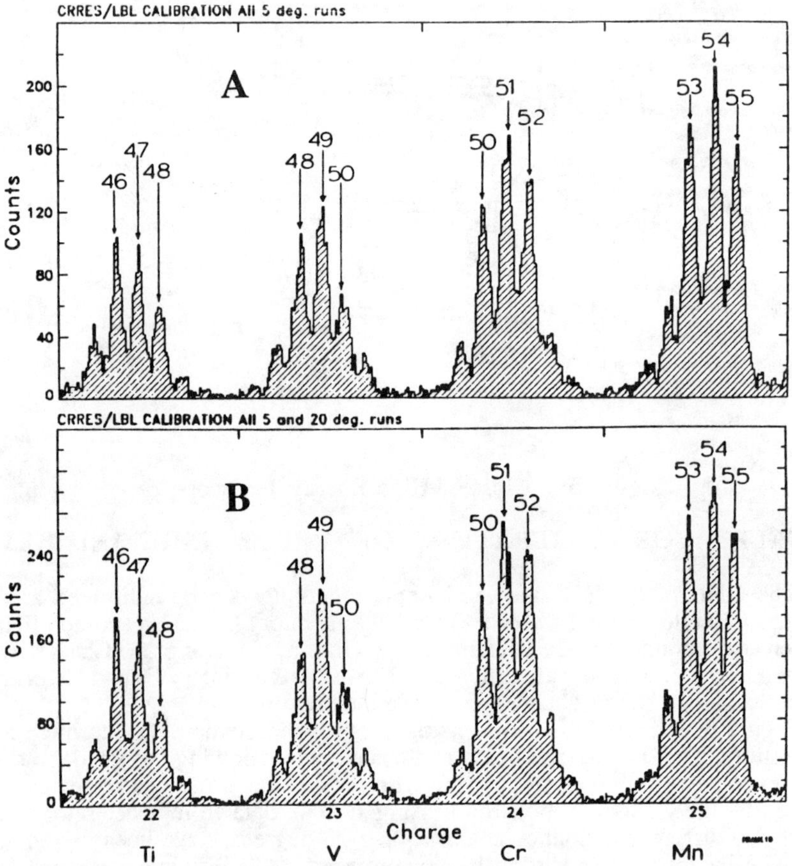

Figure 6: ONR-604 Calibration

ACKNOWLEDGEMENTS.

The ONR-604 and EHIC programs are sponsored in part by NASA Contract 5-30117, by the Office of Naval Research Contract N-00014-87-K-0334 and by the U.S. Air Force Space Test Program. "Ulysses" HET is sponsored in part by Jet Propulsion Laboratory Contract 955432. The LEPA is sponsored by the National Research Council (Canada)

REFERENCES.

1. M. S. Gusenhoven, E. G. Mullen, and R. C. Sagalyn, "CRRES/SPACERAD Experiment Description", AFGL-TR-85-0017, Environmental Research Papers, No. 906, 24 January 1985.
2. J. A. Simpson et al., 1985, "The Experiment for High-Energy, Heavy Nuclei Composition (ONR-604)", CRRES/Spacerad Experiment Description, Air Force Geophysical Laboratories, ed. M.S. Gussenhoven, E.G. Mullen and R.C. Sagalyn.
3. J. A. Simpson et al., 1983, "The ISPM Cosmic-Ray and Solar Particle Investigation", The International Solar Polar Mission - Its Scientific Investigations, ESA SP-1050, ed. K.-P. Wenzel, R.G. Marsden and B. Battrick, p. 155.

SAMPEX MISSION OVERVIEW

G. M. Mason[a], D. N. Baker[b], J. B. Blake[c], L. B. Callis[d], D. C. Hamilton[a], D. Hovestadt[e], B. Klecker[e], R. A. Mewaldt[f], M. Scholer[e], E. C. Stone[f], and T. T. von Rosenvinge[b]

[a]University of Maryland, College Park, MD 20742
[b]NASA/Goddard Space Flight Center, Greenbelt, MD 20771
[c]Aerospace Corporation, Los Angeles, CA 90009
[d]NASA/Langley Research Center, Hampton, VA 23665-5225
[e]Max-Planck-Institut für extraterrestrische Physik, 8046 Garching, FRG
[f]California Institute of Technology, Pasadena, CA 91125

ABSTRACT

The Solar, Anomalous, and Magnetospheric Particle Explorer, SAMPEX, will carry out energetic particle studies of outstanding scientific questions in the fields of space plasma physics, solar physics, magnetospheric and middle atmospheric physics, and cosmic ray physics. SAMPEX will measure the electron and ion composition of energetic particle populations from ~0.4 MeV/nucleon to hundreds of MeV/nucleon from a zenith-pointing small satellite in near-polar orbit, using a coordinated set of detectors with excellent charge and mass resolution, and with higher sensitivity than previously flown instruments. While over the magnetic poles, the instruments will study the composition of anomalous cosmic rays, solar energetic particles, and galactic cosmic rays. At lower magnetic latitudes, geomagnetic cutoff effects will allow determination of the ionization state of these particles at energies much higher than can be studied from interplanetary spacecraft. At subauroral latitudes, SAMPEX will also observe precipitating relativistic magnetospheric electrons, which undergo important interactions within the middle atmosphere.

ANOMALOUS COSMIC RAYS

A major goal of the SAMPEX mission is to determine the ionization state of the Anomalous Cosmic Rays (ACRs), elements (He, C, N, O, Ne, and Ar) in the energy range below ~50 MeV/nucleon whose abundances are anomalous compared to other solar system material. For example the oxygen:carbon ratio in ACRs is >50, while in the 'universal abundances' this ratio is close to 2. A widely accepted model for ACRs makes a unique prediction: ACRs should be singly ionized. SAMPEX will give the first definitive measurements of the charge state of anomalous He and O, assuming that their flux during 1993 and 1994 has recovered as expected from low solar maximum levels.

In the case of ACR oxygen, when SAMPEX is over the geomagnetic poles it will measure the interplanetary flux levels. If this oxygen is fully stripped the geomagnetic field will prevent it from reaching the satellite when it is more than ~25° from the magnetic poles; if it is singly ionized, it will reach the spacecraft up to ~40° from the poles – a simple measurement. Even though ACR fluxes are low, the sensitive detectors on SAMPEX will collect thousands of counts/year–more than enough to discriminate between singly and fully ionized oxygen charge states.

If the ACR component is shown to be singly charged, then this component represents a direct sample of the local interstellar medium that carries important information about galactic evolution in the solar neighborhood over the time interval since the formation of the solar nebula; SAMPEX will measure the isotopic composition of this sample of matter.

PRECIPITATING RELATIVISTIC MAGNETOSPHERIC ELECTRONS

Another unique study possible with the combination of SAMPEX instruments and orbit concerns precipitating relativistic magnetospheric electrons. Observations at synchronous orbit have found $\gtrsim$ 1 MeV electron intensities that increase substantially for periods of a few days often with a recurrence period of ~27 days. Numerical models show that the multi-MeV electrons—when they precipitate—can cause a large (10-100 keV/cm^3-s) energy deposition at 40-60 km altitude in the atmosphere. This could dominate other ionization sources (such as cosmic rays and solar EUV) at these heights by orders of magnitude.

Precipitating relativistic electrons may indeed lead to substantial long-term increases in the levels of odd nitrogen compounds (NO_x) at these heights with an attendant impact on local ozone levels via the reactions $NO + O_3 \rightarrow NO_2 + O_2$, and $NO_2 + O \rightarrow NO + O_2$. It is therefore a critical problem to determine the actual intensity and spatial extent of relativistic electron precipitation. SAMPEX will determine the longitude, latitude, and local time precipitation characteristics for >1 MeV electrons required for these studies.

SOLAR ENERGETIC PARTICLES

By launching into the period shortly after sunspot maximum, SAMPEX should observe several large solar particle events; as the solar cycle winds down, it will observe numerous implusive flares such as ^{3}He-rich flares. Solar energetic particles provide a sample of ions from the solar atmosphere. Since the charge states of SEPs are determined by the plasma conditions at the acceleration site, measurements of the average charge state can be used as a diagnostic tool. For example, previous limited measurements at low energies have already revealed large (~ 40% for Fe) differences between the charge states of Si and Fe in ^{3}He-rich events and those in events with normal composition, indicating that the ^{3}He-rich flares originate in hot sites in the corona. Using geomagnetic cutoff techniques, SAMPEX will measure the mean charge states of these solar flare elements with greater statistical accuracy, and to significantly higher energies than previous missions.

The solar energetic particle composition provides a direct measure of the present solar elemental and isotopic makeup, providing crucial information for understanding the history of solar system material and adding new dimensions to the study of solar flare acceleration and propagation processes. It is almost a decade since the first high-resolution measurements of solar flare isotopes heavier than He were provided by an instrument on ISEE-3. Since that time little progress has been made because no instruments with improved collecting power have been launched, and this field has therefore remained relatively unexplored. Spectrometers on SAMPEX will resume these studies and will significantly improve on the collecting power of past instruments.

GALACTIC COSMIC RAYS

A final area of the SAMPEX scientific studies involves galactic cosmic rays, which are a directly accessible sample of matter from outside the solar system. A spectrometer on SAMPEX will carry out measurements of the *isotopic* composition of this sample of high energy matter, which contains a record of the *nuclear* history of cosmic rays including their synthesis in stars and subsequent nuclear interactions with the interstellar gas. Cosmic ray isotope observations have already revolutionized our views of both cosmic ray origin and propagation. As an example, measurements by a number of groups have shown that ^{22}Ne is a factor of at least 3 times more abundant in cosmic ray source material than in the solar system, while the abundances of ^{25}Mg, ^{26}Mg, ^{29}Si, and ^{30}Si are all enhanced by a factor of $\sim$1.5. The exposure obtained on SAMPEX will make it possible to extend the search for isotopic differences between galactic and solar cosmic ray material to several additional key elements.

Table 1. Instrument Performance

	LEICA	HILT	MAST	PET
Energy range for				
electrons	—	—	—	1-120 MeV
H	0.75-6	—	7-15	18-350 MeV
He	0.4-6	3.9-90	7-91	18-500 MeV/n
C	0.35-12	7.2-160	12-210	
Si	0.26-6	9.6-177	19-345	54-195 MeV/n
Fe	0.16-4	11.0-90	24-470	
Charge range for elements	1-28	2-28	1-28	1-28
Charge range for isotopes	2-16	2	1-28	1-2
Geometrical factor (cm^2 sr)	1.0	50	7-16	0.3-1.6

INSTRUMENTATION AND MISSION DESCRIPTION

The SAMPEX instrumentation includes a complementary set of four high resolution, high sensitivity particle detectors: the Heavy Ion Large Telescope (HILT), the Low Energy Ion Composition Analyzer (LEICA), the Mass Spectrometer Telescope (MAST), and the Proton/Electron Telescope (PET). Prototypes of LEICA and HILT flew on the Space Shuttle in August, 1989. MAST and PET were originally approved and under construction for the (cancelled) U.S. spacecraft of the International Solar Polar Mission. Thus, these new instruments are in an advanced state of development and can be readied for SAMPEX quickly and at moderate cost. Table 1 lists instrument characteristics, and Table 2 gives mission and spacecraft parameters.

The broad range of studies carried out by SAMPEX could *only* be carried out on a low altitude near-polar orbiting spacecraft such as can be flown under the Small-Class Explorer Program. The 1992 launch date will allow important solar studies to be carried out during a period of enhanced solar activity and continuing into the declining phase of the activity cycle. This time interval is not only ideal for the objectives discussed above but also would provide coverage of energetic particle fluxes in the period preceding the launch of the WIND spacecraft. SAMPEX will also provide key particle data to complement the measurements of Solar-A, GOES, and GRO during the solar maximum, as well as provide baseline coverage for Ulysses, Galileo, and the Pioneer and Voyager spacecraft. SAMPEX will also provide overlapping coverage for a number of long duration balloon flights planned as part of the Max '91 program. SAMPEX precipitating particle measurements will be compared with observations on geostationary satellites and CRRES, and also with ozone measurements from UARS.

Table 2. SAMPEX Mission Description

Orbit: 580 km circular, 82° inclination	Orbit average bitrate: 3 kpbs
Launch Date: first half 1992	Mission Duration: ≥3 years
Mass: 58kg instruments, 119kg S/C + margin	Total: 177kg
Power: 20W instruments, 55W S/C	Total: 75W
Spacecraft: to be built at GSFC	Launch site: Vandenberg AFB

Acknowledgements: This work is supported in part by NASA contract NAS5-30704 and by the German Government.

THE ADVANCED COMPOSITION EXPLORER

E. C. Stone[a], L. F. Burlaga[b], A. C. Cummings[a], W. C. Feldman[c], W. E. Frain[f], J. Geiss[d], G. Gloeckler[e], R. E. Gold[f], D. Hovestadt[g], S. M. Krimigis[f], G. M. Mason[e], D. McComas[c], R. A. Mewaldt[a], J. A. Simpson[h], T. T. von Rosenvinge[b], and M. E. Wiedenbeck[h]

a) California Institute of Technology; Pasadena, CA 91125 USA
b) Goddard Space Flight Center; Greenbelt, MD 20771 USA
c) Los Alamos National Laboratory; Los Alamos, NM 87545 USA
d) Physikalisches Institut, University of Bern; Bern, Switzerland
e) University of Maryland; College Park, MD 20742 USA
f) Applied Physics Lab, The Johns Hopkins University; Laurel, MD 20707 USA
g) Max-Planck-Institut fur Physik und Astrophysik; Munich, Federal Republic of Germany
h) University of Chicago; Chicago, IL 60637 USA

ABSTRACT

The Advanced Composition Explorer (ACE) was recently selected as one of two new Explorer-class missions to be developed for launch during the mid-1990's. ACE will observe particles of solar, interplanetary, interstellar, and galactic origins, spanning the energy range from that of the solar wind (~1 keV/nucleon) to galactic cosmic ray energies (several hundred MeV/nucleon). Definitive studies will be made of the abundance of nearly all isotopes from H to Zn ($1 \leq Z \leq 30$), with exploratory isotope studies extending to Zr ($Z = 40$). To accomplish this, the ACE payload includes six high-resolution spectrometers, each designed to provide the optimum charge, mass, or charge-state resolution in its particular energy range, and each having a geometry factor optimized for the expected flux levels, so as to provide a collecting power a factor of 10 to 1000 times greater than previous or planned experiments. The payload also includes several instruments of standard design that will monitor solar wind and magnetic field conditions and energetic H, He, and electron fluxes. We summarize here the scientific objectives, instrumentation, spacecraft, and mission approach that were defined for ACE during the Phase-A study period.

INTRODUCTION

There are a variety of energetic particle populations observed in near-Earth interplanetary space, as indicated in Figure 1. These include nuclei accelerated in the solar wind and in solar flares, in interplanetary space, in galactic cosmic ray sources, and apparently, in the case of the "anomalous" cosmic rays, at the termination shock of the solar wind. Measurements of these particles allow a direct study of the elemental and isotopic composition of three distinct samples of matter. Of these, the anomalous cosmic rays are thought to be a sample of the very local interstellar medium, while galactic cosmic rays are a second contemporary sample

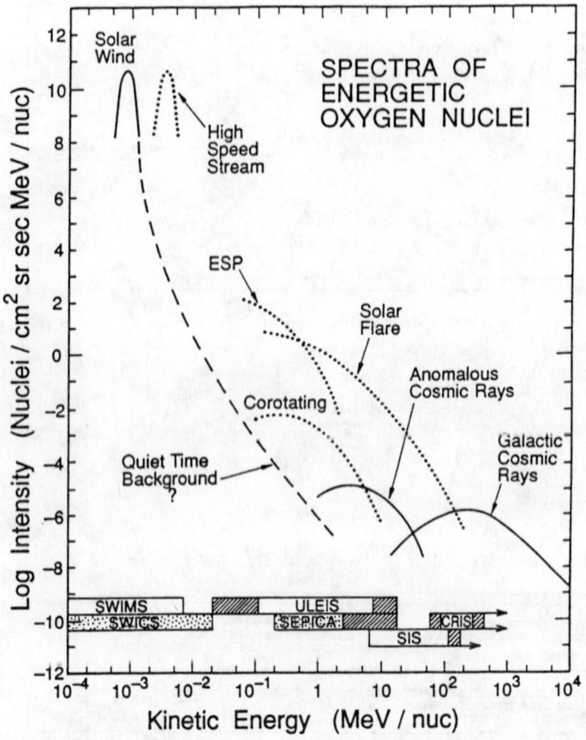

Figure 1: Typical energy spectra of energetic ^{16}O nuclei resulting from various particle populations. The solid curves represent "steady-state" components, while the dotted curves are for "transient" phenomena. Also shown is a postulated quiet-time flux of supra-thermal solar particles (dashed curve). The energy ranges of the various ACE instruments are indicated for resolution of isotopes (blank), elements only (cross-hatched), and ionic charge-states (stippled). Other nuclear species generally tend to have spectra that are similarly shaped to that of oxygen (but of varying intensity) when plotted as a function of energy/nucleon.

from somewhat more remote regions of the galaxy. The solar wind and solar energetic particles represent an older sample of interstellar matter that has been stored in the Sun for the last 4.6 billion years.

ACE will compare the composition of these distinct samples of matter by performing comprehensive and coordinated determinations of the elemental and isotopic composition of energetic nuclei accelerated on the Sun, in interplanetary space, and from galactic sources. These observations will span six decades in energy, as indicated in Figure 2, extending from solar wind to galactic cosmic ray energies, and will cover the element range from $_1H$ to $_{40}Zr$. The comparison of these samples of matter will be used to study the origin and subsequent evolution of both solar system and galactic material by isolating the effects of fundamental processes that include nucleosynthesis, charged and neutral-particle separation, bulk plasma acceleration, and the acceleration of suprathermal and high energy particles.

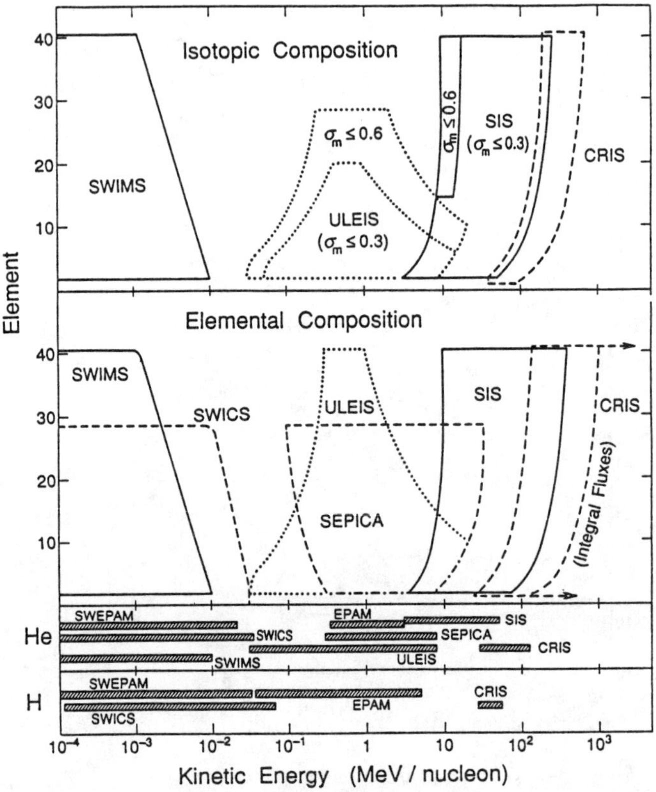

Figure 2: A plot of the energy/nucleon range over which the various ACE sensors will measure the elemental and isotopic composition of nuclei from He to Zr (Z = 2 to 40). In the top panel, boundaries for a mass resolution of $\sigma_M \lesssim 0.3$ and $\sigma_M \lesssim 0.6$ amu are indicated (note that many studies, e.g., ^{22}Ne/^{20}Ne, do not require resolution of adjacent isotopes). The H and He panels at the bottom indicate the coverage for these elements. Note that a 400 km/sec solar wind velocity corresponds to $\sim 8 \times 10^{-4}$ MeV/nucleon.

SCIENTIFIC OBJECTIVES

The observations to be performed on ACE will allow the investigation of a wide range of fundamental problems in the following major areas:

1) The Elemental and Isotopic Composition of Matter

A major objective is the accurate and comprehensive determination of the elemental and isotopic composition of the various samples of "source material" from which nuclei are accelerated. Thus, using ACE measurements we will:

- Generate a set of solar *isotopic* abundances based on *direct* sampling of solar material.
- Determine the coronal *elemental and isotopic* composition with greatly improved accuracy.

- Establish the pattern of isotopic differences between galactic cosmic ray and solar system matter.
- Measure the elemental and isotopic abundances of interstellar and interplanetary "pick-up ions".
- Determine the isotopic composition of the "anomalous cosmic ray component", thought to represent a sample of the local interstellar medium.

2) Origin of the Elements and Subsequent Evolutionary Processing

Isotopic "anomalies" in meteorites indicate that the solar system was not homogeneous when formed, while other data suggest that the solar composition continues to evolve. Similarly, the galaxy is neither uniform in space nor constant in time due to continuous stellar nucleosynthesis. Using measurements from ACE we will:

- Search for additional differences between the isotopic composition of solar and meteoritic material.
- Determine the contributions of solar wind and solar flare nuclei to lunar and meteoritic material, and to planetary atmospheres and magnetospheres.
- Determine the dominant nucleosynthetic processes that contribute to cosmic ray source material.
- Determine whether cosmic rays are a sample of freshly synthesized material (e.g., from supernovae), or of the contemporary interstellar medium.
- Search for isotopic patterns in solar and galactic material as a test of galactic evolution models.

3) Formation of the Solar Corona and Acceleration of the Solar Wind

Solar energetic particle, solar wind, and spectroscopic observations show that the *elemental* composition of the corona is differentiated from that of the photosphere, although the processes by which this occurs, and by which the solar wind is subsequently accelerated, are poorly understood. The detailed composition and charge-state data provided by ACE will allow us to:

- Isolate the dominant coronal formation processes by comparing a broad range of coronal and photospheric abundances.
- Study plasma conditions at the source of the solar wind and the solar energetic particles by measuring and comparing the charge states of these two populations.
- Study solar wind acceleration processes and any charge or mass-dependent fractionation in various types of solar wind flows.

4) Particle Acceleration and Transport in Nature

Particle acceleration is ubiquitous in nature and is one of the fundamental problems of space plasma astrophysics. The unique data set obtained by ACE measurements will enable us to:

- Make direct measurements of charge and/or mass-dependent fractionation during solar flare and interplanetary acceleration.
- Constrain solar flare and interplanetary acceleration models with charge, mass, and spectral data spanning up to five decades in energy.
- Test theoretical models for ^{3}He-rich flares and solar γ-ray events.
- Measure cosmic ray acceleration and propagation time scales using radioactive clocks.
- Test whether the "anomalous cosmic rays" are a singly-ionized sample of the neutral interstellar gas by directly measuring their charge state.

INSTRUMENT SYSTEM

The scientific objectives described above imply a need for coordinated high-precision measurements of the isotopic, elemental, and ionic charge state composition of energetic nuclei over a broad energy range, made with excellent time resolution in order to account for dynamical processes that affect the composition. The ACE sensor systems employ the optimum technology for each energy regime.

The sensor systems for the ACE payload are:

SWIMS - A Solar Wind Ion Mass Spectrometer that uses time-of-flight (TOF) through a specially tailored electric field to achieve unprecedented mass resolution at solar wind energies.

SWICS - A Solar Wind Ion Composition Spectrometer which is the Ulysses flight spare of the GLG[1] instrument and which uses electrostatic deflection, post-acceleration, TOF, and energy measurement to determine the elemental and ionic-charge composition of the solar wind.

ULEIS - An Ultra Low Energy Isotope Spectrometer that combines TOF with energy measurement to achieve excellent mass resolution for particles with energies ~20 keV/nuc to 10 MeV/nuc.

SEPICA - A Solar Energetic Particle Ionic Charge Analyzer that uses electrostatic deflection, differential energy loss (dE/dX), and residual energy measurement (E') to obtain the ionic charge state and nuclear charge of energetic ions above ~0.2 MeV/nuc.

SIS - A Solar Isotope Spectrometer that uses trajectory measuring detectors, dE/dX, and E' measurements to determine the elemental and isotopic composition of solar, anomalous, and interplanetary particles from He to Zn ($Z = 2$ to 30) over the energy range from ~8 to 150 MeV/nuc.

CRIS - A Cosmic Ray Isotope Spectrometer that uses trajectory measurements, dE/dX, and E' measurements to provide high-statistics measurements of all stable and long-lived isotopes of galactic cosmic ray nuclei from He to Zn ($Z = 2$ to 30) over the energy range from ~100 to 600 MeV/nuc.

SWEPAM - A Solar Wind Electron, Proton, and Alpha-particle Monitor which is the Ulysses flight spare of the BAM[2] instrument and which will provide baseline

Table 1: Instrument System Summary					
Instrument	Wt. (kg)	Nom. Power (W)	Data Rate (bits/s)	Technique	Typical Energy (MeV/nuc)
CRIS	20.6	10.6	462	dE/dX × E	~300
SIS	19.5	18.1	2000	dE/dX × E	~30
ULEIS	17.0	15.1	1000	TOF × E	~1
SEPICA	16.0	6.0	600	dE/dX × E × E/Q	~1
SWICS	5.5	4.0	500	TOF × E × E/Q	~0.001
SWIMS	6.8	5.0	505	TOF thru special E-field	~0.001
SWEPAM	6.7	5.4	1000	Electrostatic Analyzer	~0.001
SWIMS/SWICS/ SWEPAM DPU	3.0	3.0	-	-	-
EPAM	5.0	5.0	160	dE/dX × E	~0.3
MAG	4.7	4.0	300	Triaxial fluxgate	-

monitoring of solar wind electron and ion fluxes.

EPAM - An energetic Electron, Proton, and Alpha-particle Monitor which is the Ulysses flight spare of the LAN[3] instrument and which will provide baseline monitoring of low-energy particle fluxes.

MAG - A Magnetometer which is the Voyager Proof Test Model[4] and which will provide baseline monitoring of the interplanetary magnetic field.

Table 1 provides a summary of some of the important instrument parameters.

There are previous and planned space flight instruments which have and will be providing definitive studies of the most abundant isotopes, exploratory studies (<100 particles) of some of the less abundant nuclides, and surveys of the composition of the solar wind. We summarize here a comparison of the capabilities of these earlier instruments with those planned for ACE. In comparing energetic particle instruments of comparable resolution it should be remembered that for resolved isotope measurements it is the *collecting power* of the instrument that determines the statistical accuracy to which a given isotope ratio can be measured. It is also the collecting power that determines the minimum time scale on which the composition of transient events can be characterized. These collecting power estimates take into account all essential features of instrument design, including the geometry factor, energy range, immunity to "pile-up" and "accidental coincidences", and telemetry bit rate.

Solar Wind Ion Mass Spectrometers: As indicated in Table 2, our present knowledge of the isotopic composition of the solar wind is very limited. SWIMS will extend solar-wind spectroscopy to a wide range of additional elements and isotopes.

Solar and Interplanetary Isotope Spectrometers: Figure 3 compares the characteristics of ULEIS and SIS with other instruments designed to measure solar

Table 2: Solar Wind Composition Instruments

Element	Time Period of Measurements						
	1960	1965	1970	1975	1980	1985	ACE/ SWIMS
H	E	E	E	E	E	E	E
He		E	I	E	I	I	I
C						E	I
N						E	I
O			E	E	E	E	I
Ne			I		E	E	I
Na							I
Mg						E	I
Al							I
Si			E	E	E	E	I
P							I
S						E	I
Cl							I
Ar			I				I
K							I
Ca							I
Ti							I
V							I
Cr							I
Mn							I
Fe			E	E	E	E	I
Co							I
Ni							I
Technique Used:	Faraday Cup (F.C.)	F.C., E/Q	E/Q Foil	E/Q	M/Q	E/Q-A TOF-E	E/Q-TOF $V = X^2$

E: Indicated elements or element groups are resolved.
I: Listed isotopes of indicated elements are resolved.

and interplanetary isotopes with a mass resolution of $\leq$0.25 amu at Si. ULEIS will provide the first comprehensive isotope measurements below a few MeV/nuc.

Solar Energetic Particle Charge Analyzers: Through the use of increased electrostatic deflection voltages and the addition of a time-of-flight system with position sensing, SEPICA will achieve a factor of ~25 increase in collecting power over previous or planned instruments. At the same time, the charge-state resolution near 1 MeV/nuc will be greatly improved.

Galactic Cosmic Ray Spectrometers: Figure 4 compares the proposed SIS and CRIS instruments with those of other investigations. Note that CRIS will provide the necessary order of magnitude increase in collecting power to provide definitive measurements of rare isotopes, while SIS will extend these measurements to lower energy, also with very large collecting power. Although a balloon-borne instrument may have 5 to 10 times the collecting power of CRIS, it would have a duty cycle of $<10^{-2}$ (i.e., 2 days/yr). The superconducting magnet facility planned for the Space Station (Astromag) will be limited to energies $>$2 GeV/nuc by the geomagnetic cutoff. Thus, ACE and Astromag observations will complement each other.

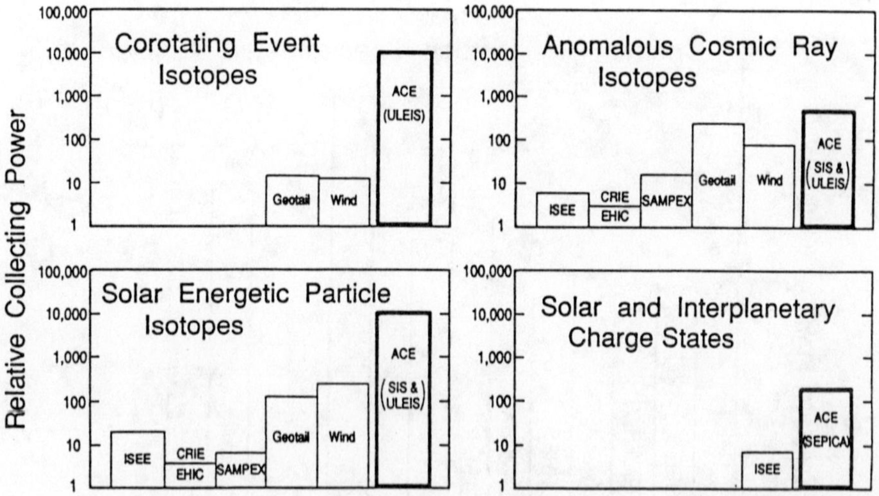

Figure 3: A comparison of the collecting power of previous and planned spectrometers designed to measure the isotopic composition and charge states of solar and interplanetary particles. Note that the collecting power scale is logarithmic.

SPACECRAFT, MISSION, and DATA ANALYSIS APPROACH

To minimize cost, the proposed ACE spacecraft is based on designs that evolved from the AMPTE/CCE and SIMPL programs. The AMPTE/CCE spacecraft was designed as a Sun-pointing mounting bus for scientific instruments, built by The Johns Hopkins University/Applied Physics Laboratory, and successfully launched in August 1984. The SIMPL study defined a spacecraft at the Earth/Sun L_1 point that would provide an early warning for solar-terrestrial disturbances.

The ACE spacecraft bus provides mounting platforms, electrical power, attitude control, thermal control, and data handling for the 9 scientific instruments. The spacecraft body consists of a two deck, irregular octagon, 78 inches across the points and about 4 feet high. Most of the instruments and spacecraft electronics are mounted on the upper deck. The lower deck supports a hydrazine propulsion system that provides for both orbit insertion and maintenance. Electrical power is supplied by four solar panels and an 8 A-hr battery. RF communication is achieved with a 60-cm high-gain, parabolic dish antenna and four low-gain quadrifilar helical antennas.

A Multi-purpose Launch Vehicle (MLV) with the capability of a Delta II 7920 will deliver the ACE spacecraft to the vicinity of the Earth-Sun libration point, L_1. The spacecraft propulsion system will maneuver ACE so that an orbit of 10 degrees with respect to an Earth-Sun line will be established. Periodic thrust maneuvers and spin axis torquing will be required over the mission life. The

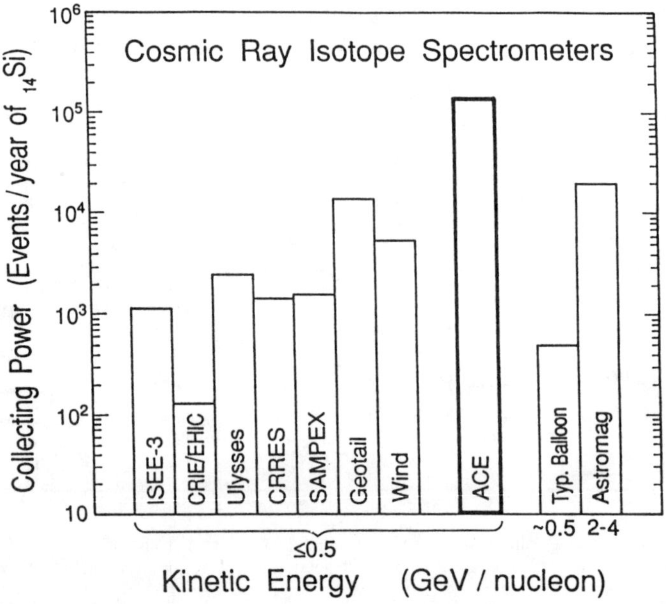

Figure 4: A comparison of the collecting power of previous and planned cosmic ray isotope spectrometers with that of ACE.

spacecraft attitude will be set to provide spin axis pointing toward the Sun to within ±20°. Data will be collected on tape and will be played back once per day to the 26-meter antennas of the Deep Space Network. The data are then transmitted to the ACE ground station for analysis and subsequent distribution.

Correlation of the diverse measurements is a key component of the ACE science objectives. Consequently, the initial science processing of the telemetry stream will be accomplished at a single central Production Data Center (PDC), which will convert the data to calibrated science measurements. This data set will be made available to nine Analysis Data Sites (ADS), located at each of the local Investigator Institutions, where comprehensive local science analysis will take place. In addition, each Investigator Team will supply a set of algorithms that will enable the other Investigators to display summary data in a standard format to facilitate correlation amongst the various data sets. A subset of the calibrated science measurements will be made available to the National Space Science Data Center in parallel. Finally, ACE has the capability to transmit in real time a set of monitoring data to provide a continuous monitor of solar and interplanetary conditions.

SUMMARY

In summary, ACE will use advanced spectrometers with large collecting power and unprecedented resolution to make possible a major advance in our knowledge of the elemental and isotopic composition of solar, interplanetary, and interstellar particles over a wide energy range. The ACE data plan will allow correlations of the various data sets in order to address outstanding scientific problems in the areas of the origin of the elements, the formation of the solar corona, and particle acceleration and transport in nature.

ACKNOWLEDGEMENTS

The ACE Phase-A study could not have been completed without the contributions of many individuals, including exceptional efforts from many scientists, engineers, and staff at the various Investigator Institutions. D. Mehoke of JHU/APL was the systems engineer for the spacecraft study team. He and the other members of the spacecraft study team at JHU/APL deserve thanks for their response to the tasks given them. We are grateful to R. Farquhar of GSFC and D. Dunham of Computer Sciences Corp. for providing trajectory calculations for ACE. The ACE Technical Officers at GFSC were E. Mercanti and T. Karras and their aid in providing liaison with the GSFC Project Office was appreciated. We are thankful for the support of S. Nylund of JHU/APL who was largely responsible for formulating the ACE Data Plan. We also appreciate the assistance of G. Floyd of GSFC for helping to formulate the ground data capture and mission operations and control systems. Our efforts in defining the telecom parameters for ACE benefited from discussions with W. L. Martin of JPL. Finally, the ACE Phase-A report achieved its final form due to the considerable effort and expertise of L. Sartain, C. Silva, and K. Gary, all of Caltech. The ACE Phase-A Study was supported by NASA Contract NAS5-30340.

REFERENCES

1. Gloeckler, G., J. Geiss, H. Balsiger, L. A. Fisk, F. Gliem, F. M. Ipavich, K. W. Ogilvie, W. Stüdemann, and B. Wilken, *ESA Special Publication SP-1050,* 77 (1983).

2. Bame, S. J., J. P. Glore, D. J. McComas, K. R. Moore, J. C. Chavez, T. J. Ellis, G. R. Peterson, J. H. Temple, and F. J. Wymer, *ESA Special Publication SP-1050,* 49 (1983).

3. Lanzerotti, L. J., R. E. Gold, L. A. Anderson, T. P. Armstrong, R. P. Lin, S. M. Krimigis, M. Pick, E. C. Roelof, E. T. Sarris, G. M. Simnett, and W. E. Frain, *ESA Special Publication SP-1050,* 143 (1983).

4. Behannon, K. W., M. H. Acuña, L. F. Burlaga, R. P. Lepping, N. F. Ness, and F. M. Neubauer, *Space Sci. Rev.,* **21,** 235, (1977).

POsitron Electron Magnet Spectrometer (POEMS) for the Eos Mission

Paul Evenson,[1] John P. Wefel,[2] Simon Swordy,[3] Robert Streitmatter,[4] Michael Salamon,[5] Louis Barbier,[4] T. Gregory Guzik,[2] Karen P. Magee-Sauer,[1] John W. Mitchell,[2] Jonathan Ormes,[4] Reuven Ramaty,[4] and Donald Reames[4]

[1]Bartol Research Institute, University of Delaware, Newark, DE 19716 USA
[2]Department of Physics and Astronomy, Louisiana State University Baton Rouge, LA 70803 USA
[3]Department of Physics, University of Chicago, Chicago, IL 60637 USA
[4]Laboratory for High Energy Astrophysics, Goddard Space Flight Center Greenbelt, MD 20771 USA
[5]Department of Physics, University of Utah Salt Lake City, UT 84112 USA

ABSTRACT

The POsitron Electron Magnet Spectrometer (POEMS) has been selected to enter definition phase study for flight on one of the Earth Observing System (Eos) polar platforms. Following launch, which could come as early as 1997, POEMS will measure the critical positron and electron components of the cosmic radiation and utilize this information to trace processes occurring within our geospace environment, in solar flares, in the solar wind, and elsewhere in the galaxy. In addition, POEMS will measure heavier charged particles to complement the electron/positron measurements. With POEMS data we will investigate the origin (primary or secondary) of galactic positrons, study the charge sign dependence of solar modulation over a large fraction of a solar magnetic cycle, measure positron and neutral particle emission from solar flares, and monitor the temporal variations of the charged particle intensities and energy spectra in the Eos orbit. Two orthogonal charged particle telescopes are used, each terminating in a shared bismuth germanate (BGO) detector array that also serves as a calorimeter for neutral particles, specifically neutrons and gamma-rays from solar flares.

INTRODUCTION At the present time the particle astrophysics community is concerned with development of the generation of spacecraft instruments that will carry research into the next century. One key element of this program will be Astromag, a large magnetic spectrometer that will take advantage of observing opportunities presented by the manned Space Station. However Astromag alone will not be able to answer all important physics questions. In a low inclination orbit, the manned Space Station will be at a high rigidity geomagnetic cutoff precluding study of low energy particles. Some important data can be obtained from balloon platforms, but these are

limited by weight capability, launch location, flight duration, and atmospheric background. Valuable data will also be provided by such missions as WIND, CRIE, ACE, SAMPEX, and HNC. A critical link between these other missions, measurement of low energy (5 MeV to 5 GeV) positrons, will be provided by the POEMS investigation, now entering Phase B definition study for flight on the first NASA/Eos polar platform, scheduled for launch in 1997. POEMS is expected to return continuous data for a period of 15 years after launch. The positron spectrum itself and its time variation are quite important scientifically. Combined with simultaneous measurement of electrons, protons, alpha particles and higher-Z elements systematic measurement of the positron component will provide a potent tool for understanding a number of the most important problems in cosmic, solar and heliospheric physics. In this report we briefly summarize the scientific objectives of POEMS and sketch the design and performance of the instrument.

PRIMARY POSITRONS A gamma-ray line at 0.511 MeV from $e^+ \leftrightarrow e^-$ annihilation has been observed from a variety of astrophysical sites – the galactic center region, the interstellar medium, gamma-ray bursters and solar flares. Production of positrons in some sources of the 0.511 MeV gamma radiation is calculated to exceed the production rate of secondary positrons in cosmic ray interactions by nearly two orders of magnitude (Ramaty and Lingenfelter, 1979). There are several possible astrophysical sources of cosmic ray positrons. First, positrons result from interactions of accelerated particles (Kozlovsky, Lingenfelter, and Ramaty, 1987). Cosmic ray interactions in the interstellar medium must be the source of some of the positrons observed at Earth (Protheroe, 1982), generally termed secondary positrons. High energy particles almost certainly exist in energetic compact objects such as Cyg X-3, Her X-1 or pulsars. Positrons produced in such sources and escaping into the interstellar medium would be primary particles with respect to the use of cosmic rays to understand astrophysical processed in the galaxy. Positrons also emerge from the decay of nucleosynthesis products (such as ^{56}Co) that power the light curves of supernova like the recent SN1987A. Radionuclides ^{56}Co, ^{44}Ti, ^{22}Na and ^{26}Al, produced in Type I and Type II supernova explosions, are expected to be important sources. Although the initial energy of such radioactive decay products is low when compared to the energy of typical cosmic rays such positrons form a unique sample of material injected at energies sufficient to enter directly into many proposed cosmic ray acceleration processes. Finally, positrons may originate from direct pair production in a high energy environment such as pulsars (Harding and Ramaty, 1987).

PARTICLE CONFINEMENT IN THE GALAXY Electrons and positrons are unique tracers of the confinement and transport of cosmic rays due to their low rest mass and electromagnetic interaction with the interstellar medium (Webber, 1983). Other widely used tracers of particle "history" involve secondary to primary ratios such as $\bar{p}/p$, ^{2}H/^{4}He, ^{3}He/^{4}He, B/C, F/Ne, sub-Fe/Fe $(21 \leq Z \leq 25)/(Z = 26)$, V/Fe or Sc/Fe, The numerator in these ratios is a secondary particle produced by nuclear interactions of primary nuclei such as those in the denominator. These ratios, therefore, measure the propagation history of different components in the cosmic radiation. Both positrons and electrons are produced as secondary particles in nuclear interactions (primarily by $\pi^\pm \rightarrow \mu^\pm \rightarrow e^\pm$ but for e^- it is established that there are also primary sources (Cummings, Stone and Vogt, 1973). POEMS data may confirm the secondary nature of all positrons, or they may define an energy range in which secondaries dominate. Stephens and Golden (1987) and Webber (1987) have summarized available data and have shown that a propagation model can be developed that predicts an anti-proton flux and an e^+ abundance consistent with low energy results. However, this new prediction falls well above the measured positron fraction around 1 GeV, so there is in fact currently no satisfactory model which explains all of the data.

SOLAR MODULATION Modulation, the variation on short timescales of the intensity of galactic cosmic radiation observed at earth, is caused by the flow of solar wind plasma containing trapped magnetic fields outward from the sun (Parker, 1965; Jokipii, 1971). A spherically symmetric model of solar modulation, the "Standard Model", has been developed (Urch and Gleeson, 1972; Fisk, 1979) which explains the gross features of the modulation process in an equilibrium situation that neglects off-diagonal terms in the diffusion tensor, particularly particle drifts due to the gradient and curvature of the magnetic field. Predictions of this model are independent of the particle charge sign. The algebraic sign of particle charge may affect modulation by either or both of two presently known mechanisms: gradient and curvature drift (Jokipii and Levy, 1977) and net magnetic helicity of scattering turbulence (Bieber, Evenson, and Matthaeus, 1987). Both processes are expected to have manifestations related to the reversal of solar magnetic polarity at eleven year intervals. For both particle drifts and helical scattering, reversal of the solar magnetic polarity has the same effect as reversing the charge sign of the particles. Proper study of charge sign dependence of modulation requires comparison of fluxes of oppositely charged particles having the same mass and energy so that velocity and rigidity are identical. Detailed spectra of e^+ and e^- (over long time periods and thus as a function of level of modulation) will allow this analysis to be made. Combined with simultaneous proton and helium measurements from POEMS and other instruments, POEMS electron and positron data will permit the modulation process to be analyzed in detail and the role of particle drifts and/or helical scattering to be determined quantitatively.

SOLAR FLARES Solar flares are an important source of information on the dynamics of the solar atmosphere as well as on the particle acceleration process, which may have important similarities to energetic particle acceleration in various plasma/field environments elsewhere in the galaxy. By correlating observations of solar energetic particles (SEP's) with neutron and gamma-ray measurements, the time dependence of particle acceleration and release can be studied. The large BGO shower counter in POEMS provides a sensitivity to gammas and neutrons comparable to the Solar Maximum Mission (SMM) gamma ray spectrometer at high energy and about 1/3 the sensitivity of SMM at low energy. POEMS will provide simultaneous solar charged particle and gamma ray observations over almost two solar cycles.

Flares producing gamma rays tend to be the electron rich. Extending this type of correlation analysis to a larger sample of flares is one of the goals of the POEMS experiment. More interesting, however, is the question of whether such electron rich gamma-ray flares have a detectable positron fraction. Production of neutrons and gammas in these flares shows that SEP's interact in the solar atmosphere. Among the products of such interactions are radioactive nuclei (*e.g.* ^{11}C) which decay by e^+ emission (typically producing 0.1-1 MeV positrons). These e^+ can be accelerated and emitted into interplanetary space. Low energy positrons (0.1-0.5 MeV) may have been seen from the impulsive phase of the 3 June 82 flare (Kirsch, Keppler and Richter, 1985) but instruments to study higher energy e^+ have not been available. If POEMS can detect a significant positron fraction in flares, it would set new constraints on acceleration time scales, the mixing of material in the "flare regions" and the possible topology of the magnetic field lines.

GEOSPACE STUDIES Data obtained by POEMS can be interpreted to yield valuable information on the Geospace environment. The main POEMS spectrometer will provide a long duration monitor of the geomagnetic trapped electron population at energies above 5 MeV during a wide range of solar conditions, providing information on the time constants for electron enhancements following magnetic storms, as well as resultant energy spectra and pitch angle distributions. POEMS has the unique ability to determine whether there is a significant positron component in the geomagnetic trapped particle population and, if there is, to measure its characteristics over a substantial period of time. Specifically, POEMS will look for evidence that positrons emitted from nuclear reactors are accelerated so as to become a separate belt population. Additionally, the particles observed by POEMS will be used in various ways to explore the geomagnetic field and to study its variability on many timescales.

ATMOSPHERIC PHYSICS AND CHEMISTRY Particle spectra and intensities measured by POEMS can be converted into a complete, daily, three dimensional, global map of the input of ionizing radiation into the atmosphere at altitudes below 25 kilometers. This will be directly useful in monitoring hazards to aircraft operation. Perhaps more important, very little is presently known about the influence of ionizing radiation on physical processes in the atmosphere (such as cloud formation) and on the chemistry of trace components (such as oxides of nitrogen and ozone). The long term, stable dataset provided by POEMS will be important to both real time and retrospective studies of these effects.

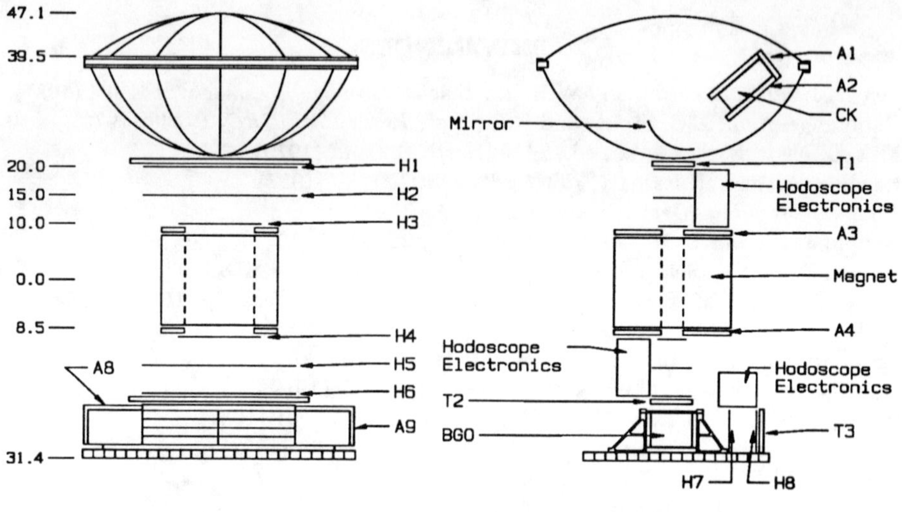

Figure 1: *Two schematic views of the POEMS sensor. Reference dimensions are given in centimeters, in a coordinate system centered on the permanent magnet.*

INSTRUMENT DESIGN SUMMARY Cosmic positrons have eluded precise observation primarily for operational rather than strictly technological reasons. It is in this respect that the Eos mission uniquely provides a perfect opportunity to conduct this investigation. Relatively massive detectors are required to isolate the small fluxes of positrons and electrons from the large flux of cosmic protons and to distinguish the positrons from the electrons. Fundamental physical principles dictate the use of a shower calorimeter and possibly a gas Cherenkov detector to separate the electrons and positrons from the protons, and bending in a magnetic field to separate the electrons from the positrons. Approximately half the 125 kg proposed mass of POEMS is devoted to these components. It has to date been impractical to include such a massive detector on typical interplanetary spacecraft, however it is of modest size compared to the available Eos polar platform payload. Although this mass is also modest for a balloon payload, atmospheric secondary positrons prevent truly definitive positron measurements from balloons below approximately 1 GeV.

POEMS, shown schematically in Figure 1, retains four basic design elements of earlier investigations (permanent magnet, trajectory measuring system, gas Cherenkov counter, and shower counter) however each is designed for improved performance. The POEMS magnet is an advanced design using rare-earth materials with virtually no stray field. The trajectory system will be constructed from solid state silicon strip detectors. These detectors will offer excellent position resolution and yet require no consumables. The sealed gas Cherenkov counter will use a non-scintillating gas (ethylene) and photomultipliers which are shielded against accidental triggers using plastic scintillators in anticoincidence. The shower counter will be composed of four layers of Bismuth Germanate (BGO) scintillator, each approximately one radiation length thick. Comparing shower starts in the different layers will allow direct, in flight determinations of the effective interaction cross sections needed to make background corrections.

REFERENCES

Bieber, J.W., P. Evenson, and W.H. Matthaeus, *Geophys. Res. Lett.,* **14**, 864 (1987).
Cummings, A.C., E.C. Stone, and R.E. Vogt, *Proc. 13th ICRC,* **1**, 340 (1973).
Fisk, L. A., in Parker *et al.,* ed., p179, North Holland (1979).
Harding, A., and R. Ramaty, *Proc. 20th ICRC,* **2**, 92 (1987).
Jokipii, J. R., *Rev. Geophys. Space Phys.,* **9**, 27 (1971).
Jokipii, J. R., and E. H. Levy, *Astrophys. J.,* **213**, L85 (1977).
Kirsch, E., E. Keppler and K. Richter, *Proc. 19th ICRC,* **4**, 158 (1985).
Kozlovsky, B., R. E. Lingenfelter, and R. Ramaty, *Astrophys. J.,* **316**, 801 (1987).
Parker, E. N., *Planet. Space Sci.,* **13**, 9 (1965).
Protheroe, R. J., *Astrophys. J.,* **254**, 391 (1982).
Ramaty, R. and R. E. Lingenfelter, *Nature,* **278**, 127 (1979).
Stephens, S. A., and R. L. Golden, *Space Sci. Rev.,* **46**, 31 (1987).
Urch, I. H. and L. J. Gleeson, *Astro. Space Sci.,* **17**, 426 (1972).
Webber, W. R., in Shapiro, ed., p83, Reidel (1983).
Webber, W. R., *Proc. 20th ICRC,* **8**, 65 (1987).

HEAVY NUCLEUS COLLECTOR FOR SPACE STATION

P. B. Price
Physics Department, University of California, Berkeley, CA 94720

ABSTRACT

Two opportunities exist in the next decade for measuring the charge distribution of ultraheavy cosmic rays with unprecedented statistics and elemental resolution. Groups at Berkeley, University of Michigan, and University of Utah will expose an array of track-recording phosphate glass plates ~16 m^2 in area for five years at an orbital inclination of 28.5^o on Space Station Freedom. In addition, negotiations are underway between Berkeley and Soviet colleagues to expose 1 to 3 m^2 of glass for 2 years at an inclination of 51.6^o on Space Station Mir. The elemental resolution for the two sets of experiments is expected to be ~0.25 charge unit. If negotiations are successful, the Mir experiment will be launched in 1991 and will collect about six times as many events with $Z > 50$ as were studied on HEAO-3 and about one-fifth as many events as will be collected on Freedom.

INTRODUCTION

HEAO-3[1] and Ariel-6[2] experiments have provided interesting data on the abundances of ultraheavy cosmic rays with $Z > 30$. Each instrument collected between 300 and 400 events with $Z > 50$. On HEAO-3 the charge resolution was adequate to detect even-Z peaks up to $Z = 60$. One possible actinide event was detected on HEAO-3 and perhaps two or three on Ariel-6. The next study of ultraheavy cosmic rays must obtain much higher statistics and must attain charge resolution sufficient to resolve both odd-Z and even-Z nuclei. In its original conception, the Heavy Nucleus Collector (HNC) would have used large arrays of track-recording plastic films on a reflight of the Long-Duration Exposure Facility (LDEF), with an event thermometer by means of which the temperature-dependence of the response of the plastic films could be corrected. The Challenger accident dashed hopes for this version of HNC. Since the accident, we have developed a new class of detectors made of phosphate glass, which have several advantages over the plastic films.[3] The version of HNC that will fly as an attached payload on Freedom will not require an event thermometer and will require the analysis of only 16 layers of glass, in contrast to the ~120 layers of plastic for the old HNC. The reduction in the number of dE/dx detectors is a consequence of the much higher charge resolution of glass than of plastic.

ASTROPHYSICAL GOALS AND EXPECTED NUMBER OF EVENTS

With the ability to measure both odd-Z and even-Z elements with good statistics, a number of astrophysical goals can be achieved. First and foremost is the goal of determining the source of the ultraheavy cosmic rays. Neutron capture is the dominant process of nucleosynthesis for the elements beyond the iron peak. The analyses of HEAO-3 and Ariel-6 data showed that the abundances of the even-Z elements 32 to 60 are consistent with a cosmic ray source having a composition similar to that of the solar system, but subject to source fractionation correlated with the first ionization potential of each element. For $Z > 60$ statistics and resolution were limited, such that elements had to be analyzed by groups. The conclusion was that the abundances of both the platinum group, $74 \leq Z \leq 80$, and of the largely secondary elements $62 \leq Z \leq 73$, are enhanced relative to those in a propagated solar system source. These results suggest that there is an enhancement of the r-process contribution to the source nuclei in the charge region $Z > 60$. Over the entire region of charge with $Z > 32$, standard leaky box models of propagation account satisfactorily for secondary production. Statistics and resolution were inadequate to draw conclusions about the actinide elements, $Z \geq 90$. Thus, although much has been learned from the HEAO-3 and Ariel-6 experiments, details of the nature of the source of ultraheavy cosmic rays are still obscure.

If the goal of collection of at least 40 actinide events can be achieved, it should be possible to estimate the age of the ultraheavy cosmic rays since they were synthesized. This is done by comparing the relative numbers of Th, U, and transuranic nuclides, which span halflives from $\sim 10^6$ to $\sim 10^{10}$ years. For example, if the cosmic rays were simply swept-up interstellar matter uncontaminated by supernova debris since early in the history of the Galaxy, the Th/U ratio would be ~4, as is the case for the solar system material today. If the cosmic rays were made up of equal parts of old material and supernova debris added within the last $\sim 10^7$ years, the Th/U ratio would be <0.4, and a number of transuranic cosmic ray nuclei would be detected.

In addition to the astrophysical goals, one can use HNC to search for admittedly improbable particles such as long-lived superheavy elements (Z ~ 110), magnetic monopoles, and quark nuggets.

The table below compares the number of nuclei in various charge intervals seen in the HEAO-3 experiment with the number expected for a one-year Freedom exposure and a two-year Mir exposure. If the decision is made to leave the HNC detectors on the

Table 1. Expected Number of Ultraheavy Cosmic Rays

Z	seen on HEAO-3	estimated, 28.5° 16 m^2 for 1 yr (NASA Freedom)	est., 51.6° 3.2 m^2, 2 yr (Soviet Mir)
49-50	50	126	95
51-52	50	140	119
53-54	34	147	78
55-56	54	319	171
57-58	34	142	62
59-60	14	113	56
62-69	48	450	414
70-73	15	180	259
74-80	36	570	734
81-83	8	160	218
>88	1?	15	18
Total no.	340	2362	2224

[HEAO-3 + Ariel-6 data, and dependence of aperture on ionization rate, were used to predict fluxes at solar min.; at solar max., fluxes will be lower.]

Freedom for a full five years, the numbers in column 3 can be multiplied by five. Thus, in round numbers, a Mir experiment with 6 m^2 yr would improve the statistics by a factor ~6 over the HEAO-3 data, and a Freedom experiment with 80 m^2 yr would lead to a factor ~35 improvement. Either of these possibilities would lead to good statistics in the collection of odd-Z and actinide elements, with resolution sufficient to provide firm charge identification.

PHOSPHATE GLASS DETECTION SYSTEM

Figure 1, from ref. 4, shows the unprecedented charge resolution exhibited by phosphate glass. The main peak is that of a beam of gold (Z = 79) nuclei of energy 800 MeV/N produced at the LBL Bevalac. The other peaks are due to projectile fragmentation of gold. The charge standard deviation is ~0.06 e. An extraordinary feature of these detectors is that instantaneous charge states are recorded at each surface of a stack, and pickup and stripping of single electrons are unambiguously seen.

Ref. 4 gives the composition and properties of BP-1, the phosphate glass that we developed and that will be used on the HNC. Aside from its remarkable charge resolution, two noteworthy features are that its response is independent of the ambient atmosphere and far less sensitive to ambient temperature than is the response of the plastic detectors originally planned for HNC.

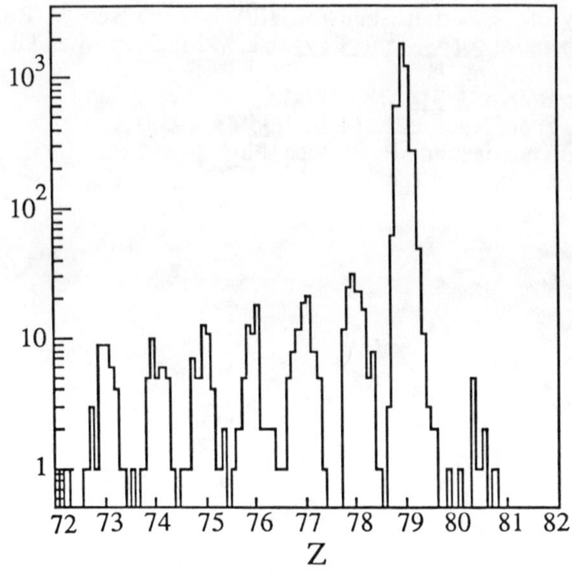

Figure 1. Charge distribution of fragments produced in fragmentation of 800 MeV/N Au projectiles at LBL Bevalac. Standard deviation is 0.06 e.

TWO POSSIBLE SPACE STATION EXPERIMENTS

At this writing, NASA has given its provisional approval of HNC for early deployment as an attached payload on Freedom. The glass plates will be processed and analyzed with automated scanning and measuring systems at the Universities of California (Berkeley) and Michigan. Charge, Z, and velocity, βc, are determined from the value of the ratio Z/β inferred from the ratio of the track etch rate to the general etch rate, v_T/v_G, at the point of entrance of the particle into the stack, and from the rate of change of Z/β as the particle slows. G. Tarle' will be in charge of the Michigan effort, and M. H. Salamon (University of Utah) will participate in data analysis. The requirement of a space exposure of several years will put the beginning of data analysis several years into the next millenium.

Soviet physicists at the Soviet Space Research Institute and at Moscow State University are interested in collaborating with the Berkeley group (together with D. O'Sullivan of the Dublin Institute of Advanced Studies) in a scaled-down version of HNC that could be deployed by Cosmonauts on the outside wall of the Mir, which will remain in space until toward the end of 1993. Details and a definite commitment have yet to be worked out, but the prospect of having the first really high-resolution data on ultraheavy cosmic rays a decade before the Freedom experiment is complete is very attractive.

REFERENCES

1. For a summary of the current status of HEAO-3 data, see W. Robert Binns, in Genesis and Propagation of Cosmic Rays, ed. M.M. Shapiro and J. P. Wefel (D. Reidel, New York, 1988), p. 71.
2. P. H. Fowler et al., Ap. J. 314, 739 (1987).
3. Shicheng Wang et al., Nucl. Instr. Meth. B35 43 (1988).
4. G. Gerbier, Ren Guoxiao, and P. B. Price, Phys. Rev. Lett. 60, 2258 (1988).

THE ASTROMAG FACILITY

George F. Smoot*
Lawrence Berkeley Laboratory and Space Sciences Laboratory
University of California, Berkeley CA 94720

ABSTRACT

The USA and Italy are planning to place a large superconducting magnet facility, Astromag, on the Space Station Freedom in the late 1990's. Astromag has been designed as primarily a cosmic ray astrophysics facility, but it could also be used for any investigation which can make use of a large volume, high intensity magnetic field in a microgravity or space environment. This paper describes the facility and mentions the first set of astrophysical problems to be investigated.

INTRODUCTION

In 1985 NASA formed the Superconducting Magnet Facility Definition Study Team to explore the possibility of accomplishing a broad range of particle astrophysics objectives with a superconducting magnet facility. The team named the facility Astromag and issued an Interim Report in August 1986 and a final report in May 1988. Those reports detailed the scientific rationale for the facility, outlined the approach to be followed, and emphasized the Space Station Freedom (SSF) as the preferred space platform for the observations. From the start scientists representing a number of countries have been involved in the study activities. In particular, the Agenzia Spaziale Italiana (ASI) has suggested that Astromag become a bilateral project, and National Laboratory for High Energy Physics (KEK) in Japan has expressed interest in contributing to the design and development of the superconducting coils.

In mid-1988, NASA issued an Announcement of Opportunity (AO) for Attached Payloads to conduct scientific investigations during the assembly phase of the Space Station Freedom. The May 1988 report convinced NASA to identify Astromag as one of two facilities which could be accommodated during this phase of the Freedom's development. Potential investigators were instructed to assume that a superconducting magnet facility would be available and to propose how it might be used to accomplish their science objectives. Since the Freedom interfaces were not established, only the critical interfaces between Astromag and the possible experiments were specified in the AO. In July 1989 NASA selected three experiments for first phase of Astromag. Isotope studies will be made by LISA[1], A Large Isotope Spectrometer for Astromag, for which J. F. Ormes (Goddard Space Flight Center) is the Principal Investigator; WiZard[2], a collaboration headed by R. L. Golden (New Mexico State University) and P. Picozza (U. di Roma), will search for antimatter in the cosmic radiations; and the experiment SCINATT[3], will study Spectra, Composition, and Interactions of Nuclei Above 10 TeV, for which T. A. Parnell (Marshall Space Flight Center) is the Principal Investigator with T. Ogata (University of Tokyo). Meanwhile Goddard Space Flight Center (GSFC), designated as the development center for Astromag, has made considerable progress towards

---

* Representing the Astromag Facility Definition Team

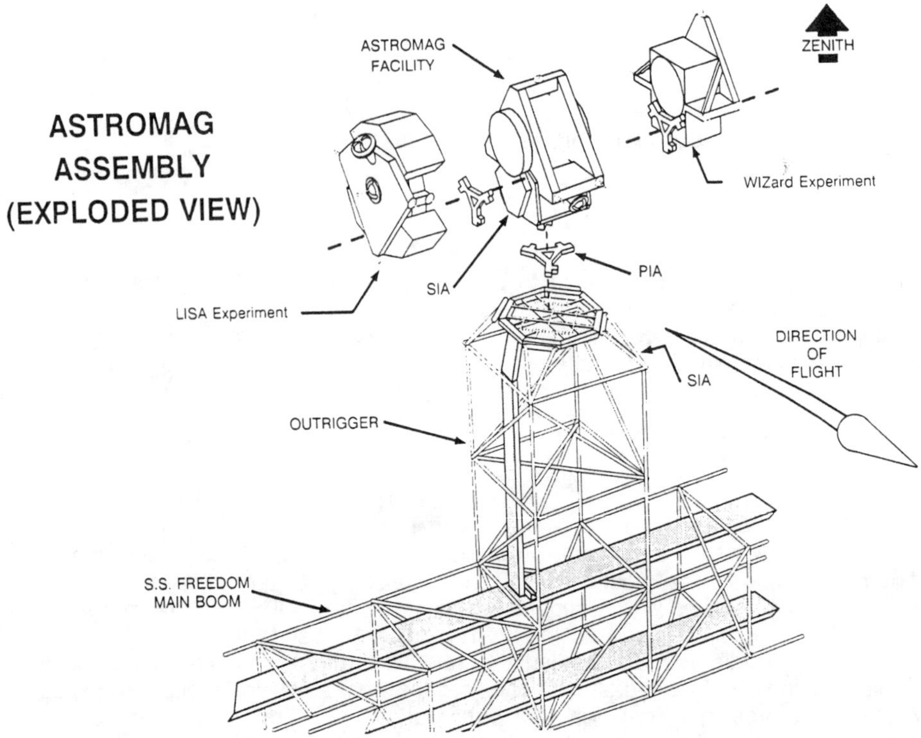

Figure 1. One conceptual drawing of Astromag as it will be attached to the Space Station Freedom.

---

definition of the facility and the accommodations on the Space Station Freedom.[4] Figure 1 illustrates a concept as to how Astromag could be adapted to SSF.

SCIENTIFIC OBJECTIVES

Astromag's primary scientific goal is the study of cosmic rays, which are a suprathermal gas of energetic charged particles, still of unknown origin, that pervade the galaxy and perhaps the space beyond. Cosmic rays provide a unique sample of matter from far outside our solar system, and are important for the dynamics and evolution of the galaxy, and the interpretation of measurements in the electromagnetic spectrum. The studies of cosmic ray origin, acceleration, and propagation are certain to yield important new understandings of the ubiquitous phenomenon of relativistic plasmas in space. In addition this sample of matter will be searched with unprecendented sensitivity (by 3 orders of magnitude) for antimatter and other exotic particles. While the results of such a search are highly speculative, they should be carried as far as is practical. If antimatter were to be found in the cosmic rays, our ideas about fundamental physical processes in the Universe would be revolutionized.

Cosmic rays have been studied from spacecraft for many years, but earlier detectors have been severely limited in size, which restricts the energy range of their investigations. Also, there are many interesting questions which are beyond the capability of current spacecraft instruments, despite their long exposure times. Larger instruments have been flown on balloons and have given us much information on higher energy cosmic rays. However, the short duration of balloon flights has restricted the investigation of the rarer particles. Clearly the next step is to obtain longer exposure times for larger cosmic ray experiments.

The proposed Astromag facility is very well suited to take this next step. Astromag will permit a series of complementary long-duration, large cosmic ray experiments to be flown, each tuned to explore different aspects of high-energy particle astrophysics. These experiments will be capable of measuring the composition and spectra of energetic charged particles with unprecedented accuracy and will address a wide range of fundamental questions.

Additional scientific goals have been identified in a number of related fields which can utilize the large volume of high magnetic field provided by Astromag. These include investigation of fundamental plasma processes in the environs of Astromag and investigations of celestial gamma rays. A brief overview of the proposed Astromag facility is presented below.

THE MAGNET

The requirements for the design of the Astromag facility were based on the "Strawman payload" described by the Astromag Study Team in its Final Report issued in May 1988. The Study Team consisted of cosmic ray and plasma physicists, as well as engineers familiar with magnets, superfluid helium and space flight dewar technologies needed to accomplish the objectives. This team defined a baseline core facility consisting of the coils and dewar.

In order to accomplish the cosmic ray objectives previously discussed, a mass spectrometer is required. The first such device was invented by Francis W. Aston, a student of J. J. Thompson. Aston was awarded the Nobel Prize for chemistry in 1922 for the identification of isotopes with his device. For galactic cosmic ray studies the spectrometer must be capable of determining the momentum per unit charge (rigidity) and sign of the charge of highly relativistic particles. The approach selected is to have a long-lived facility on the Space Station Freedom consisting of a set of coils which produce an intense magnetic field. The field map is shown in Figure 2. The coils are wound using niobium-titanium wire which, when cooled by liquid helium, becomes superconducting (i.e., the electrical resistance becomes zero) so that the current and the associated magnetic field persists indefinitely. A spectrometer requires instrumentation which tracks the particles through the magnetic field measuring their bending. The radius of curvature for the bent tracks is inversely proportional to their rigidity and the strength of the field. Tracking detectors required for the spectrometers will be provided by experimenters.

The high field regions and the lower far field surrounding Astromag can also accommodate instrumentation to conduct other kinds of experiments. For example, surrounding Astromag is a containerless plasma which can be probed to study a variety of phenomena under conditions different from those in ground-based laboratories and different from those so far encountered in space. Plasma experiments which might require different magnetic field intensities could be done in a campaign mode during the three or four anticipated annual charge/discharge cycles.

Astromag will pave the way for future applications in space which require a combination of low gravity and/or high vacuum and high magnetic fields. Examples which come to mind are magnetic resonance imaging experiments on the physiological effects of low gravity on living systems, crystal growth applications, and possibly even variable thrust plasma driven rockets, inductive energy storage and multi-megawatt power generation.

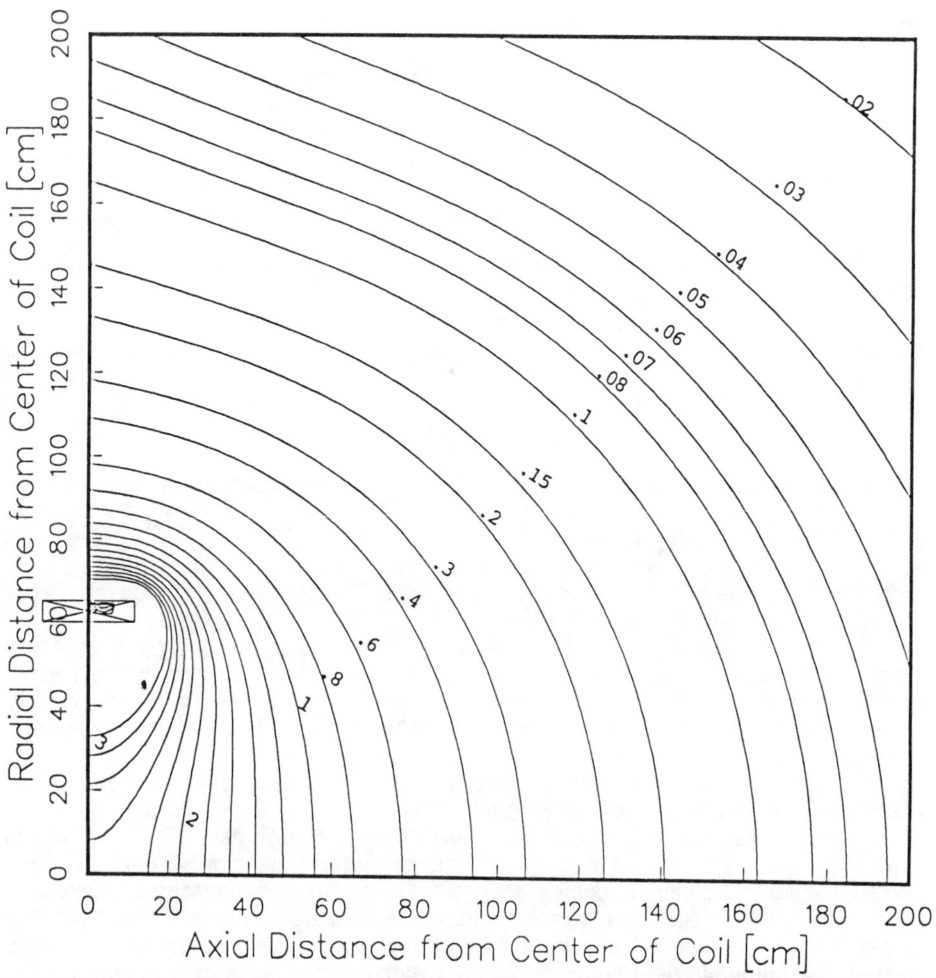

Figure 2. Magnetic field intensity from double coil configuration. This view represents one quadrant of the full field, which is symmetric about the central coil axes and the vertical axis between the two coils. The intensity indicated is in Tesla.

The proposed concept is to have the core facility located on the Space Station Freedom. The program baseline calls for the facility to be capable of 2 to 4 years of operation without resupply of liquid helium; after which time the facility would be returned to earth for refurnishing, resupply, and required modifications. Experiment apparatus would be brought up to the Freedom and attached to Astromag for a period of time sufficient to undertake the required observations and then be replaced by alternate instrumentation for different investigations. The core facility would be analogous to the mirrors in a telescope facility, and the experiments would be exchangeable like the focal plane instrumentation. Access to the Space Station Mobile Servicer and Flight Telerobotic Servicer (FTS) as well as the STS Remote Manipulator System would provide the ability to assemble and service the facility and experiments on orbit. These facilities make the Space Station Freedom the ideal spacecraft for the Astromag investigations.

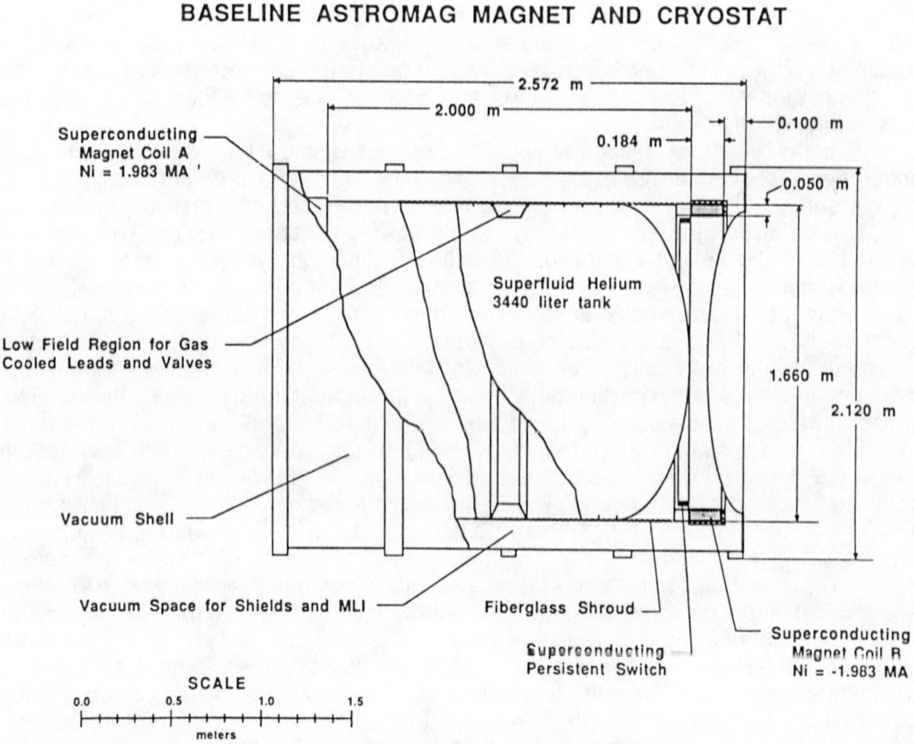

Figure 3. Schematic view of superconducting coils and dewar.

Coils and Dewar

At the heart of the Astromag facility is the superconducting magnet and its supporting cryogenic system. This system is comprised of two superconducting coils, a persistent switch, gas-cooled detachable current leads, and a superfluid helium circulation system, all housed inside a superfluid helium dewar as shown in Figure 3.

The baseline superconducting coils are envisioned to be made of copper or aluminum-stabilized niobium-titanium (NbTi) superconducting wire impregnated with epoxy for structural rigidity and wound on an aluminum bobbin. They would have a 1.7 meter outer diameter and be separated by approximately 2 meters. The coil cross section would be a rectangle 184 mm wide and 50 mm along the coil axis.

Each coil would have 2448 turns and carry a current of about 810 amps. At full current the stored magnetic energy in both coils would be approximately 11 megajoules. The peak field within the coils would be less than 7.0 tesla. The field decreases rapidly outside the facility, falling to approximately the Earth's magnetic field 15 meters from the center of the magnet system.

The coils would be identical but have opposite polarity so that the dipole moments cancel. The net dipole moment is determined by the manufacturing differences between the coils, minimizing the torque induced on the Space Station Freedom when the magnet is on. In addition, the coils would be connected in series to ensure that the current in each coil would be the same at all times.

Since the coils are superconducting, the circulating current is persistent (decays less than 1% per year) so that the magnetic field is maintained without continuous application of power. The coils can be charged or discharged by attaching the gas-cooled electrical current leads and opening the coil series circuit using a persistent switch.

In the event of a spontaneous or induced quench (in which the coil wire undergoes a conversion from the superconducting state to a normal resistive state), the current would decay in a two-second time period because of resistive losses and the 11 megajoules of magnetic energy would go into heating the coils. The coils are isolated from the helium except for the helium circulation system. If they were not isolated, that energy would vaporize all the liquid helium in the dewar. The connection of the two coils in series is particularly important during a quench, because it will keep the net dipole moment equal to zero as the field decays simultaneously in both coils. The coils are designed to quench together so that the stored magnetic energy is distributed evenly throughout and between them. The temperature of the isolated coils would rise to about 120 K and they would have to be re-cooled to operating temperature by helium gas flowing through the helium circulation system. Using this scheme, it is estimated that the system can recover from a quench in 24 to 48 hours. The dewar and magnet can survive a quench at any fill level. It should be noted that a spontaneous quench is an extremely unlikely event.

The superfluid helium dewar consists of: a cryogen tank capable of holding about 3500 liters of liquid helium; vapor-cooled shields and multilayer insulation to intercept conductive and radiative heat loads; a vacuum shell; and associated plumbing, valves, and instrumentation. The pressure vessel would be approximately 2.6 meters long and 2.1 meters in diameter. The cryogen tank vapor-cooled shields would be supported from the pressure vessel by a low heat leak composite support structure. In turn, the magnet coils would be connected to the cryogen tank by a fiberglass epoxy shroud. The shroud and cryogen tank must be capable of withstanding launch loads as well as the intercoil magnetic force, which could be as large as 25 metric tons in tension at the full design current.

The magnet and dewar will be designed for a number of charges and discharges over the lifetime of the magnet system so the magnet can be turned off as required by other Space Station Freedom operations.

In order to provide liquid helium (LHe) replenishment every 2 or 4 years, two approaches are possible. One is to return the empty dewar to Earth for refilling, the second is on-station resupply via a LHe tanker. While the baseline calls for ground resupply, on-orbit resupply will be studied during Phase B.

ASTROMAG SAFETY ISSUES

The safety issues raised by Astromag in conjunction with its launch on the STS and its operation on the Space Station Freedom do not appear to present any intractable problems. The most serious hazards are those associated with the presence of a large dewar filled with liquid helium, the large magnetic field and the high rate of change of field in the event of a transition of the magnet from the superconducting to the normal state.

Whenever Astromag is within the atmosphere, either on the ground or during the launch phase, loss of the vacuum integrity could result in a high heat load to, and possible rupture of, the cryogen tank. Both the cryogen tank and the main shell include pressure relief devices to relieve pressure under worst case conditions.

Astromag will be designed to ensure that the thermal and pressure effects of the resulting high effluent on the payload bay do not pose a hazard to the shuttle.

An over-pressure in the cryogen tank on orbit could result from inadequate venting capacity, incorrect valve operation, or freezing air in the vent line. Care must also be taken to design the dewar to withstand the forces and pressure rise induced by eddy current heating in the event of a quench.

There are also safety concerns related to the two large superconducting coils and their associated magnetic fields. The large intra- and intercoil forces which could induce structural failure will be analyzed and compared to those due to launch, flight, and landing environments. The worst case loads will be used in the structural design. Structural loads induced by a quench must also be considered. A rapidly collapsing magnetic field will induce eddy currents in the dewar and coil support structure. Preliminary calculations show that the pressure rise in the cryogen tank due to eddy current heating would be small. However, the structural loads due to the interaction between the eddy currents and the magnetic field are large and must be accounted for in the dewar and magnet design.

In the event of a quench, the collapsing magnetic field can induce voltages on nearby structure and instrumentation, but they fall rapidly with distance. The distance from the main boom and relatively long quench time provide general protection. A few meters from Astromag the quench-induced voltages expected for a typical circuit are in the nanovolt to microvolt range. The forces and eddy currents induced in structural elements are similarly small.

Limits have been established by magnetic resonance imaging facilities and by national laboratories for both AC and DC magnetic field levels where people live and work. We anticipate that the Space Station will make a determination and establish limits for astronauts. These limits will be used to establish an EVA avoidance zone around Astromag which we estimate to be between 5 and 8 meters from Astromag's exterior surface. The Astromag field will be shut off for astronaut activity inside this region.

Experiments using the Astromag Facility will also be designed to meet safety standards and to be protected against B and dB/dt effects.

FIRST PHASE EXPERIMENTS

As mentioned in the introduction, science investigations for Astromag have been selected by NASA Headquarters. The first three instruments that will fly with Astromag are WiZard, which will measure spectra of electrons, positrons, antiprotons and other low Z cosmic ray components and carry out an extensive search for antimatter helium; LISA, which will measure isotopic composition of the cosmic rays, search for heavier antinuclei and carry out detailed spectral observations of high Z elements, $3 < Z < 31$; and SCINATT, an emulsion experiment to measure composition and spectra above 10^{14} eV and search for evidence of quark-gluon plasma phases of matter.

WiZard consists of 32 planes of tracking detectors (either MWPC's or drift chambers) 1.5 meters in height, a 10 layer carbon fiber/MWPC Transition Radiation Detector (TRD), and an electromagnetic calorimeter. The instrument is designed to measure rigidities up to 1 TeV with a geometry factor between 0.1 - 0.4 m^2sr. WiZard will have a sensitivity for antiprotons of better than 1 part in 10^8 extending in energy up to 300 GeV. Antihelium, the most likely form of primordial antimatter, will be detectable to 1 part in 10^8. While these will be the main goals of WiZard, it will also be able to study relativistic solar flare particles, relativistic particle anisotropies, and even perform as an all sky gamma ray survey telescope, with energy resolution of 1% between 1 GeV and 1 TeV. A more complete description of this experiment can be found in Dr. R. Golden's paper on WiZard.

The second instrument on Astromag will be a Large Isotope Spectrometer for Astromag (LISA). This instrument will have a 0.05 m^2sr geometry factor and an isotope resolution of 0.3 amu for $3 < Z < 29$. LISA uses 5 scintillating optical fiber arrays for trajectory information (these planes are 50 cm in width and vary in length from 92 cm to 172 cm.) These 100 micron diameter fibers are viewed by 34 CID cameras and provide position information to within 70 microns. Two layers of silica aerogel with indices of refraction of 1.03 and 1.04 provide velocity measurements of the cosmic ray nuclei. With LISA elemental cosmic ray abundances will be measured from 2 to 500 GeV/nucleon, and isotope identification will be made from 2.5 to 4 GeV/nucleon. More details of the LISA experiment are in Dr. R. Binn's paper.

The SCINATT experiment consists of 4 stacks of passive detectors (emulsions, CR 39 plastics, and X-ray films) to be flown for 3 months. Two of these stacks are optimized for measuring the sign and transverse momenta of the secondary particles from an interaction vertex. These stacks are 52 cm in height and weigh 176 kg. The other stacks are more traditional emulsion chambers, 14 cm high and weighing 165 kg. The total exposure factor (assuming 3 months on orbit) is 3.76 x 107 m^2sr sec. This experiment will address the energy regime above a TeV in energy, which neither of the other experiments is capable of doing. Some of the most interesting things to be studied are high energy density events where a phase transition to a quark gluon plasma may be seen. With the given exposure factor, more than 500 events with energy densities exceeding 2 GeV/fm^3 should be seen, making this search extremely interesting. More about the SCIN-MAGIC experiment can be found in Dr. T. Parnell's paper.

PLASMA DRAG

The Astromag field interacts with the ambient plasma in a complex manner, but there is no significant drag on the Space Station. Assuming an ionospheric neutral density of 10^8 atoms/cm^3, the largest drag would be caused by neutral atmospheric atoms and molecules incident on a perfectly reflecting (5m X 5m) solid object. This is a non-plasma interaction and the drag force is ~0.007 N.

The only plasma effect which is large enough to consider is the charge exchange of ionospheric neutrals on an enhanced plasma build-up on closed field lines (which can be controlled by collecting the resulting ions on a conducting surface near the facility). Nevertheless, for a worst case where the trapped plasma density is limited by interchange instability the drag is only 0.003 N. All other plasma effects (Alfven wings, etc.) are an order of magnitude smaller. Thus for the configuration proposed, on a 10 meter outrigger, the total neutral plus plasma drag torque would be < 0.10 N-m (in pitch).

CONCLUSIONS

The past few decades have seen great advances in all aspects of physics and astronomy, spanning the very small, the very large, the periodic and the chaotic. Mounted upon the Space Station Freedom, Astromag and its detector complements will provide a new platform on which we can stand and survey the universe. The technical challenges to be overcome in getting there are formidable. We have tried to demonstrate in this brief paper that those challenges are not insurmountable and that the knowledge to be gained from Astromag will be well worth the investment.

ACKNOWLEDGEMENTS:

The original Astromag Facility Definition Team who's report led to the Astromag Facility being in the SSF AO were: T. Bowen, R. Golden, W. Hibbard, M. Israel, W. V. Jones, R. Mewaldt, D. Muller, J. Ormes, G. Smoot, and M. Weidenbeck. The foreign participants were: G. Auriemma, G. Basini, P. Carlson, P. Goret, J. Nishmura, I. Rasmussen, M. Simon, P. Spillantini, S. A. Stephens, A. Wolfendale, A. Yamamoto. The subcommittee members included: J. Adams, W. R. Binns, S. Castles, R. Dixon, P. Evenson, M. Green, R. Kephart, P. Meyer, R. Nieman, J. O'Connor, D. Reames, R. Streitmatter, J. Sullivan, and J. Wefel. Many others contributed.

Recognition is due to the Astromag Technical Advisory Team and associated members for the work done on the Astromag design. The members of the team were G. Basini, M. Green, R. Ibba, P. Marston, J. Ormes, J. Phenix, G. Smoot, P. Spillantini, J. Sullivan, A. Yamamoto. Many others participated in the work including G. Anikis, S. Breon, R. Boyle and S. Levin. Louis Barbier and Eric Christian were heavily involved in this condensation of that work.

REFERENCES

1. Binns et al., this volume.
2. Golden et al., this volume.
3. Parnell et al., this volume.
4. Phenix et al., Astromag Phase-A Study Final Report, 1989.

WIZARD - AN EXPERIMENT TO MEASURE THE COSMIC RAYS INCLUDING ANTI-PROTONS, POSITRONS, NUCLEI AND TO CONDUCT A SEARCH FOR PRIMORDIAL ANTIMATTER

WiZard Science Team
presented by R. L. Golden, Particle Astrophysics Lab., New Mexico State University, NM 88003

ABSTRACT

The WiZard experiment will utilize the Astromag magnet facility onboard space station Freedom to explore the composition and energy spectra of low-Z cosmic rays. Particular emphasis will be placed on a search for primordial antimatter and measurement of antiproton and positron fluxes at energies up to 400 GeV. In this paper we present the scientific goals and rationale; we describe the experimental method and summarize the present status of the WiZard project.

INTRODUCTION

The WiZard science team is made up of members with extensive experience in related balloon and accelerator technologies. A list of member institutions is provided in Table 1. In this paper we will review each of the primary objectives, the current state of observations and the expected results from the WiZard experiment.

OBJECTIVES

Search for Antimatter: Detection of antimatter of primary origin in cosmic rays would be a discovery of fundamental significance. Cosmic ray searches that have been made so far have yielded only upper limits of $1:10^4$ for heavy nuclei ($Z>2$) and several parts in 10^5 for helium[1,2]. In looking for primordial antimatter in cosmic rays, one hopes to find particles from distant regions of the universe that have been accelerated to high energies and have traveled to our solar system. One might think of looking for antiprotons, the antimatter equivalent of the most numerous nuclei in our part of the universe. Unfortunately, antiprotons are not good indicators for the existence of primordial antimatter. This is because they can be produced by collisions of the cosmic rays with the interstellar medium. However, antihelium cannot be produced in observable amounts by such interactions. Next to hydrogen, helium is the most abundant element in the universe. Thus, if there is primordial antimatter, antihelium is the most likely form to be detected in cosmic rays.

TABLE 1.
Member Institutions of the WiZard Science Team

U.S.	ITALY	GERMANY
NMSU	INFN	University of Siegen
Boston U.	Universita di Bari	
GSFC	Universita di Firenze	
Indiana U.	Universita di Perugia	
LSU	U. Di Roma - Tor Vergata	
NRL	Universita di Trieste	
UCB		

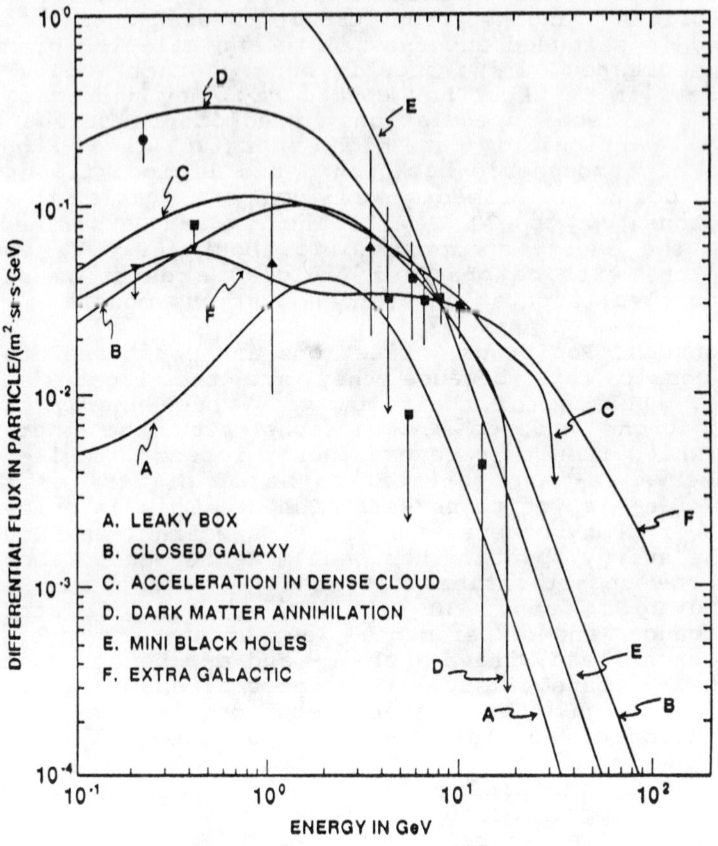

Figure 1. Antiproton Spectra.
The current data on antiproton fluxes are shown along with theoretical predictions. The shear number of theoretical curves is indicative of the interest in antiproton observations. The WiZard experiment proposes to make detailed antiproton observations in the rigidity range 4-400 GeV.

The WiZard experiment will be capable of detecting antihelium with a sensitivity of 1 part in 10^8. This level is sufficient to guarantee that a significant number of nuclei from beyond our galaxy have been sampled. Whether or not antihelium is observed, new insights will gained in the role of CP violations in the early universe.

Antiproton Spectrum: Antiprotons have been observed in the cosmic rays since 1979[3,4]. One feature of the observed flux is that, at high energies, it appears to be 3 times that expected if the antiprotons are produced by interactions of the cosmic rays with the interstellar medium. The overabundance of antiprotons has led to speculations about their origin ranging from models where antiprotons are produced in shrouded supernovae, to annihilation of Majorana fermions created during the big bang. Other models explain the excess in terms of how cosmic rays may propagate in the Galaxy. These speculations predict different energy spectra, particularly at high energies (see Figure 1). Because of atmospheric background and limited flight times, balloon borne experiments will only be able to measure antiprotons up to 20 GeV. The WiZard experiment will measure the energy spectrum up to about 400 GeV. Over 10^5 antiprotons will be observed yielding a detailed spectrum, allowing clear separation of these various models.

Electrons and Positrons: Electrons and positrons are unique among cosmic rays because they are the lightest charged leptons. Because of their low mass, high-energy electrons and positrons undergo interactions with the interstellar medium which result in severe energy losses . While most of the observed e- are believed to be of primary origin, the origin of e+ is yet to be established. This is evident from the number of hypotheses for the e+ spectra (see Figure 2).

The energy spectrum has been measured up to about 1 TeV with large uncertainties[5,6,7] and the e+ spectrum has been measured up to about 30 GeV[8,9,10,11] but the statistical significance and experimental agreement are poor. The WiZard experiment will measure e- and e+ spectra up to 1 TeV and 400 GeV respectively. The spectral shapes will provide information on the (i)acceleration of e- and the distribution of acceleration sites on time scales of about 10^5 years, (ii)cosmic ray lifetime and the physical conditions in the confinement volume, (iii)magnitude of re-acceleration by interstellar shock waves, and (iv)origin and propagation of e+ in relationship with antiprotons.

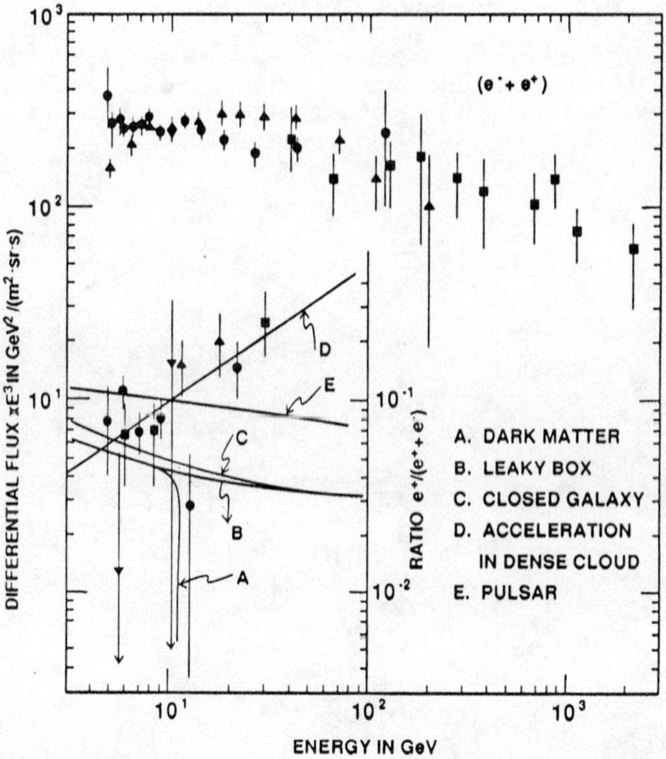

Figure 2. Electron and Positron Spectra.
While data for electrons exists up to 1000 GeV, positron data is very meager. The source mechanism for positrons is not known yet. A number of possible theoretical predictions are shown. The WiZard experiment will measure positrons to about 400 GeV.

Summary of Additional Objectives: In addition to the primary investigations, there are a number of additional studies which we summarize briefly:

-Composition and Spectra of Nuclei (Hydrogen -> Oxygen) from 4 GeV - 1000 GeV.
-Study of Solar Flare Spectra.
-Solar Terrestrial Relationships. Study variations in composition and spectra correlated with orbital location, solar rotation, diurnal variations etc.
-Galactic Anisotropies. Study of sidereal variations.
-Geomagnetic Effects.
-Spectra of Heavier Nuclei.
-Gamma-ray Astronomy.

INSTRUMENT DESCRIPTION

The WiZard experiment utilizes an array of particle detectors mounted on one end of the Astromag magnet system. Figure 3 illustrates the detectors proposed for the WiZard experiment. In this section we describe the basic concepts in the WiZard design.

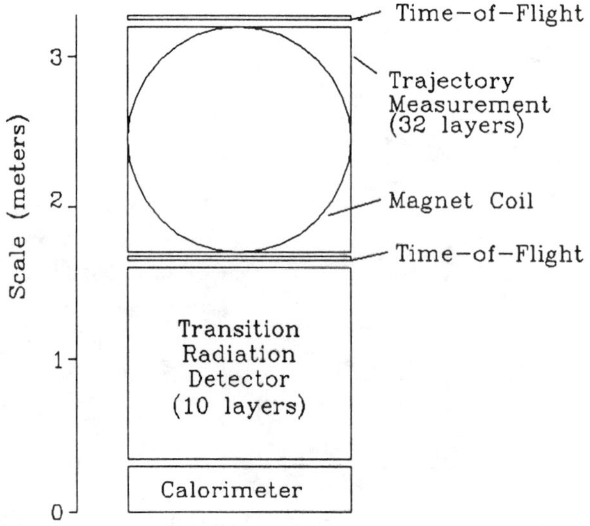

Figure 3. WiZard Instrument Assembly.
The detector elements of the WiZard experiment are shown in schematic form. Note the location of the experiment relative to the magnet coil.

Starting from the top, the instrument includes a pair of time-of-flight (TOF) detectors, a tracking system and another pair of TOF detectors. The tracking system is an instrumented volume which allows an almost continuous measurement of a particles trajectory as it traverses the magnetic field produced by the Astromag coil. The TOF detectors measure the time required to traverse from the top plane to the bottom. Both the TOF and tracking system make measurements of the particles charge as it traverses the system. Taken as a unit, the TOF and tracking systems are all that is required for the search for primordial antimatter. The TOF tells the direction of flight and the magnitude of the charge. The tracking system allows determination of the particles rigidity and sign of charge. Great care has been taken in the design of this system to minimize the mass that must be traversed by particles being measured. Any extra mass introduced here results in increased chances of the particle undergoing interactions which increase the difficulty of interpreting the event. To measure positron and antiproton fluxes, additional detectors are required. For example, an electron and an antiproton would be indistinguishable using the TOF and tracking systems described above. The transition radiation detector (TRD) and the calorimeter provide means for separately identifying light particles (e.g. electrons and positrons) from heavier particles (protons and antiprotons).

The TRD detector is comprised of 10 layers of carbon fiber filaments separated by 10 Xe filled multiwire proportional counters. The Calorimeter is an array of 25 planes of silicon detectors sandwiched between tungsten sheets. The total thickness of the calorimeter corresponds to 10 radiation lengths.

The total instrument TOF, tracking system, TRD and Calorimeter , constitute an array of particle detectors capable of identifying electrons, positrons, antiprotons, nuclei and antinuclei over a wide range of energies. We have chosen types of detectors that allow cross-checks in observational technique and we have designed each detector so that it contains internal redundancy (for purposes of reliability AND resolution). The total instrument weight (including all shuttle and space station interfaces) is 2593 Kg and the projected power requirement is 950 watts.

STATUS

Definition programs are funded and underway in Italy and Germany. US funding is expected in early '90. Activities in the US are currently being carried out through symbiotic relationships with our SRT programs. On a detector-by-detector basis, the status of the WiZard detector is as follows:

TOF - Prototype operational on New Mexico State Universities Balloon Borne Magnet Facility (BBMF).

Tracking Detectors - Development and definition activities are underway at 3 institutions. Accelerator tests are scheduled in December at CERN.

TRD - The first series of tests have been completed at CERN. A balloon prototype will be constructed beginning in Jan. 1990.

Calorimeter - Silicon for the first tests has been order. The first test run is scheduled for June 1990.

Basic Infastructure - Operational prototype in operation (BBMF).

CONCLUSIONS

Although not funded in the US yet WiZard is already a very active program. The WiZard instrument is based on existent technology and makes effective use of the space station and its Astromag facility. WiZard will be able to improve the sensitivity of the search for primordial antimatter by more than 3 orders of magnitude. The observed energy range and numbers of observed antiprotons and positrons will also be increased by 2-3 orders of magnitude. The observations done with WiZard can be expected to bring new understanding in many areas of space physics and cosmology including the role of antimatter in the universe and acceleration and containment mechanisms for cosmic rays.

REFERENCES

1) Smoot, G. F., A. Buffington and C. D. Orth, Phys. Rev. Lett., **35**, 258 (1975).

2) Badhwar, G. D., R. L. Golden, J. L. Lacy, J. E. Zipse, R. R. Daniel and S. A. Stephens, Nature, **274**, 137 (1978).

3) Golden, R. L., S. Horan, B. Mauger, G. D. Badhwar, J. L. Lacy, S. A. Stephens, R. R. Daniel and J. E. Zipse, Phys. Rev. Lett., **43**, 1196 (1979).

4) Bogomolov, E. A., N. D. Lubyanaya, V. A. Romanov, S. V. Stepanov, M. S. Shulakova, Proc. 16th Int. Cosmic Ray Conf., **1**, 330 (1979).

5) Nishimura, J., M. Fuji, T. Taira, E. Aizu, Y. Nomura, T. Kobayashi, K. Niu, A. Nishio, R. L. Golden, T. A. Koss, J. J. Lord, R. J. Wilkes, Ap. J., **238**, 394 (1980).

6) Golden, R. L., S. A. Stephens, B. G. Mauger, G. D. Badhwar, R. R. Daniel, S. Horan, J. L. Lacy and J. E. Zipse, Astr. Ap., **188**, 145 (1987).

7) Tang, T. T., Ap. J., **278**, 881 (1984).

8) Faneslow, J. L., R. C. Hartman, R. H. Hildebrand, P. Meyer, Ap. J., **158**, 771 (1969).

9) Buffington, A., C. D. Orth and G. P. Smoot, Ap. J., **199**, 699)1975). et al 1975

10) Golden R. L., S. A. Stephens, B. G. Mauger, G. D. Badhwar, R. R. Daniel, S. Horan, J. L. Lacy and J. E. Zipse, Astr. Ap., **188**, 145 (1987).

11) Muller, D., and K. K. Tang, Ap. J., **312**, 183 (1987).

LARGE ISOTOPE SPECTROMETER FOR ASTROMAG

W. R. Binns, J. Klarmann, and M.H. Israel
Department of Physics and McDonnell Center for the Space Sciences, Washington University, St. Louis, Mo. 63130 USA

T. L. Garrard, R. A. Mewaldt and E. C. Stone
California Institute of Technology, Pasadena Ca., 91125 USA

J. F. Ormes and R. E. Streitmatter
NASA/Goddard Space Flight Center, Greenbelt, Md. 20771 USA

I.L. Rasmussen
Danish Space Res. Institute, Lundtoftevej 7, DK-2800 Lyngby, Denmark

M. E. Wiedenbeck
The University of Chicago, Enrico Fermi Institute, 933 E. 56th St., Chicago Ill. 60637 USA

ABSTRACT

The Large Isotope Spectrometer for Astromag (LISA) is an experiment designed to measure the isotopic composition and energy spectra of cosmic rays for elements extending from beryllium through zinc. The overall objectives of this investigation are to study the origin and evolution of galactic matter; the acceleration, transport, and time scales of cosmic rays in the galaxy; and search for heavy antinuclei in the cosmic radiation.

To achieve these objectives the LISA experiment will make the first identifications of individual heavy cosmic ray isotopes in the energy range from about 2.5 to 4 GeV/n where relativistic time dilation effects enhance the abundances of radioactive clocks and where the effects of solar modulation and cross-section variations are minimized. It will extend high resolution measurements of individual element abundances and their energy spectra to energies of nearly 1 TeV/n, and has the potential for discovering heavy anti-nuclei which could not have been formed except in extra-galactic sources.

EXPERIMENT OBJECTIVES AND RATIONALE

1. Elemental and Isotopic composition of CR sources--It has recently been shown that the matter from which cosmic rays originate has a distinctly different isotopic composition than typical solar system matter for several isotopes[1-3]. In particular, ^{22}Ne is at least three times more abundant in cosmic ray source material, while the abundances of the neutron rich isotopes of Mg and Si are enhanced by a factor of about

1.5. For each of the 5 isotopes for which the source composition has been determined to an accuracy of 30% or better, there is evidence for a difference between cosmic ray and solar system material. This implies a difference between the nucleosynthesis of cosmic ray and solar system matter, and has stimulated a number of theoretical suggestions as to how such a difference may have occurred. Of these, the most quantitative are the "supermetallicity" model[4] which postulates that cosmic rays are produced in "metal" rich regions of the galaxy, and the Wolf-Rayet model[5] which assumes that a fraction of the cosmic rays originate as surface mass loss of Helium burning products of WR stars.

The results from LISA will allow us to test these and other models by extending the present observations to a variety of other nuclei. The number of individual isotopes that we expect to detect and resolve with resolution $\leq$0.3 amu in two years of data collection is shown in Fig. 1. For isotopes such as ^{57}Fe and ^{58}Fe for which secondary contributions are small, the derived source abundances relative to ^{56}Fe will be limited by statistics, and are ~6% and ~15% respectively. Measurement of these and other species including ^{13}C, ^{18}O, ^{29,30}Si, ^{38}Ar, ^{34}S, and ^{54}Fe should make it possible to constrain and discriminate between models.

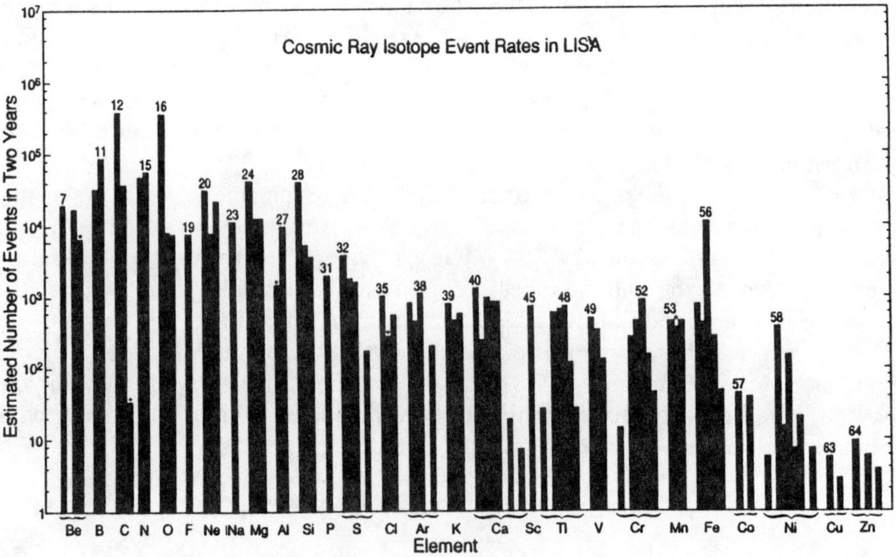

Fig. 1--Expected yields of isotopes from Be to Ni that LISA will observe with excellent mass resolution (<0.3 amu). The estimates include the effects of the geomagnetic field averaged over the orbit, nuclear interactions in instrument, and 2 years of data. Selected radioactive species which can decay in flight are marked with a star.

The abundance of isotopes such as ^{13}C and ^{18}O in cosmic rays is also an important test of galactic evolution models. Recent optical measurements[6] indicate a factor of 2 enhancement of these isotopes in the local galactic region. High precision measurements of the cosmic ray abundances, coupled with recent improved knowledge of spallation production cross-sections, should enable us to differentiate between the Wolf-Rayet, solar system, and local galactic values.

Cosmic rays with energies of ~100 GeV/n have apparently traversed only about 1 g/cm^2 of material so that their sources become almost "bare". Thus their source composition can be determined with greater precision than is possible at lower energies[7]. For example, in the iron "secondary" region, a measurement of the ratio of Cr (partly primary) to the (Sc+Ti+V) abundances (mostly secondary) as a function of energy should give a reasonably precise measurement of the Cr source abundance. Other rare source elements where this approach should work include N, F, Na, Al, K, P, Ar, Ca, and Mn.

Accurate measurements of particle energy spectra over a wide energy interval (LISA covers 2 1/2 decades in energy/nucleon) are also important tools for identifying particle acceleration mechanisms. This is potentially important since special sources such as Wolf-Rayet stars may have atypical energy spectra. Table 1 summarizes the expected yields of representative elements.

Number of Events per Two Years		
Element	>10 GeV/n	>100 GeV/n
B	66,000	600
O	600,000	12,000
Ne	270,000	4,000
Ca	15,000	200
Fe	100,000	2,000

2. *Cosmic Ray Acceleration Time Scales*--The fundamental question of the time scale of the acceleration of cosmic ray nuclei will be addressed by measuring radioactive isotopes (and their daughters) that decay only by electron capture. Examples of such isotopes are ^{56}Ni, ^{57}Co, and ^{59}Ni (shown in Fig. 2) which have half-lives ranging from a few days to 10^5 years.

3. Cosmic Ray Transport and Reacceleration-- At energies greater than ~30 GeV/n, element ratios such as B/C and (Sc+Ti+V+Cr)/Fe are sensitive to propagation and may contain information relevant to acceleration. These ratios should distinguish between the energy dependent leaky box, reacceleration, and the "closed galaxy" models.

High energy isotope measurements can also add a new dimension to studies of cosmic ray propagation using radioactive isotopes such as ^{10}Be, ^{14}C, ^{26}Al, ^{36}Cl, and ^{54}Mn as natural clocks to determine the time constants associated with the storage of cosmic rays in the galaxy. Fig. 3 shows how the abundances of ^{10}Be and ^{26}Al are expected to depend on energy and on the average density of material in the cosmic ray confinement region. The LISA measurements will be made at energies

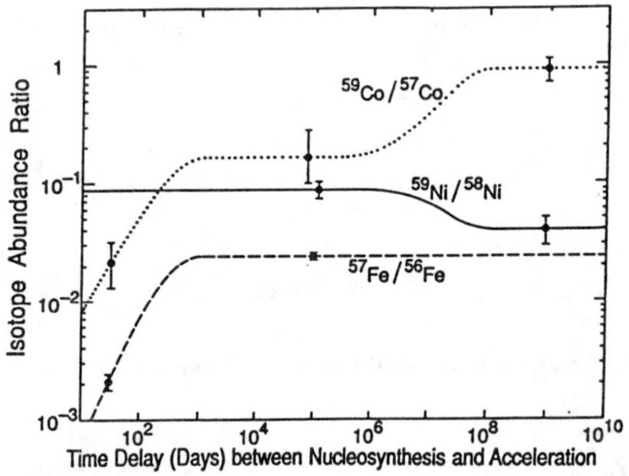

Fig. 2--Expected Fe, Co, and Ni isotope ratios as a function of the time delay between nucleosynthesis and cosmic ray acceleration. Measurement uncertainties expected from LISA are shown.

corresponding to Lorentz factors of $\gamma = 3.6$ to 5.5, where the effective half-lives of these clocks are several times greater than at low energy. Comparison of precise measurements by LISA with the improved measurements expected at 0.1 to 0.2 GeV/n from missions in the 1990's will provide sensitive tests of models which predict the energy dependence of isotope abundance ratios based on a postulated distribution of matter in the confinement volume.

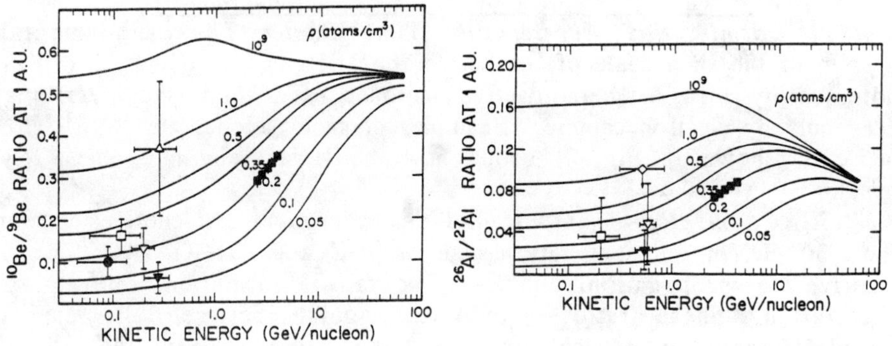

Fig.3--^{10}Be/^{9}Be and ^{26}A/^{27}Al ratios vs. energy(calculations by Guzik and Wefel[8]) parameterized by the density of the propagation region. Simulated LISA results (solid squares; statistical error bars slightly larger than the solid squares) are indicated assuming $\rho = 0.2/cm^3$.

4. Cosmological Antimatter--LISA is also capable of searching for heavy antinuclei with Z>3 having energies >2 GeV/n. In 2 years of operation LISA should collect about 7 x 10^6 heavy nuclei giving an upper limit on the antimatter/matter ratio of <4 x 10^{-7}, assuming that no antimatter candidates are observed. This is an improvement of about 2 orders of magnitude over present limits.

THE LISA INSTRUMENT

The geometric arrangement of the LISA detectors is shown in Fig. 4. LISA consists of a central core for tracking particles through the magnetic field, and an outer ring of detectors for determining velocity, charge, and time-of-flight (TOF). Measurements of particle trajectories in the magnetic field are made using a scintillating optical fiber trajectory (SOFT) detector consisting of five planes (H1 through H5), each providing two-dimensional coordinates with an rms resolution of 70μm or better. The outer SOFT planes (H1 and H5) are each constructed as four segments of a polygon, as are the other sensor elements. Two aerogel Cherenkov counters (C1 and C2, or C3 and C4), one Pilot 425 Cherenkov counter (T1 or T2), and a set of scintillators (S1 or S2) are combined in each of eight modules.

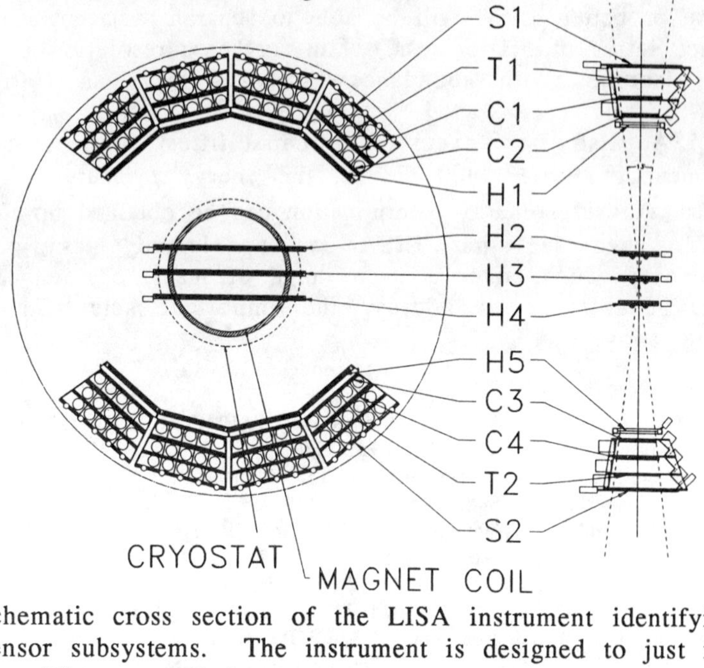

Fig. 4--Schematic cross section of the LISA instrument identifying the various sensor subsystems. The instrument is designed to just fill the shuttle bay (diameter=4.5m).

For the particles to be measured, the Cherenkov response of the Pilot 425 is nearly saturated and provides a direct measure of the atomic number, Z, after pathlength and mapping corrections are applied. Using the track positions measured in the SOFT planes we fit the particle trajectory through the magnetic field and determine the magnetic rigidity (momentum per unit charge) as well as the incidence angle and position. For particles of energy ≤10 GeV/n, the aerogel Cherenkov counters measure the particle's momentum per nucleon or velocity. Combining these three measurements allows us to determine the mass, charge, and energy of each nucleus traversing the LISA instrument. The TOF scintillators determine the direction in which the particle traversed the detector (up or down), thus identifying antinuclei, identifying the arrival direction through the geomagnetic field and the corresponding rigidity cutoff, and serving as a consistency check. More details of the detection method used in LISA are given in Ref. 9.

The expected mass resolution including contributions from the trajectory measurement errors and multiple scattering, as well as contributions due to uncertainties in the measured Cherenkov emission, has been calculated to lie in the range of 0.17 to 0.30amu for Fe nuclei with kinetic energy between about 2.5 and 3.7 GeV/n. For lighter nuclei this resolution is improved. For energies where the mass resolution is 0.30 amu or better, LISA will be able to separate adjacent isotopes with abundance ratios of 5:1 or less, while for mass resolution of 0.25 amu adjacent isotopes with abundance ratios as large as 100:1 can be studied.

LISA also has exceptional capabilities for energy spectra measurements. For E ≤300 GeV/n, the energy measurement precision is ≤20%. A useful energy determination can be obtained up to 1TeV/n.

Thus we see that LISA is capable of making precision measurements which cover a wide range of mass, charge, and energy, and will enable us to address the important scientific objectives described above.

References

1. R.A. Mewaldt, "Cosmic Abundances of Matter", AIP Conf. Proc. **183,** 124 (1988); and Rev. Geophys. and Sp. Phys., **21**, 295 (1983).
2. J.A. Simpson, Ann. Rev. Nucl. Part. Sci. **33**, 323 (1983).
3. M. E. Wiedenbeck, Adv. Space Res. **4**, 15, (1984).
4. S.E. Woosley and T.A. Weaver, Ap. J., **243**, 561 (1981).
5. M. Casse, and J.A. Paul, Ap. J., **258**, 860 (1982).
6. I. Hawkins and M. Jura, Ap.J., **317**, 926 (1987).
7. W.R. Binns, T.L. Garrard, M.H. Israel, M.D. Jones, M.P. Kamionkowski, J. Klarmann, E.C. Stone, and C.J. Waddington, Ap. J., **324**,1106 (1988).
8. T.G. Guzik and J.P. Wefel, Private Communication (1986).
9. R.A. Mewaldt, Proc. of "Workshop on Scientific Uses of the Space Station in Cosmic Ray and High Energy Gamma-Ray Physics", J. Klarmann, Ed., 140, (1986).

SPECTRA, COMPOSITION, AND INTERACTIONS OF NUCLEI WITH MAGNET INTERACTION CHAMBERS

T. A. Parnell[i], T. H. Burnett[j], M. C. Cherry[h], S. Dake[b], J. H. Derrickson[i], W. F. Fountain[i], M. Fuki[d], J. C. Gregory[g], T. Hayashi[g], R. Holynski[k], J. Iwai[j], A. Jurak[k], J. J. Lord[j], O. Miyamura[c], K. Niwa[m], H. Oda[b], T. Ogata[a], F. E. Roberts[i], T. Shibata[e], S. C. Strausz[j], T. Tabuki[i], T. Taira[l], Y. Takahashi[g], T. Tominaga[h], J. W. Watts[i], J. P. Wefel[h], B. Wilczynska[k], H. Wilczynski[k], R. J. Wilkes[j], W. Wolter[k], T. Wosiek[k], A. Yamamoto[n], H. Yokomi[o], T. Yuda[a]

[a]ICRR, University of Tokyo; [b]Kobe University, Kobe; [c]Osaka University, Osaka; [d]Matsusho Gaukuen Junior College, Matsumoto; [e]Aoyama Gakuin University, Tokyo; [g]University of Alabama, Huntsville; [h]Louisiana State University, Baton Rouge; [i]NASA Marshall Space Flight Center, Huntsville; [j]University of Washington; [k]INP, Krakow; [l]Kanagaw University; [m]Nagoya University; [n]KEK, Tsukuba; [o]Tezukayama University, Nara

ABSTRACT

Emulsion chambers will be flown in the Astromag Facility to measure the cosmic ray composition (H through Fe) and spectra to 10^{15} eV total energy and to definitively study the characteristics of nucleus-nucleus interactions above 10^{12} eV/n. Two configurations of emulsion chambers will be flown in the SCIN/MAGIC experiment. One chamber has an emulsion target and a calorimeter similar to those recently flown on balloons for composition and spectra measurements. The other has an identical calorimeter and a low-density target section optimized for performing rigidity measurements on charged particles produced in interactions. The transverse momenta of charged and neutral mesons, direct hadronic pairs from resonance decays and interference effects, and possible charge clustering in high density states of matter will be studied. The magnetic field allows more definitive tests of heavy nucleus interaction models at high energy, and enhances searches for evidence of nuclear matter phase change such as a quark-gluon plasma or chiral symmetry. These interaction studies are important for energy measurements below and above 10^{15} eV using ionization calorimetry, including air shower experiments.

INTRODUCTION AND SCIENTIFIC OBJECTIVES

From data and from models of propagation and acceleration, it is expected that the cosmic ray composition will change significantly between 10^{14} to 10^{16} eV total energy. The shock acceleration model which explains the spectrum ($\propto E^{\approx 2.7}$) below 1 TeV/n, may fail to power the cosmic rays above 10^{14} eV[1]. However, air shower experiments have shown a small intensity enhancement in the spectrum above 10^{14} eV, followed by a sudden steepening (the "knee") at ~3 x 10^{15} eV (Figure 1). The steepening is generally understood as the decline of galactic confinement, whereas the enhancement (if any) is

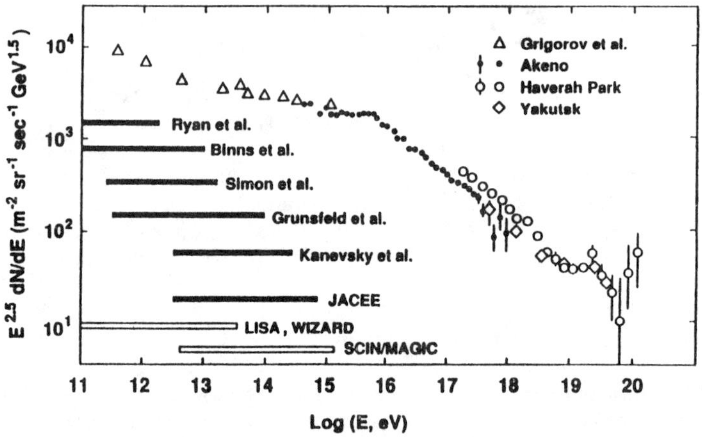

Fig. 1. The all-particle cosmic ray spectrum compiled from several measurements, and energy ranges of various experiments.

speculated to be due to particular sources and acceleration mechanisms. The decline of galactic confinement in a rigidity dependent manner would also change the composition below and above the "knee".[2] The leaky box model of cosmic ray propagation fits the spectra data reasonably well below 10^{11} eV/n where adequate data exists. Predictions of the models generally diverge at higher energies, particularly for secondary components.

Figure 2 shows the status of some directly measured composition and spectra data in the region above 1 TeV/n, contributed by balloon borne and space experiments[3,4,5]. Although data are meager at present, there are indications of changes. The Intercosmos[6] and JACEE[4] groups report on increase in the relative abundance of helium above 5 TeV/n. The JACEE data indicates (with nebulous statistics) an increase in intensity from carbon to sulfur above 10 TeV/n. The Spacelab 2 results[5] indicate the spectrum of iron is flatter and Ne, Mg, Si are steeper than anticipated above 1 TeV/n. More direct data is obviously needed to resolve the experimental ambiguities and questions presented by the models.

The Spectra, Composition, and Interactions of Nuclei (SCIN) experiment in the Astromag Facility will directly measure the composition of cosmic rays with conventional emulsion chambers, and with Magnet Interaction Chambers (MAGIC). The statistical accuracy that may be achieved in a 3 month exposure of 0.8 m^2 of conventional (EMCAL) and 0.8 m^2 of MAGIC chambers are shown in Figure 2 and Table I. The main goal of the SCIN/MAGIC composition measurements is to determine whether multiple processes accelerate cosmic rays up to 10^{15} eV. Sufficient data on primary electrons and secondary nuclei will be gathered above 10^{12} eV/n to distinguish between some propagation models.

Emulsion chamber experiments used to gather the highest energy direct data (Figure 2) have also yielded novel and interesting

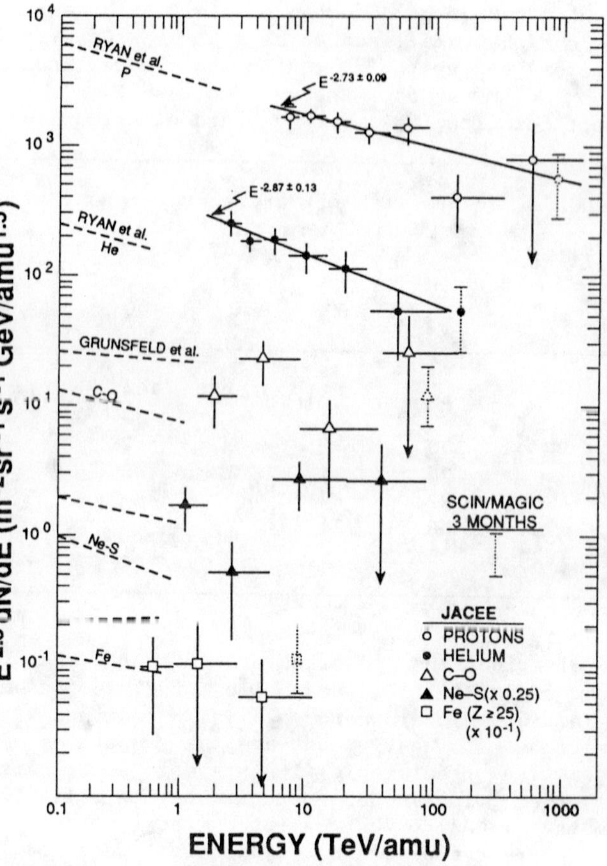

Fig. 2. Composition and spectra data from Ryan et al. (balloons), JACEE (balloons)[4], and Grunsfeld et al. (Spacelab 2). The SCIN/MAGIC symbols indicate expected statistics, not intensities.

observations of nuclear interactions through the study of the produced charged and neutral particles. Those observations include a larger than expected abundance of closely spaced particles (pairs)[7], large scale angular structure of produced particles[8], heavy nucleus interactions with very high energy density and transverse momenta[9], and possible observations of bottom quark meson decay[10]. The study of cosmic ray interactions is essential to the use of ionization calorimetry for primary energy measurement (and air shower observations) but much of additional interest derives from quantum chromodynamics (QCD) theory and associated calculations. Calculations using QCD predict a phase transition of hadronic matter into a deconfined quark and gluon plasma (QGP) at high energy density (a few GeV/fm^3), which existed in the early stage of the big bang. Recently, heavy ion interaction experiments at CERN (O^{16}, S^{32} at 200 GeV/n) have failed to reach that energy density[11], and have measured

Table I Expected number of detected events. A three month exposure of eight chambers (each 40 cm x 50 cm area with a 12 rl calorimeter) and up to 75° zenith angle acceptance are assumed (3.76×10^7 m^2 sr s). Energy spectra $\propto E^{-2.7}$ are assumed, except for iron (Fe#) $\propto E^{-2.5}$, and electrons (e) $\propto E^{-3.3}$.

Primary Particle	Number of events with primary energy > Eo (TeV/n) or [Total Energy>10^{14} eV]* E_0: 1	5	10	50	100	500	[*]
p			1.1E4	620	184	10	[184]
He			1580	98	30	2	[320]
C - O		320	102	6	2		[174]
Ne - S		120	36	2			[174]
Fe	360	22	8				[134]
Fe#	780	68	24	2			[320]
e	1220	30	6				

multiplicity distributions and transverse momenta consistant with superposition models[12,13,20]. Meanwhile, studies of cosmic ray interactions at much higher energy > 1 TeV/n continue with two dominant questions. To what extent are the properties of heavy nucleus interactions predictable with standard superposition models[12,13] so that primary energy spectra may be reliably derived from calorimetry[14]; and to what extent do the rare central collisions of heavy nuclei exhibit a different behavior (perhaps QGP)?

Measurements on interaction products with properly configured emulsion chambers in a magnetic field can significantly enhance the cosmic ray interaction studies. Previous emulsion chamber measurements of the average transverse momenta ($\langle P_T \rangle$) of produced π° (via measurement of individual gamma rays in the calorimeter) indicate a significant increase in average momenta above collision energy densities of ~2 GeV/fm^3 (Figure 3)[9]. Such behavior of $\langle P_T \rangle$ is anticipated if a quark qluon plasma is formed in the collisions[15]. The MAGIC chambers will measure $\langle P_T \rangle$ for both neutral and charged mesons. For a 10 TeV/amu primary these measurements will cover the target fragmentation and central regions of rapidity. For an exposure of 3.76×10^6 m^2 sr s approximately 100 central interactions with energy density exceeding 2 GeV/fm^3 will be measured.

By measurement of charge-signs, detailed study can also be made of hadronic pairs previously observed as overabundant in N-N collisions in several emulsion experiments[7,16]. Like sign pairs are understood as Bose-Einstein interference particles and can be used to measure the source radius of pionization by second order interferometry[17,18]. Unlike sign pairs are decay particles and can be

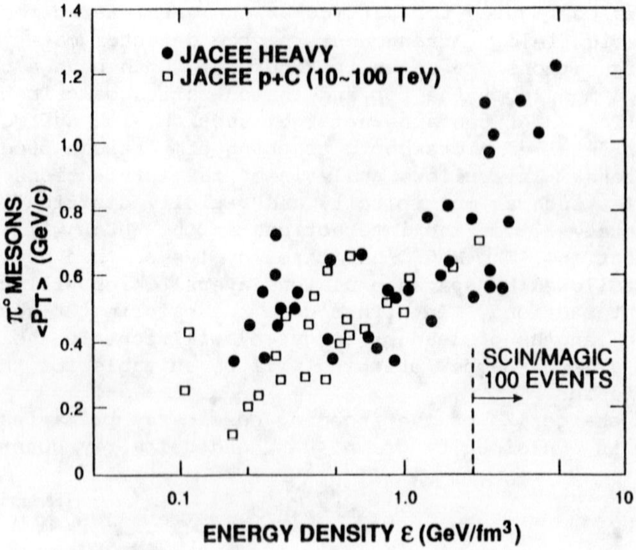

Fig. 3. The average transverse momenta of produced neutral mesons as a function of energy density during collision. This sample of proton and heavy nucleus interactions is from the JACEE data[9].

analyzed in terms of Q values to determine their possible origins, such as in resonance mesons. For example, if a chiral-symmetry phase were produced in heavy ion collisions, the resonance mesons η, σ, ρ, and A_1 might have a reduced mass[19], which would be revealed through a closer spacing of the observed pions. If this phase were produced the dynamical mass of nucleons and mesons would be reduced to zero, leaving the bare mass of the constituent quarks (a few MeV). An understanding of the origin of the majority of mass in the universe depends on this transition.

The emulsion grain density for tracks with small rapidities ($0.5 < \eta < 1.5$) will be measured and when coupled with the momentum data, will allow the identification of produced particles (for determination of the K/π ratio, etc. in the target fragmentation region).

In high multiplicity events the distribution of charge-signs on a target plane will be studied. If charge clustering (non-poisson behavior) is observed, it may be due to a chiral symmetry phase (where charge conjugation does not hold[19]), and to multiple Bose-Einstein interference.

INSTRUMENT DESCRIPTION

Two configurations of emulsion chambers will be used; one is a conventional emulsion-calorimeter (EMCAL)[14] for measuring cosmic ray composition and general characteristics of interactions. The other

(MAGIC) is configured for particular nuclear interaction studies in the magnetic field. Arrangements of the detector materials for the two configurations are shown in Figure 4. Both have a charge identification module at the top and an ionization calorimeter at the bottom. The EMCAL contains a target section with sufficient emulsions for detailed microscopic tracking of particle produced in interactions. This allows analysis of the interactions for general observables such as multiplicity and rapidity distributions. Approximately 30% of the interactions in the EMCAL will be in the target section. The MAGIC has three cycles of thin heavy element targets followed by spaced emulsion layers, which will contain 10% of the interactions. In either case the calorimeter section (12 radiation lengths of lead or tungsten) will contain the most interactions, a large number of which will be suitable for the composition studies.

The charge (Z) of the incoming cosmic ray primaries will be measured in emulsions by grain, gap, and delta ray counting in thick

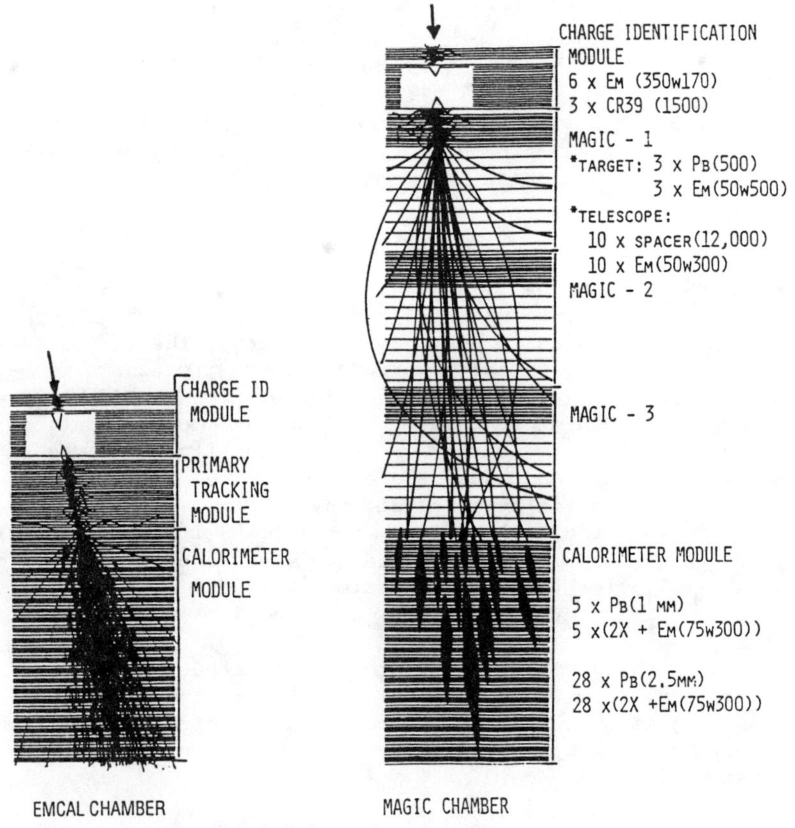

Fig. 4. Schematic of the two configurations of emulsion chambers in the SCIN/MAGIC experiment.

(≈200-300 μm)emulsions. The overall accuracy of the charge identification with emulsions is expected to be $\sigma = \pm 0.6$ e (e: unit of charge) for iron group nuclei, increasing to 0.1 e for protons. Better resolution ($\lesssim 0.3$e) is expected by CR39 etch-pit measurements for heavy nuclei ($Z \lesssim 20$ for most incident angles).

The ability to measure energies of electron-photon cascades is one of the most important functions of the instrument. Following an interaction above or in the top of the calorimeter a fraction of the total primary energy (5 - 25%) released, depending on impact parameters and atomic mass numbers of the colliding nuclei, will be deposited in the calorimeter in the form of photon and electron energy. Multiple measurements of electron number in each cascade yields the cascade energy (E_γ). Monte Carlo calculations for interactions in various parts of the chamber and for various primary nuclei are performed to describe the detection efficiency, energy resolution, and deconvolution of measured photon energy (ΣE_γ) to primary energy. The measurement techniques of ΣE_γ and conversion to primary cosmic ray energy spectra have been established[14].

The target section of the MAGIC consists of thin target layers (e.g. 3 x 300 μm lead and thin emulsions) followed by ~10 layers of thin (~50 μm double coated on 300 μm substrates) emulsions spaced 4-12 mm apart. This arrangement allows produced charged particles from interactions in the target to be tracked in the emulsion layers without serious degradation of rigidity measurements from multiple scattering. Calculations using the arrangement of Figure 4 in a 2 Tesla field and a position resolution of 2 micrometers indicates that produced charged meson momentum can be measured to 600 GeV/c. The momentum accuracy that has previously been achieved by the MAGIC technique is shown in Figure 5. This accelerator experiment[20] consisted of a target section (in a 1.8 Tesla field) similar to that planned for the SCIN/MAGIC experiment. The curvature (R) fitting of measured track coordinates was made with an accuracy of $\Delta R/R = (1.7 + P/2\ \text{GeV/c})\ \%$, where P denotes the momentum of a measured track. The momentum range and accuracy will be improved in SCIN/MAGIC (Figure 5) by optimizing the spacing of tracking emulsion layers and increasing the magnetic field.

Figure 6 shows two assemblies of EMCAL and MAGIC chambers placed in the ASTROMAG Facility. Each assembly is independent so exposure is not necessarily simultaneous. Each assembly consists of two MAGIC and two EMCAL chambers, with a mass of 1200 kg exclusive of the shuttle carrier. Thermal control to ±1 °C is achieved by external insulation, thermal radiators, and electrical heaters. Temperature, chamber pressures, and magnetic field will be monitored. Exposure in Astromag will be determined by background track considerations and an exposure of 3 months is desired. At presently planned shuttle altitudes, the background track density from trapped protons (in the South Atlantic Anomaly) will be $\lesssim 3 \times 10^6/\text{cm}^2$ for 3 months. The trapped protons will be encountered only in the "South Atlantic Anomaly" and their directional properties will considerably reduce their effect on emulsion chamber measurements.

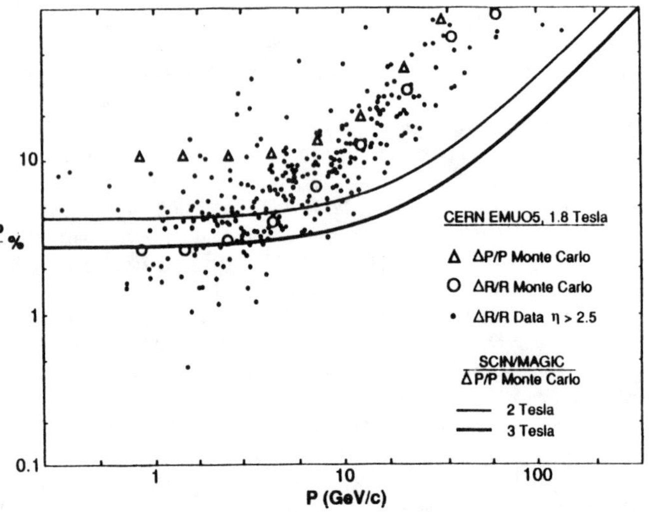

Fig. 5. The Monte Carlo predictions of momentum accuracy (delta's and solid lines) and curvature (open circles) for an accelerator experiment (20) and the SCIN/MAGIC. Data are charge particle curvature measurements (20) from (O^{16} + Pb, 200 GeV/n).

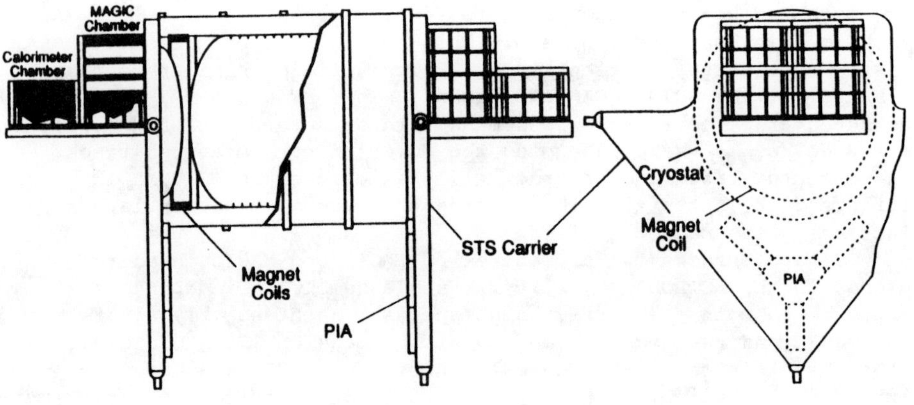

Fig. 6. The SCIN/MAGIC experiment in the Astromag Facility.

CONCLUSION

The combination of the high magnetic field of Astromag and the large area and aperture of the SCIN/MAGIC detectors will significantly extend the frontiers of high energy interaction and nuclear matter studies, and cosmic ray physics. This experiment will make direct measurements of interactions, energy spectra, and composition in the barely explored energy region of 10^{13-15} eV, where changes in spectra due to acceleration and propagation are expected and different states of matter in nucleus-nucleus collisions are predicted.

The SCIN/MAGIC experiments utilize proven emulsion chamber techniques. The EMCAL techniques have been developed in the last 30 years in multiple balloon borne experiments, and have produced the highest energy direct data. MAGIC targets have been used in accelerator experiments to produce new data on interactions. Further experiments with MAGIC chambers at accelerators and on balloons are planned by various collaborations, and simulations and design studies are in progress to refine the experimental apparatus and data analysis techniques for the SCIN/MAGIC experiments.

REFERENCES

(1) P.O. Lagage & C.J. Cesarsky, Astron. & Astrophys. 125, 249-257 (1983).

(2) A.M. Hillas, Ann. Rev. Astron. and Astrophys. 22, 425 (1984).

(3) M.J. Ryan, J.F. Ormes, & V.K. Balasubrahmanian, Phys. Rev. Lett. 28 985 (1972).

(4) T.H. Burnett et al., Astrophys. J. Lett, in press (1990) and T.A. Parnell et al., Adv. in Space Res. V9, 12, 45 (1989).

(5) J.M. Grunsfeld, J.L. Heureux, P. Meyer, D. Muller, and S.P. Swordy, Ap. J. Lett. 327, L31 (1988).

(6) I.P. Ivanenko et al., JETP Letters 48, 510 (1988).

(7) T.H. Burnett et al., Proc. Int. Cosmic Ray Conf., La Jolla 6, 168 (1985).

(8) T.H. Burnett et al., Phys. Rev. D 35, 824 (1987).

(9) T.H. Burnett et al., Phys. Rev. Lett. 57:26, 3249 (1986).

10) B. Wilczynska et al., INP (Cracow, Poland) Report 1447/PH (1989), Phys. Rev. submitted (1990).

(11) K. Kajantie, Nucl. Phys. A 498, 355 (1989); G. Baym, P. Braun-Munzinger, S. Nagamiya (eds.) Nuc. Phys. A498-1 (1989).

(12) H. Sumiyoshi et al., Z. Phys., C23, 391 (1984).

(13) B. Anderson et al., Phys. Scripta, 34, 451 (1986).

(14) T.H. Burnett et al., NIM A251, 583 (1986).

(15) E.V. Shuryak, Phys. Reports, 61C, 71 (1980).

(16) P. Freier, AIP 49, 87 (1979).

(17) A. Bamberger et al., Phys. Lett. B 203, 320 (1988); H. Satz, H.J. Specht, & R. Stock (eds.) Z. Phys. C38, 1 (1989).

(18) S.S. Padula & M. Gyulassy, Nucl. Phys. A 498, 555 (1989).

(19) C.E. DeTar & J.B. Kogut, Phys. Rev. Lett. 59, 599 (1987).

(20) Y. Takahashi et al., Nucl. Phys. A498, 529c (1989); J. Phys. Soc. Jpn. Supple. 58, 746 (1989).

Relevant Planning/Studies

THE SOLAR PROBE MISSION

W.C. Feldman[a], J. Anderson[b], J.D. Bohlin[c], L.F. Burlaga[d], R. Farquhar[d], G. Gloeckler[e], B.E. Goldstein[b], J.W, Harvey[f], T.E. Holzer[g], W.V. Jones[c], P.J. Kellogg[h], S.M. Krimigis[i], M.R. Kundu[e], A.J. Lazarus[j], M.M. Mellott[c], E.N. Parker[k], R. Rosner[k], G.J. Rottman[l], J.A. Slavin[d], S.T. Suess[m], B.T. Tsurutani[b], R.T. Woo[b], and R.D. Zwickl[n]

[a]Los Alamos National Laboratory; Los Alamos, NM 87545
[b]Jet Propulsion Laboratory/Caltech; Pasadena, CA 91109
[c]NASA/Headquarters; Washington, DC 20546
[d]NASA/Goddard Space Flight Center; Greenbelt, MD 20771
[e]University of Maryland; College Park, MD 20742
[f]NOAO/National Solar Observatory; Tucson, AZ 85726
[g]NCAR/High Altitude Observatory; Boulder, CO 80307
[h]University of Minnesota; Minneapolis, MN 55455
[i]JHU/Applied Physics Laboratory; Laurel, MD 20707
[j]Massachusetts Institute of Technology; Cambridge, MA 02138
[k]University of Chicago; Chicago, IL 60637
[l]University of Colorado; Boulder, CO 80309
[m]NASA/Marshall Space Flight Center; Huntsville, AL 35812
[n]NOAA/Space Environment Laboratory; Boulder, CO 80303

ABSTRACT

The Solar Probe will deliver a 133.5 kg science payload into a 4 R_S perihelion solar polar orbit (with the first perihelion passage in 2004) to explore *in situ* one of the last frontiers in the solar system - the solar corona. This mission is both affordable and technologically feasible. Using a payload of 12 (predominantly particles and fields) scientific experiments, it will be possible to answer many long-standing, fundamental problems concerning the structure and dynamics of the outer solar atmosphere, including the acceleration, storage, and transport of energetic particles near the Sun and in the inner (< 65 R_S) heliosphere.

INTRODUCTION

While much has been learned over the past several decades from *in situ* observations beyond 0.3 AU and remote-sensing of solar-atmospheric photons, many ambiguities remain about the mechanisms that heat the solar corona, shape its structure, and accelerate the solar wind and solar particles, due to a lack of *in situ* observations close to the Sun. It is now technologically feasible to fly a comprehensive payload of sophisticated experiments as close as 4 solar radii of Sun center, allowing the resolution of the many long-standing fundamental problems concerning the structure and dynamics of the outer solar atmosphere.

The general realization of the scientific possibilities provided by a solar probe had gained widespread acceptance by the scientific community by 1978 and has been documented in several reports of studies and workshops held since. Table 1 details the sequence of events leading to the present concept of a Solar Probe Mission.

Table I **Chronology of Events Leading to the Present Concept for a Solar Probe Mission**

SCIENTIFIC IMPORTANCE OF SOLAR PROBE GAINS WIDESPREAD ACCEPTANCE

- **Symposium at CIT in 1978**
- **"A Close Up View of the Sun" (JPL Publ. 78-70, 1978)**

SOLAR PROBE MISSION INCLUDED IN SCIENTIFIC PLANNING DOCUMENTS OF NRC

- **"A Strategy for Space Astronomy and Astrophysics for the 1980's" (1979)**
- **"Solar-System Space Physics in the 1980s: A Research Strategy" (1980)**
- **"An Implementation Plan for Priorities in Solar System Space Physics (1985)**
- **"Space Science in the Twenty First Century: Imperatives for the Decade 1995 to 2015" (Volume on Solar and Space Physics, 1980)**

NASA FUNDS INTENSIVE STUDY OF MISSION CONCEPTS AT JPL

- **Three major disciplines**
  Gravity and Relativity
  Particles and Fields
  Imaging
- **"Starprobe, Scientific Rationale, A Report of the Ad Hoc Working Groups" (JPL Publ. 82-49, 1982)**

SSB COMMITTEES RECOMMEND A TECHNICALLY FEASIBLE AND AFFORDABLE SOLAR PROBE MISSION

- **Predominantly Particle and Fields Investigations**
  "To measure density, velocity, and composition of the thermal Solar Wind plasma, as well as the magnetic fields, plasma waves, and Energetic Particles present in this unexplored region of the heliosphere"

A recent report[1] of the Solar Probe Science Study Team, commissioned by the Space Physics Division of NASA, in response to several National Research Council planning documents strongly promoting a Solar Polar Mission, 1) details the salient scientific problems, 2) places them in historical and scientific perspective, 3) assembles a minimum integrated payload of current-day technology scientific experiments capable of resolving many of the open issues, and 4) presents an affordable mission concept and a spacecraft design that is capable of supporting the necessary scientific measurements.

This paper summarizes the scientific objectives, measurement requirements, mission design, and science instrument and spacecraft descriptions, presented in that report, emphasizing results expected from suprathermal and energetic particles measurements.

SCIENTIFIC OBJECTIVES

The Sun has long been an object of intense interest and scientific study for two fundamental reasons. First, the Sun is the very special star that has nurtured and sustained life on Earth for aeons, strongly affecting human events through subtle control of terrestrial climate. The second fundamental reason for studying the Sun is simply that it is a mysterious object. Basically, it is the classical star, a self gravitating sphere of gas heated by steady nuclear fusion in its core. But a closer look shows only about a third of the expected neutrino emission, oscillation modes whose frequencies disagree with predictions, a 22-year magnetic cycle that sometimes disappears for a century (the Maunder Minimum) or operates at an ultra high level for a century (the Medieval Maximum), and an outer atmosphere enigmatically heated to 10^6 K that emits copius X-rays from its denser regions, with the tenuous regions continually expanding to fill interplanetary space with the supersonic magnetized solar wind. The general rule is that the magnetic fields emerging through the surface of the Sun enhance the heating, and the UV and X-ray emission from the outer atmosphere of the Sun varies widely - sometimes explosively (flares) - so that the terrestrial environment is bathed in a highly variable flux of hard radiation. No less important is the observed variation of the total luminosity of the Sun over a period of several years, well correlated with the level of magnetic activity. This raises the question of how much the luminosity of the Sun increased during the Medieval Maximum and declined during the Maunder Minimum. It is a curious fact that the climate in the northern temperate zone of Earth was conspicuously warmer and colder, respectively, during those periods.

The major scientific objectives of the Solar Probe Mission are listed in Table 2. These goals were formulated to fill the gaps in our current understanding of the extended solar corona. They include the investigation of coronal heating mechanisms and energy transport, acceleration of the solar wind, and of the transient corona, including the acceleration of solar energetic particles. Specifically, they address the physical processes that help shape the many different forms of the solar corona, drive the solar wind, and accelerate a small percentage of solar particles to very high energies.

A wide range of energetic particle populations that escape from the Sun are commonly observed throughout interplanetary space. Although they attain highest intensities after solar flares, they are present in measurable numbers at all times (see Figure 1). Measured energy and angular flux distributions (beyond 0.3 AU) have suggested 1) particle storage and/or delayed release near the Sun, 2) modification of flux distributions by coronal and interplanetary transport processes, 3) local, shock-associated acceleration in interplanetary space, and 4) a mix of acceleration mechanisms that generate and release the population near the Sun.

Most energetic charged particles are accelerated within closed magnetic regions and so are not directly observable in interplanetary space. We infer their characteristics indirectly through the secondary radiation they generate, consisting of hard X rays, microwaves, Type II radio emission, Type IV radio emission, gamma rays, and neutrons. This indirect method of looking at particle populations trapped near the sun is strongly model-dependent and therefore subject to essential ambiguities.

Table II **Solar Probe Scientific Objectives**

Explore *in situ* the last major unprobed region of our solar system: the inner heliosphere between 4 and 65 R_S

- **Strong scientific reasons**
- **Technology exists now**
- **Mission costs are reasonable**

Major Scientific Objectives

- **Coronal Structure**
  Large scale structure
  Small scale strucure
  Time variations
- **Coronal Heating**
  Energy transport in the lower atmosphere
- **Solar Wind Acceleration**
- **Plasma Turbulence within the Coronal Envelope**
- **Acceleration, Storage and Transport of Energetic Particles**
- **Sources, Sinks and Dynamics of Interplanetary Dust**

Much of what we know about the acceleration mechanisms that generate solar energetic particles comes from observations in interplanetary space beyond 0.3 AU. However, the use of these measurements to infer acceleration mechanisms and physical conditions within coronal acceleration regions is also subject to essential ambiguities. Specifically, acceleration efficiency cannot be inferred from interplanetary observations because most energetic particles remain confined to the Sun on closed magnetic loops. Furthermore, flux distributions of acclerated particles that manage to escape are significantly modified in interplanetary propagation by velocity dispersion (caused by energy-dependent diffusion) and by convection and deceleration in the expanding solar wind. These modifications are so strong, especially at energies below 1 MeV/nuc, that no decisive discrimination among the different coronal acceleration models is possible using interplanetary data obtained beyond 0.3 AU. Only measurements close to the Sun, especially at lower energies, which will be made by Solar Probe, will provide the necessary discrimination between competing models.

Suprathermal and energetic particles observed within the coronal envelope with Solar Probe can be used to probe the plasma and wave conditions at lower altitudes where coronal heating and particle acceleration are occurring. More specifically energetic electrons observed with Solar Probe as close as 4 R_S can be used to probe the solar corona at lower altitudes to provide definitive information on density profiles and magnetic field topology. Measurements of charge states of suprathermal and energetic ions will give a clear indication of the origin for these particles and will identify the regions where acceleration takes place. Composition and energy spectra of suprathermal and energetic (> 10s of MeV) ions measured close to the Sun will provide the most unbiased information on solar acceleration processes and should settle the question

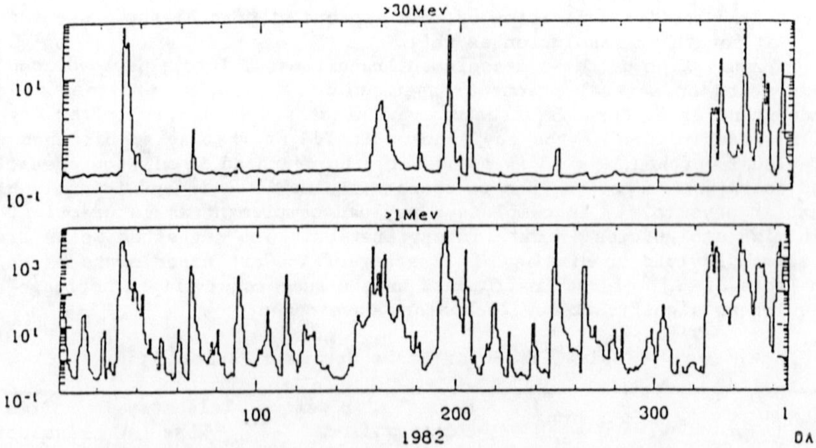

Fig. 1 One year of daily averaged proton integral fluxes for two energy levels measured by the Johns Hopkins University Applied Physics Laboratory (JHU/APL) energetic particle instrument on IMP-8 during 1982. Most of the activity at low energies is produced by variable interplanetary propagation and interplanetary acceleration conditions.

of energetic particle storage and/or continuous acceleration near the Sun. Measurements of the radial and azimuthal intensity gradients of both the suprathermal and energetic ions in the inner (< 65 R_S) solar system during times of solar active periods will give us crucial information on particle transport near the Sun and the escape of solar particles into interplanetary space. Gradient measurements with Solar Probe during quiet times should establish the origin of the steady suprathermal component and its role in the dynamics of the inner heliosphere. And finally, observation of galactic cosmic rays over the solar poles will determine modulation effects in an essentially radial field configuration.

MEASUREMENT REQUIREMENTS AND SCIENCE PAYLOAD

The recurring theme throughout the discussion of science objectives presented above was that although we have been able to establish useful limits on mechanisms that heat the corona, accelerate energetic particles, and drive both the steady state and transient solar wind, essential ambiguities remain that prevent a sufficient level of understanding to allow application to general astrophysical systems. These ambiguities stem primarily from the difficulty in correcting for transport from the Sun to the closest solar distance yet sampled using *in situ* detection techniques, about 0.3 AU. Therefore, a first priority for the measurement objectives of a Solar Probe is to fully define the particle and field, spatial and temporal structure of that part of the solar corona that can be directly sampled, i.e., $R > 4\ R_S$. However, many of the scientific objectives identified require definition of some aspects of coronal structure inside of 4 R_S for their resolution as well.

Translation of these measurement requirements into a payload complement rests on several practical assumptions. Clearly, progress in our understanding of coronal structure and processes will result from even a partial definition of the environment inside of that solar distance not previously probed, $R < 65\ R_S$. However, the overhead in mission operation and cost just to get to, and to operate inside 65 R_S is sufficiently high that it pays to fly as complete a payload complement as is practical to minimize ambiguities in data interpretation. The suggested Solar Probe Strawman Payload consisting of 12 state-of-the-art experiments is shown in Table 3. All of the instruments have a successful flight heritage and require no significant new technology development.

Table III **Solar Probe Strawman Payload**

	Mass kg	Power W	Telemetry kbps	Time Resolution
Fast Plasma	8	8	5	5 ms
3 D Ions	12.5	17	10	1 s
3 D Electrons	3.5	3	5	1 s
Magnetometer	5.5	6	2.5	40 ms
Plasma wave	19	11	15	5 ms
Thermal Composition	12	8	5	1 s
Suprathermal Comp.	13	9	10	1 s
Medium Energy Particles	10	8	10	1 s
High Energy Particles	10	8	1	1 s
Coronal Spectral Imager	25	15	5	30 min
Neutron	10	5	1	10 s
Dust	5	5	0.5	10 s
TOTAL	133.5	103	70	

The **plasma package** consists of three sensors required to define the radial dependence of the bulk flow beyond 4 R_S. However, it also contributes to 1) a definition of the nature of the local plasma turbulence through measurement of ion and electron velocity distribution functions, 2) a definition of the magnetic topology through measurement of suprathermal electron angular anisotropies, 3) separately resolving spatial from temporal variations through measurements of currents and through cross correlations between measured bulk flow moment, and vector magnetic field variations, 4) a definition of the suprathermal ion and electron populations, and 5) an understanding of mass fractionation processes through measurements of alpha-particle velocity distribution functions.

The **magnetometer experiment** measures the vector d.c. magnetic field beyond 4 R_S, which is crucial for defining the bulk structure of the solar corona. This includes both the magnetic topology of the corona as well as its total energy content. It is also essential, combined with measurements of the bulk plasma parameters, in defining scale lengths, frequencies and speeds that characterize the ambient medium, and in defining the nature of coronal MHD turbulence.

The primary role of the **plasma wave experiment** is to define the nature of coronal plasma turbulence thereby helping to identify local heating mechanisms and define the local transport characteristics of the medium. The radio wave portion of this experiment will also help detect and identify nonlocal transient processes induced by solar activity as well as provide a 3-D overview of coronal turbulence and filamentary structure using the X-band transmission channel to Earth-based receivers. This last type of information will provide an essential link to our present understanding of coronal structure, processes, and resultant acceleration, since our most direct present information stems from many years of measurements of radio scintillations.

The **thermal and suprathermal ion composition** package, consisting of two sensors will characterize the local thermal and suprathermal ion populations, essential for determining how and where solar wind heavy ions are accelerated. These measurements, extending from 15 eV/Q to ~ 200 keV/Q, will provide 1) precise composition and charge state measurements, 2) profiles of the number densities, velocities, and temperatures of heavy ion species, and will probe 3) mass fractionation processes through definition of the total thermal and suprathermal heavy-ion population, and 4) the coronal temperature profile inside of 4 R_S through measurement of heavy-ion charge-state distributions.

The package of two **energetic particle sensors** characterizes the local suprathermal and energetic particle populations and provides a nonlocal probe of 1) the occurrence of transient energy releases through detection of resulting accelerated particles, 2) magnetic topology through measurement of time-dependent, energy-angle flux distributions, 3) coronal MHD turbulence through inferences of the efficiency of particle acceleration by the first- and second-order Fermi mechanisms, and 4) local spatial coronal filaments through detection of non field-aligned angular anisotropies.

The imaging experiment, a **coronal spectral imager**, provides an essential 3-D overview of coronal structure to define the global context and its time variability, within which the local measurements by the other Solar Probe experiments fit. Additionally it helps define 1) the presence of fine-scale filamentation, in 3-D, through spatial and/or time-dependent variations in EUV-line and white-light intensities, 2) the MHD component of coronal turbulence through measurement of the broadening of selected resonance lines, 3) the global bulk velocity structure through measurement of the centroid wavelength shifts of selected resonance lines, 4) the structure of the neutral exosphere, and 5) corona dust shells.

The combined **neutron/gamma-ray experiment** provides a nonlocal probe of solar transients to help define the true occurrence rate of solar transients, information regarding ion acceleration mechanisms, and hence, concomitant plasma structures and turbulence generated by the transient energy releases. This type of probing may be our only way to view the smaller energy-release events, such as microflares, because these events are expected to remain totally confined to closed magnetic structures that do not extend to heights that overlap possible Solar Probe trajectories.

Finally, the **dust experiment** will measure the meteoroid flux and mass distribution, as well as velocity and trajectory distribution throughout the inner solar system between 4 and 1100 R_S. These measurements are essential for determining the relative importance of the several sources and sinks of the zodiacal dust cloud as well as the mechanisms that regulate the dynamics of individual dust grains during their lifetime.

MISSION DESIGN

A mission concept that is both affordable and feasible using current technology requires a ΔV maneuver at initial aphelion near 2 AU followed by an Earth swingby that adds sufficient energy to the spacecraft to enable a Jupiter gravity assist trajectory. The final heliocentric orbit passes over both solar poles and reaches perihelion over the solar equator at 4 R_S. A polar trajectory was chosen for two reasons. First, it maximizes the number of different coronal structures likely to be intercepted during perihelion passage, and second, it maximizes the real-time telemetry rate because the spacecraft will always be in view from the Earth. A 4 R_S perihelion was chosen because it falls well inside the known turbulence envelope of the Sun, yet presents a safe distance using currently available heat-shield technology.

A Solar Probe ΔV-Earth Jupiter Gravity Assist trajectory for a September 1998 launch is shown in Figure 2. The added capability of the ΔV-EJGA mission option allows delivery of the entire strawman payload given in Table 3, to a perihelion of 4 R_S with a liberal mass margin. The first perihelion passage would take place in September 2002 and the period of the final heliocentric polar orbit would be 4.2 years. Several options for reducing this orbital period to 2.5 and perhaps 2 years, are being considered.

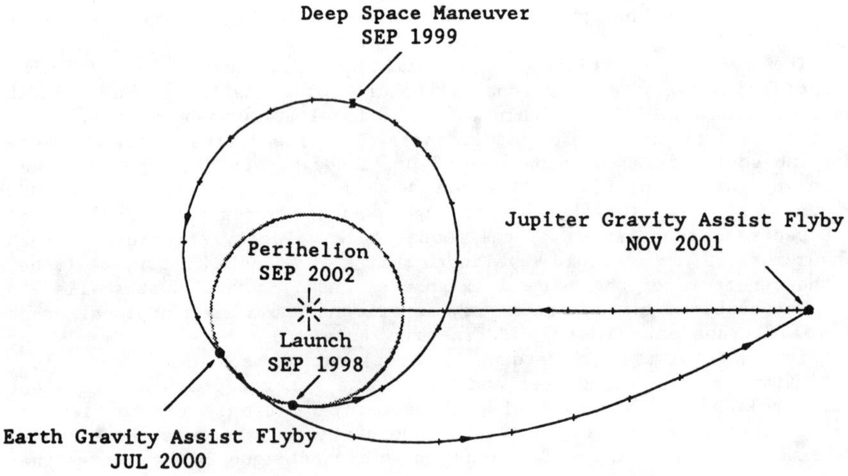

Fig. 2 Solar Probe ΔV - Earth Jupiter Gravity Assist Trajectory.

SPACECRAFT AND INSTRUMENT ACCOMMODATION

The primary technical driving requirements of the Solar Probe mission include delivery of a science payload totalling 133.5 kg at 103 watts that generates a 70 kbit/s data stream, into a 4 R_S perihelion solar polar orbit. Auxiliary, but important support requirements include 1) a 60 RPM, rotating instrument spin platform, 2) a retractable magnetometer boom, 3) two survivable or retractable (one long, one short) electric dipole antennae, 4) a 6.5 Gbit data storage medium, and 5) capability to shorten the final orbital period to 2.5 years.

A concept drawing of the Solar Probe spacecraft is shown in Figure 3. The spacecraft is 3-axis stabilized with a large carbon-carbon conical heat shield employed to shield the payload section during the perihelion encounter. Most of the science instruments are mounted on a spinning platform that is extended from the bus portion of the spacecraft. A liquid kick stage to provide the ΔV is needed for the ΔV-EJGA trajectory.

Most of the science payload will be on an extendable spinning platform allowing nearly hemispherical viewing away from the shield. Instruments most sensitive to radiation will be placed on the far side of this platform to maximimize the distance to the RTGs. Exceptions will be the Fast Plasma Experiment, the Coronal Spectral Imager, the Plasma Wave Experiment, and the Magnetometer. The Magnetometer will be placed on an extendable, zenith-pointing mast; the other three will be located near the electronics bus.

The current design of the Solar Probe spacecraft calls for the liquid stage kick motor to be used for all deep space ΔV maneuvers, and carried until just after the Earth flyby. The science platform and magnetometer mast can, therefore, not be extended until the liquid stage is jettisoned. The magnetometer boom will need to be retracted for thermal reasons during the perihelion phase of the mission ($\sim< 10$ R_S). The plasma wave electric antennae will either be composed of a refractory material that will permit exposure during the perihelion phase, or will need to be retracted during this period. The science platform will be held close to the edge of the umbra in order to optimize sensor fields of view during the early perihelion phase. This will be accomplished by yawing the spacecraft to tip the science platform as close to the edge of the umbra as possible.

Three RTGs composed of ten modules each will provide 530 watts of beginning-of-mission power (end-of-mission power of about 430 watts). The RTGs are oriented such that the axial end is facing the science platform to minimize the RTG radiation effects. The distance from the end of the RTGs to the center of the science platform will be about 3 meters at perihelion.

The telecom subsystem will receive and transmit an X-band signal via a 1 m High Gain Antenna (HGA), mounted in the area between the spacecraft bus and science platform. The real time telemetry rate at perihelion will be 70 kbits/s. There is also a Low Gain Antenna (LGA) that will be used for communications with Earth immediately after launch and as a backup for the HGA. The transmitted signal will be generated by a 5.6 watt X-band solid state amplifier similar to that designed for Mariner Mark II. In addition to the real time telemetry, the Digital Tape Recorder (DTR) will have a data capacity of 6.5 X 10^9 bits.

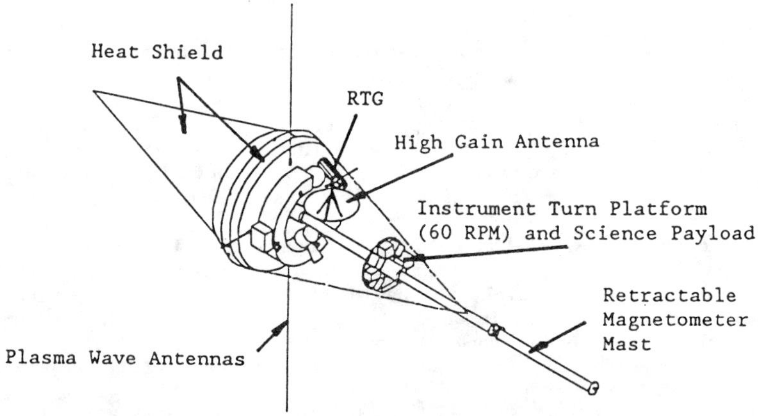

Fig. 3 Solar Probe Spacecraft, deployed configuration.

The spacecraft propulsion subsystem composed of ten 20 N thrusters (Isp = 2254 N-s/kg) will be able to supply 250 m/s ΔV. The spacecraft propulsion subsystem could allow an orbital final period of as low as 2.5 years. A liquid propulsion module that will supply an additional 2200 m/s ΔV is also included.

The payload bay will be maintained at a temperature of 25° to 30° C by a set of carbon-carbon heat shields. The primary shield will be conical in shape and will be pointed toward the Sun during the encounter phase. The mass loss rate from the shield at perihelion shall be less than 2.5 mg/s. In addition to the primary shields, two secondary shields will be employed to provide IR shielding. The primary cone will have an included angle of 30° and at perihelion, will create a 30° conical umbra containing the payload (see Fig. 3). The shield has a diameter of 4 m and a length of 7 m.

Acknowledgements. The Solar Probe Study Team acknowledges the extensive and essential engineering support provided by JPL personnel under the direction of James Randolph.

REFERENCES

1. Feldman, W.C., M. Mellott, B. Tsurutani, B. Goldstein, R. Farquhar, J. Anderson, D. Bohlin, L. Burlaga, G. Gloeckler, J. Harvey, T. Holzer, V. Jones, P. Kellogg, S.M. Krimigis, M. Kundu, A. Lazarus, E. Parker, R. Rosner, G. Rottman, J. Slavin, S. Suess, R. Woo, and R. Zwickl, Solar Probe - Scientific Rationale and Mission Concept; A report of the 1989 Solar Probe Science Study Team, JPL D-6797, (1989).

THE MERCURY DUAL ORBITER MISSION

D. N. Baker and J. A. Slavin
NASA/Goddard Space Flight Center, Greenbelt, MD 20771

ABSTRACT

The Mercury Orbiter (MeO) will carry out a full range of particles, fields, and planetary imaging science at Mercury. Present mission plans call for a launch in 1999 with a flight time of about 4.5 years. By means of multiple Venus and Mercury gravitational assists, the mission can be accomplished with present U.S. launch vehicles and a very large payload can be placed in orbit around Mercury. The dual-spacecraft concept will permit outstanding scientific study of solar cosmic rays and the solar wind throughout the inner heliosphere from 0.3 AU to 1.0 AU. Modest enhancements to the planned magnetospheric instruments and utilization of onboard solar instruments will permit unique investigation of solar particle acceleration and transport with the MeO spacecraft.

HISTORICAL BACKGROUND

Mariner 10 was the first spacecraft to fly by the planet Mercury. In all there were three close encounters in 1974 and 1975. Mariner 10 results were very exciting: they revealed that Mercury possesses a strong intrinsic magnetic field, that it has a well-developed, remarkably Earth-like magnetosphere, and that there appears to be activity near Mercury that is very much like magnetospheric substorms at Earth. How such a small, slowly rotating planet can have such a strong magnetic field is a puzzle.[1] When and how Mercury formed is one of the outstanding issues of solar system evolution.[2]

The brief glimpses of Mercury afforded by Mariner 10 suggested the need for complete imaging of its surface and for more thorough documentation of its magnetic and plasma environment. To achieve these goals, an orbiter would be required. Studies were undertaken by the Jet Propulsion Laboratory (JPL) in 1976-78 to define a Mercury orbiter mission. In order to place significant payloads in orbit around the planet, it was concluded that very advanced propulsion systems (e.g., solar electric or solar sail) would be required. Moreover, it was concluded that thermal design problems would be extremely formidable for a large, 3-axis stabilized spacecraft at low altitudes over the hot (600 K) dayside surface of Mercury. These technology drivers made a Mercury orbiter mission appear out of reach.

In 1985, a class of multiple Venus and Mercury gravitational assist trajectories were discovered[3] that permitted insertion of large payload masses (~ 2000 kg) at Mercury. These trajectories required long cruise durations (Fig. 1) but they used present conventional U.S. launch capabilities of the Titan IV class. This new mission delivery system opened the exciting possibility of a very capable, moderate cost mission to Mercury. In 1988, the NASA Space

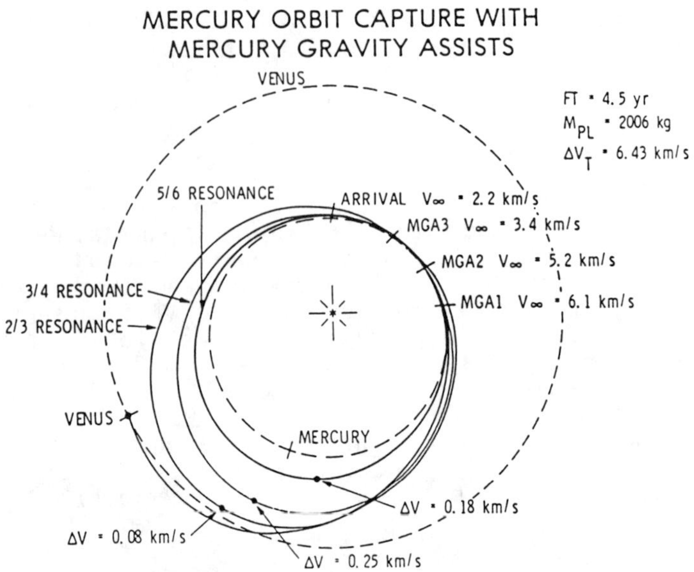

Figure 1. MeO gravity assist trajectory (total flight time = 4.5 year).

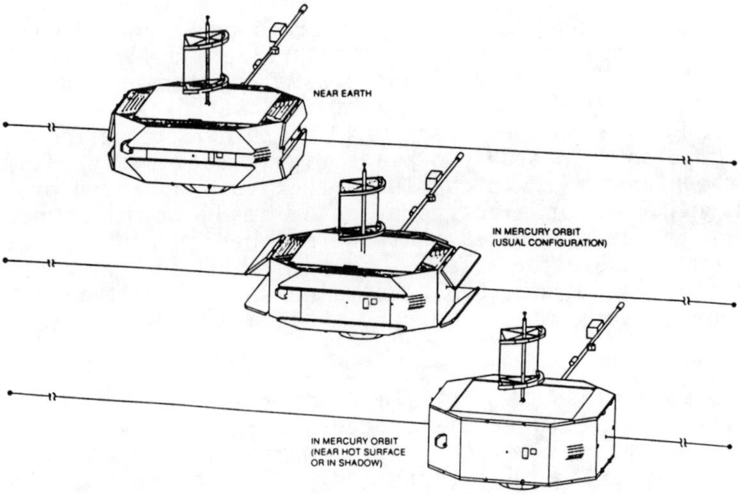

Figure 2. MeO spacecraft with solar panels in three positions.

Physics (Code ES) and Planetary Exploration (Code EL) Divisions initiated a pre-Phase A study of a possible Mercury orbiter mission.

STUDY TEAM RESULTS

The Mercury Orbiter Science Working Team (SWT) explored and refined the rationale for a return to Mercury. It was concluded that there are outstanding opportunities to compare and contrast the magnetosphere of Mercury with those of Earth and the other planets. The principal space physics objectives of the MeO mission would be to: (1) map the detailed plasma and magnetic structure (in three dimensions) of Mercury's small-scale magnetosphere; (2) determine the key similarities (and differences) during magnetospheric substorms at Earth (with its dense ionosphere) and at Mercury (with its virtually absent ionosphere); (3) examine the rate and nature of extraction of energy by Mercury's magnetosphere from the denser, more variable solar wind at 0.3-0.5 AU; (4) investigate heliosphere structure and dynamics within 0.5 AU of the sun; and (5) take advantage of the closeness of Mercury to the sun to measure solar neutrons, gamma rays, and solar energetic particles.

With light-weight, cleverly designed spin scan systems, it is possible to achieve several basic planetology goals at Mercury. The surface science objectives of MeO are to: (1) accomplish the global surface mapping which was only 40-percent completed by Mariner 10; (2) obtain global geochemical terrain maps of the occurrence of Fe, Th, K, Ti, Al, Mg, Si and other elements in the Mercury regolith; (3) determine the intrinsic magnetic field with sufficient detail to detect magnetic anomalies and (it is hoped) understand the field origin; and (4) measure the Mercury gravitational field in detail (including associated anomalies).

To meet these science objectives, the Mercury Orbiter SWT identified a strawman payload consisting of 10 instruments. These instruments, all of which are based upon mature technologies, are listed in Table 1.

Spacecraft engineering and mission design studies were carried out at JPL in parallel with the SWT activities (Fig. 2). The findings of the engineering study show that two spin-stabilized spacecraft can be launched from a single Titan IV launch vehicle. These spacecraft[4] can have over 1500 kg of propellant and can have dry payload weights approaching 1000 kg. The resulting total ΔV capability of several thousand m/s will allow a wide range of orbital maneuvers. Thus, the studies show that the comprehensive particle and field experiments along with the very capable planetology instrument complement can be placed in highly elliptical orbits at Mercury and can survive without costly sun shields or active cooling systems.

Table 1. Mercury Orbiter Strawman Payload

	Power (w)	Mass (kg)
DC Electric Field Analyzer	6.0	14.6
Energetic Particle Detector	15.0	15.0
Fast Electron Analyzer	5.0	4.0
Fast Ion Analyzer	5.0	4.0
Gamma/X-Ray Spectrometer	14.3	17.0
Ion Composition Plasma Analyzer	12.0	10.0
Line-Scan Imaging System	11.0	5.1
Magnetometer	5.5	5.3
Optimized Solar Wind Analyzer	10.0	10.0
Radio/Plasma Wave Analyzer	6.5	4.6
Solar Neutron Analyzer	10.0	10.0

SOLAR AND HELIOSPHERIC SCIENCE

The MeO study team recognized quite clearly the unique opportunity afforded during the MeO cruise phase for solar physics and heliospheric science studies. This comprises study of the large-scale structure of the solar wind in the inner heliosphere, including solar wind streams, coronal mass ejections, and the inner part of the heliospheric current sheet. MeO also offers unique opportunities to study evolving shock waves, discontinuities, and solar wind turbulence regions. Particularly fruitful opportunities exist to study solar flares and solar energetic particle. There will be rich opportunities to observe the flare particle acceleration process and to examine the transport of escaping particles from the sun.

The advantages of the MeO mission for the solar and heliospheric science include superior magnetic field orientations, lower Alfven Mach numbers, less highly evolved solar wind structures, and better connection to solar source regions. Of course, the proximity to the sun will permit unique detection possibilities for low-energy solar neutrons and nearly circular heliospheric orbits inside ~ 0.5 AU will allow nearly continuous sampling by MeO of a region only briefly visited by HELIOS.

The dual spacecraft nature of MeO will permit azimuthal, radial, and spiral (field-aligned) separations of identical instruments up to ~ $3x10^6$ km. This will permit detailed study of heliospheric fine structure and shock-associated particle acceleration.

A key to the unique role of MeO for solar energetic particle (SEP) studies is that such investigations at ~ 0.3 AU would allow determination of temporal histories without the interference of interplanetary propagation effects. Examining SEP acceleration for a broad range of event types without propagation uncertainties (such as exist at 1 AU) could eliminate many of the fundamental ambiguities that plague cosmic ray models that include diffusion, convection, and adiabatic deceleration. Present models suggest that in large flares, long injection time scales reflect an acceleration history at the sun due to strong shocks which move through the corona accelerating particles out to tens of solar radii. On MeO, often located near 60 solar radii, this picture could be readily tested.

Another important solar particle event type to study with the MeO mission would be small, impulsive flares. These are often rich in gamma-rays ^3He, and Fe. Near the sun, these events would have large fluxes and large anisotropies. Using the energetic particle sensors on MeO, in combination with the gamma-ray and neutron sensors, would give critical insights into the processes operating in impulsive flares.

SUMMARY

The Mercury Dual Orbiter Mission offers an exciting opportunity to perform unique magnetospheric science studies. It will help us to understand some of the fundamental issues of magnetospheric dynamics and solar wind-magnetosphere coupling thorough comparison and contrast of Mercury and Earth. It will also afford unique opportunities for the study of solar and heliospheric physics during its extended cruise phase. Particularly valuable studies can be performed in solar energetic particle investigations using the MeO platforms. Thus, MeO provides extensive, exciting science return for a broad cross-section of the space physics community for a very moderate cost.

REFERENCES

1. C. T. Russell, D. N. Baker, and J. A. Slavin, in Mercury (F. Vilas, C. Chapman, and M. Mathews, eds.; Univ. of Arizona, Tucson, 1988), p. 514.
2. J. S. Lewis, in Mercury, p. 651 (1988).
3. C. L. Yen, Paper AAS85-346, AAS/AIAA Astrodynamics Specialist Conf., Colorado (Aug. 1985).
4. C. L. Yen, D. H. Collins, and S. A. Meyer, Paper AAS89-195, AAS/AIAA Astrodynamics Specialist Conf., (Jan. 1989).

A REVIEW OF THE "21ST CENTURY STUDY"

Dietrich Müller
Enrico Fermi Institute and Department of Physics
The University of Chicago, Chicago, Illinois 60637

ABSTRACT

In 1984, the Space Sciences Board of the National Research Council initiated a study of the Directions of the Space Sciences in the 21st Century, or more exactly, for the approximate period 1995-2015. The study report appeared in print in 1988. In the following, we give an overview of this report, discuss the recommendations in the context of earlier strategy documents, and in contrast to the realities of implementation. We specifically discuss the status of the particle astrophysics program.

I. INTRODUCTION

Under the auspices of the Space Sciences Board of the National Research Council, a study "Space Science in the Twenty-First Century: Imperatives for the Decades 1995 to 2015" was initiated in 1984, and was completed in 1988. It is instructive to review this report in the context of other Academy reports of recent years that address related questions. These include the Astronomy and Astrophysics Surveys for the 1970's ("Greenstein report") and for the 1980's ("Field report"), and the "Strategy for Space Astronomy and Astrophysics for the 1980's" report. At present, a survey for the 1990's is in preparation under the chairmanship of John Bahcall.

The major recommendations of these reports and the status of implementation, and, in particular, the status of particle astrophysics, are summarized as follows:

(a) <u>Greenstein Report (1972):</u>

The highest priority major program is the construction of a very large radio array (VLA). For observations in space, the report recommends a series of large High Energy Astronomical Observatories (HEAO's), mainly towards x-ray and gamma-ray observations. Little mention is made of cosmic rays and high energy particles. The HEAO program was implemented in the 1970's at a much smaller level than originally planned, but HEAO-3 carried two cosmic ray detectors.

(b) <u>Strategy for Space Astronomy and Astrophysics for the 1980's:</u>

This report, published in 1979, is restricted to space missions. Major recommendations are, in order of priority:

(i) Space Telescope and Solar Polar Mission,

(ii) Gamma Ray Observatory (GRO), X-ray Telescope (XRT), and Cosmic ray Observatory (CRO).

None of these missions have been launched yet, but three are scheduled for lift-off in 1990: the Hubble Space Telescope, the Ulysses Mission (the European half of the Solar Polar Mission; the US spacecraft was cancelled), and the Gamma Ray Observatory. XRT is now in preparation under the name AXAF, but the Cosmic Ray Observatory was never implemented.

(c) Field Report (1982):

The Report of the Astronomy Survey Committee declares AXAF, the "Advanced X-ray Astrophysics Facility" as the highest priority *major program* for the 1980's. AXAF is followed in priority by a Very Long Baseline Array of radio telescopes (VLBA), a New Technology Telescope (NTT), and a large Deployable Reflector in Space (LDR). Among seven *moderate programs*, the report recommends a series of Cosmic Ray Experiments in Space. The High Energy Astrophysics Panel of the Survey Committee gives Cosmic Ray Studies the second highest priority (after AXAF).

None of the recommended space missions have been launched as yet, although AXAF now is an approved mission.

(d) Space Science in the 21st Century (1988):

This study covers all scientific disciplines that are involved in or make use of space borne investigations. In astronomy and astrophysics, the program recommended for the period 1995-2015 includes: Imaging Interferometry from space, Large Area and High Throughput Telescopes, and Astromag, a massive cosmic ray analyzer in space.

This study shall be discussed in more detail in the following.

II. "THE 21st CENTURY STUDY" - OVERVIEW:

The following is an excerpt of the preface of the report by Tom Donahue, Chairman of the Space Sciences Board at that time:

"Early in 1984, NASA asked the Space Science Board to undertake a study to determine the principal scientific issues that the disciplines of space science would face during the period from about 1995 to 2015. This request was made partly because NASA expected the Space Station to become available at the beginning of this period, and partly because the missions needed to implement research strategies previously developed by the various committees of the board should have been launched or their development under way by that time. A two year study was called for. To carry out the study the board put together task groups on earth sciences, planetary and lunar exploration, solar system space physics, astronomy and astrophysics, fundamental physics and chemistry (relativistic gravitation and microgravity sciences), and life sciences. Responsibility for the study was vested in a steering group whose members consisted of task group chairmen plus other senior representatives of the space science disciplines. To the board's good fortune, distinguished scientists from many countries other than the United States participated in this study."

The study group started their work under the aura of success of the Space Shuttle program, and of the expected availability of a large space station and its associated infrastructure in the early 1990's. The members of the group were asked to bring forward innovative and imaginative ideas, and not to be overly constrained by budgetary considerations. This climate of optimism was sadly shaken by the Challenger accident, and the final report was completed some time after this accident.

The report includes a wealth of ambitious projects which would severely challenge the budgetary and technological resources available in the foreseeable future. A list of these projects is given in the following:

EARTH SCIENCES: MISSION TO PLANET EARTH

- 5 Geostationary Satellites
- 2-6 Polar Orbiting Platforms
- Special Missions in other Orbits

PLANETARY AND LUNAR EXPLORATION

- *Mercury:* Orbiter, Lander/Sensor Network
- *Venus:* Probe, Lander/Sensor Network, Sample Return
- *Moon:* Lander/Sensor Network, Rover, Sample Return
- *Mars:* Lander/Sensor Network, Rover, Sample Return, Human Exploration
- *Jupiter:* Polar Orbiter, Atm. Probe, Io Lander
- *Saturn:* Orbiter, Atm. Probe, Ring Rendezvous, Titan Orbiter, Probe and Lander
- *Uranus:* Orbiter, Atm. Probe
- *Neptune:* Orbiter, Atm. Probe
- *Pluto:* Orbiter
- *Comets:* Coma and Nucleus Sample Return
- *Asteroids:* Rendezvous and Sample Returns

SOLAR SYSTEM SPACE PHYSICS

- High Resolution UV and X-ray Telescopes Solar Probe
- Advanced Spacecraft EOS, STO, ASO
- Missions to Outer Planets (see above)
- Interstellar Probe

ASTRONOMY AND ASTROPHYSICS:

- Imaging Interferometry
  (a) Large Space Telescope Array
  (b) Long Baseline Optical Space Interferometer
  (c) An array of Very-Long-Baseline Interferometry (VLBI) stations in space
- Large Area and High Throughput Telescopes
  (a) A large deployable reflector (LDR) for submillimeter studies
  (b) An 8- to 16-m optical space telescope
  (c) Large area telescopes for the energy range 20 keV to 2 MeV
  (d) A large Compton telescope for spectroscopy, 0.1 to 10 MeV
  (e) Large gamma-ray telescopes for energies above 2 MeV
- Astromag, a massive cosmic-ray analyzer in space

FUNDAMENTAL PHYSICS AND CHEMISTRY

- Laser Gravitational Wave Observatory in Space (LAGOS)
- Mercury Orbiter
- Precision Optical Interferometer in Space (POINTS)
- Star Probe
- Principle of Equivalence Experiment
- Large Area X-ray Detector (timing of x-ray pulsars)

MICROGRAVITY SCIENCE, AND LIFE SCIENCES

No <u>dedicated</u> missions proposed.

III. ASTRONOMY & ASTROPHYSICS TASK GROUP REPORT

The task group was asked to identify the main scientific thrusts of the field. Broad themes are:

- the origin of the universe
- the laws of physics governing the universe
- the birth of stars and planets, and the advent of life

To derive a plan for the 21st century, the task group first summarized the expected status of the field in 1995. The group assumed that the "great observatories" will be operative in 1995:

- HST will be operational with at least one, and probably two, instrument upgrades completed.
- GRO will be nearing the end of its operating lifetime with about 50 percent of the scientific instrumentation operating.
- AXAF will be in orbit and nearing the time for its first servicing.
- SIRTF will be operating with the first cryogen servicing planned for the near term.

Further it was assumed that "Moderate Missions" in orbit by 1995 include: EUVE, FUSE/LYMAN, XTE, QUASAT, ACE, HNC, CRN, Spacelab and Spartan Payloads, and that several European and Japanese missions are realized: ROSAT, ISO, ASTRO C, XMM, HIPPARCOS.

In Section IV, we will give further detail on the status and future of particle astrophysics. In the areas of "zero rest mass" Astronomy for the 21st Century, the following directions were identified:

(a) High Resolution Interferometry:

The goal is angular resolution in the region 10^{-3} arc seconds to, ultimately, 10^{-6} arc seconds. Realization of this goal would come from optical and infrared interferometric arrays, either mounted on a compensating structure, or in free flying configurations. For instance, two 8 m class diffraction limited telescopes, separated by 100 km, or arrays of radio telescopes, in orbit at baselines of $\sim$ 100 earth diameters, would yield μ *arcsec* resolution.

(b) High Throughput Instruments

This includes the Large Deployable Reflector (LDR), a 20-30 m aperture telescope for far infrared and sub-mm observations. This instrument would have to be assembled in space.

An 8-16 m telescope would be a successor for HST, about 10-20 years after HST is launched.

In the x-ray region, a Very High Throughput Facility (VHTF) should provide high sensitivity, angular, spectral, and time resolution. It would be assembled in space and provide a total effective area of $\sim 30\ m^2$. At higher energies (20-100 keV), a Hard X-Ray Imaging Facility (HXIF) might be considered. This would be use large area coded aperture arrays.

In gamma ray astronomy, the initiatives following GRO would include state-of-the-art spectroscopy in the 0.1 to 10 MeV region, the next generation Compton telescopes at medium energies (1-60 MeV), and large systems with good angular resolution at high energies, perhaps up to 10^{11} eV.

IV. THE COSMIC RAY PROGRAM

The Astronomy and Astrophysics Task Group recognized that only few of the cosmic ray observations recommended in previous reports are being implemented:

"A continuing series of observations of galactic cosmic rays at higher energies is being conducted on balloons and on the Space Shuttle. These provide important exploratory data, but definitive measurements have to await the availability of spacecraft capable of carrying very large payloads for extended periods of time (years). Objectives of such measurements are to determine: the isotopic abundances at high energies; the composition of the rare ultraheavy nuclei; the energy spectra of nuclei and electrons over a wide range; and the abundances of positrons and antiprotons. The Astronomy Survey Committee had recommended carrying out a series of such observations (*Astronomy and Astrophysics for the 1980s*), but only a few investigations are being implemented. Thus, by 1995 we will probably still be short of decisive data that could specify the character of particle acceleration in the galaxy, or that could define the nucleosynthesis processes leading to cosmic ray matter and governing evolution of the galaxy. We will probably also lack definitive data that could further our understanding of the structure and properties of the interstellar medium, and that could address some fundamental cosmological questions such as matter/antimatter symmetry in the present universe."

The Expected Status of the Field in 1995 was summarized as follows:

(a) *Interplanetary Spacecraft*: Pioneers 10 and 11 and Voyagers 1 and 2 will continue to explore the solar system at large distances, with near-earth monitoring with IMP-8 and ISTP/Wind. The Ulysses mission will have been launched.

(b) *Cosmic Ray Composition.* The Heavy Nucleus Collector (HNC) on the Long Duration Exposure Facility (LDEF) will have investigated ultra-heavy cosmic rays at low energies. A comprehensive set of particle composition measurements from solar wind energies to galactic cosmic rays will be underway with the Cosmic Ray Composition Explorer (ACE).

(c) *Space Shuttle and Space Station*: Several reflights of the Cosmic Ray Nuclei Detector (CRN) on the Space Shuttle will have measured the elemental composition of cosmic rays to energies around $10^{13} - 10^{14}$ eV. A long duration flight of CRN on a space platform or an early configuration of the Space Station will provide measurements up to $\sim 10^{15}$ eV.

(d) *Instrument Development*: An active program of instrument development will be in progress using balloons and in laboratories.

On this basis, the following programs are proposed for the period 1995-2015:

(a) *Astromag*: This will be a large superconducting magnet spectrometer facility with a succession of interchangeable detector systems. It is expected to be in orbit for $\sim$ 10-20 years. Observational goals include anti-proton and anti-matter searches, measurements of the isotopic composition and of the energy spectra of cosmic rays, and investigation of electrons and positrons.

Magnet Spectrometers had been flown on balloons before (Chicago, Berkeley, NASA-JSC, NASA-GSFC), and had been proposed for HEAO and for the Space Shuttle. The new and attractive element of the present proposal is the facility nature of Astromag. Servicing and interchangeability of instruments is made possible through the resources of the Space Station.

(b) *Interplanetary and Interstellar Measurements*: In-situ measurements of low energy cosmic rays must be made with a network of detectors at different locations in the heliosphere. Particle detectors on an interstellar probe should, for the first time, measure the low energy galactic cosmic rays beyond the heliosphere.

(c) *Large Detector Arrays in Space:*

(i) High Energy Array: A very massive detector array could provide direct cosmic ray composition measurements up to 10^{16} eV, i.e. beyond the "break" in the overall energy spectrum.

(ii) Ultraheavy Nuclei: A very large array of detectors will be assembled in space to measure the energy spectra of ultraheavy cosmic rays.

(d) *Experiments on a Polar Platform:* A polar platform is required for measurements of low energy antiprotons and positrons and to determine the isotopic composition of ultraheavy nuclei.

Among these projects, the Astromag program generated considerable interest among the Study Group participants. It has subsequently been further discussed and refined by a NASA definition team, and it has now been selected to become an attached payload to the Space Station Freedom.

V. CONCLUSION

Clearly, the goals outlined in the Study are ambitious and optimistic. Nevertheless, they do provide some perspective on possible future directions of Space Science. For the cosmic ray community, it might be reassuring that Astromag is one of the very few projects brought forward in the Study that in fact has now been selected for implementation. On the other hand, many of the more modest near term objectives remain unsupported and without flight opportunity. This includes HNC and CRN, as well as parts of the balloon and instrument development program. One should hope that further programmatic efforts, including the study in progress by the Bahcall committee, will recognize that the more mundane programs of today are necessary in order to develop the technological and human resources required for the ambitious goals of tomorrow.

Members of the Steering Group:

Thomas M. Donahue, (Chairman), Don Anderson, D. James Baker, Robert Berliner, Bernard Burke, A.G.W. Cameron, George Field, Herbert Friedman, Donald Hunten, Francis Johnson, Robert Kretsinger, Stamatios Krimigis, Eugene Levy, Frank B. McDonald, John Naugle, Joseph M. Reynolds, Frederick Scarf, Scott Swisher, David Usher, James Van Allen, Rainer Weiss

Members of the Task Group on Astronomy and Astrophysics:

Bernard Burke, (Chairman), James Roger Angel, Jacques Beckers, Andrea Dupree, Carl E. Fichtel, George Field, Riccardo Giacconi, Jonathan Grindlay, Martin Harwit, Frank Low, Frank McDonald, Dietrich Müller, Minoru Oda, Klaus Pinkau, Kenneth A. Pounds, Irwin Shapiro, Susan Wyckoff

Proposed Missions/Ideas

The Measurement of Elemental Abundances above 1 TeV/amu by a Long Duration Flight of CRN

Simon P. Swordy*

Enrico Fermi Institute, University of Chicago.
LASR, 933 E56th St, Chicago, Il 60637

We propose the flight of the Cosmic Ray Nuclei detector (CRN), for a long duration of at least 1 year. The instrument has already flown successfully on the Spacelab-2 mission, using transition radiation detectors to provide individual elemental abundances to energies of 1 TeV/amu for $4<Z<30$. The dynamic range of the instrument permits measurements to much higher energies if a longer exposure time can be facilitated. With a one year flight the number of particles collected would provide data to ≈10 TeV/amu or beyond.

Introduction

The CRN instrument was developed and constructed at the University of Chicago in the early 1980's, and exposed in a one week flight on the Spacelab-2 mission of the Space Shuttle Challenger in summer, 1985. It has provided data on the elemental composition of cosmic rays to energies of 1 TeV/amu[1]. This considerably extends the energy range covered by previous measurements on balloons or in space. However, statistical limitations which arise due to the relatively short duration of the Spacelab flight have left some of the most interesting astrophysical questions unanswered. The dynamic range of the instrument, and in particular the energy response of the transition radiation detector system allow measurements to still much higher particle energies. Statistically significant data over at least one more order of magnitude in energy would be obtained in an extended space flight of the CRN instrument of about one year duration. The instrument would then provide the elemental composition of cosmic rays in an energy region which is largely unknown at present, and which approaches the so-called 'knee' in the overall energy spectrum. These data will also provide a much needed calibration of air shower measurements which overlap with this energy range, but will explore the cosmic ray composition to the highest energies (≈10^{20}eV), that are not accessible to direct observations.

CRN on Spacelab-2

A cross section of the CRN instrument is shown in Figure 1. It consists of a counter telescope of plastic scintillators, gas Čerenkov counters, and transition radiation detectors (TRD). The instrument measures

*This paper is presented on behalf of the CRN investigators; Jacques L'Heureux, Dietrich Müller, Peter Meyer, and Simon Swordy.

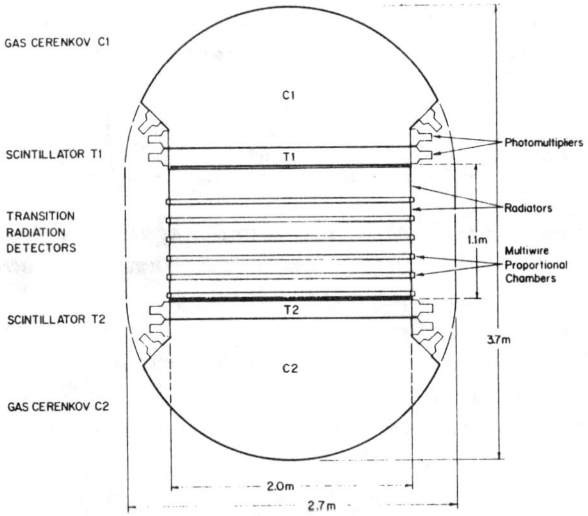

Figure 1. A cross section of CRN.

elemental energy spectra for atomic numbers, Z, in the range 4<Z<30. The energy measuring devices of CRN consist of the TRD, which responds above $\gamma \approx 400$, and two gas filled Čerenkov counters, C1 and C2, which cover the energy range $40 < \gamma < 120$. These counters are triggered by the coincidence telescope provided by the plastic scintillators T1 and T2, which are also used to measure the particle charge and the direction of traversal, (by a time-of-flight measurement). The TRD consists of layers of fiber radiator material interleaved with multiwire proportional chambers filled with a xenon gas mixture. These also serve as a hodoscope to determine the particle trajectory[2].

The instrument was flown for 8 days on the Spacelab-2 mission in July of 1985. However, the *effective* exposure time was only 78 hours, largely because the mission flight plan allowed the instrument to observe space for only a fraction of the time. The source abundances of primary nuclei, as measured by CRN on Spacelab-2, are shown in Figure 2. Here, the abundances are compared to local galactic abundances, normalized to Fe, and plotted versus the first ionization potential. We find that the abundances, for all energies, follow the pattern of depletion of elements with high first ionization potential, this pattern is well established at lower energies[3]. We also find that most source abundances do not seem to change noticeably with energy, with the possible exception of silicon whose depletion at high energy is peculiar. The flatness of the *observed* energy spectrum of iron, (see Figure 3), is likely to be a consequence of the competition between spallation and escape for iron nuclei during propagation.

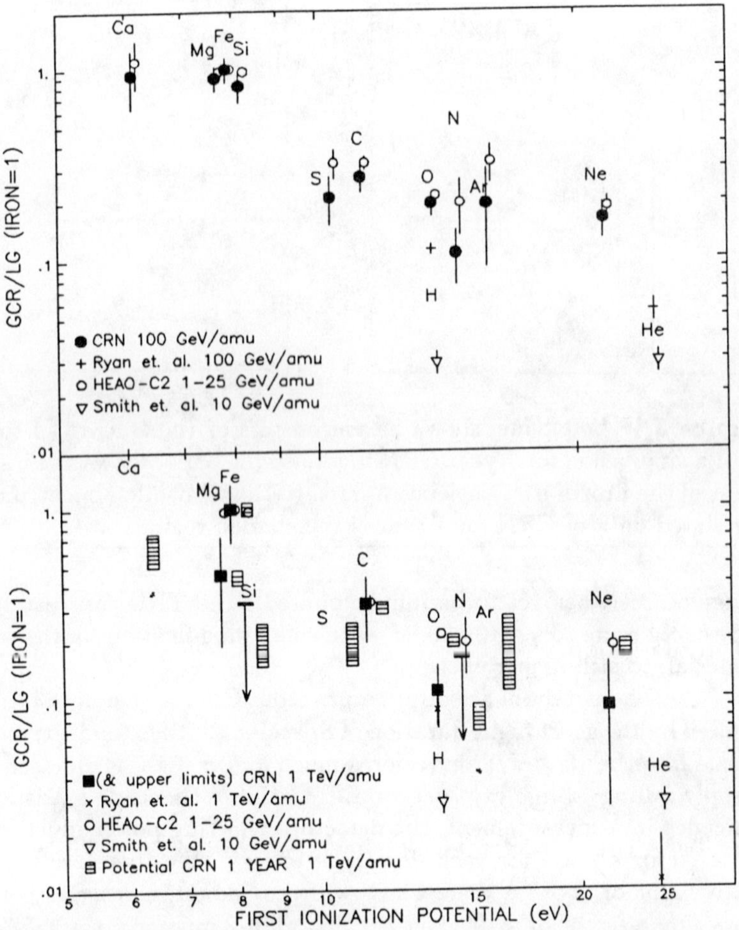

Figure 2. Galactic cosmic ray abundances relative to local galactic abundances plotted vs first ionization potential. The data from CRN on Spacelab-2 are filled symbols, (circles at 100 GeV/amu and squares at 1 TeV/amu). The lower panel shows simulated data from CRN at 1 TeV/amu for a one year exposure.

Our measurements would support a source spectrum for all species except silicon of the form $E^{-2.1\pm0.1}$ over the energy range 10 to 1000 GeV/amu, within the framework of a simple leaky box model.

A Long Duration Flight of CRN

The low particle fluxes at high energy, have limited the maximum energy for which we can make statistically meaningful measurements on the Spacelab-2 flight. For the more abundant elements this energy on Spacelab-2 is ≈1 TeV/amu, (γ ≈1000). However, the dynamic

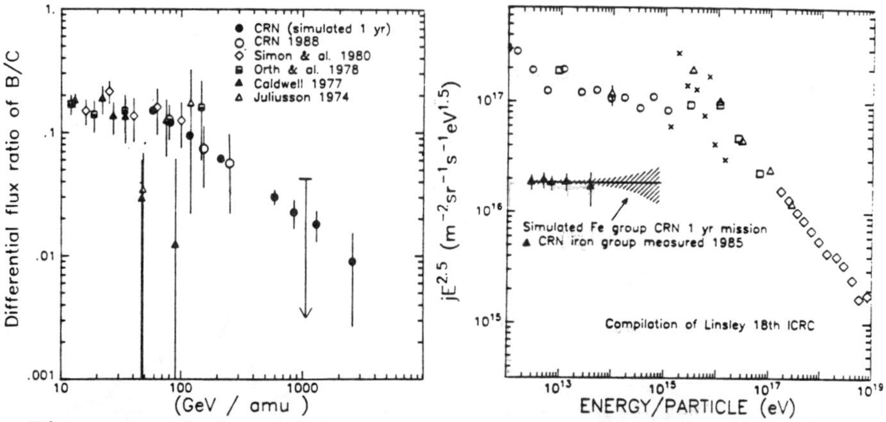

Figure 3. Left panel shows measurements of the B/C ratio together with a simulation for 1 year CRN flight. Right panel shows the measurement of the flux of iron nuclei compared to the all particle spectrum. The simulated data of CRN for 1 year is the shaded region.

range of CRN has yet to be fully exploited. The TRD can readily measure energies up to $\gamma \approx 10{,}000$, or, with small modifications in the radiator material, to still higher energies[4].

The potential scientific return from CRN is therefore at present limited by the short flight duration of Spacelab-2. To extend the measurements into the 10 TeV/amu region a much longer flight is dictated by the steeply falling cosmic ray spectrum. To achieve the same statistical significance in a measurement, the detector collecting power must increase by a factor of $\approx$50 for each decade of energy. We therefore propose a CRN flight of about a year, which will provide an increase in collecting power by a factor of $\approx$100 over the Spacelab-2 mission.

To illustrate the potential of a year's exposure of CRN we have simulated some expected results. Figure 3 shows measurements of the ratio of boron, a secondary nucleus produced in the interstellar medium, to carbon. The ratio B/C gives an indication of the pathlength traversed by the particles during propagation. Our measurements on Spacelab-2 of the flux of these nuclei as a function of energy confirm that the pathlength decreases with increasing energy, and show the mean escape length from the galaxy at 1 TeV/amu is in the range 1-2 g/cm^2. The expected quality of these data for a one year flight is also shown in Figure 3, this simulation assumes that the same energy dependence observed at low energies continues to high energies.

Figure 2 shows the expected accuracy in determination of source abundance at 1 TeV/amu for a one year flight. The statistics are clearly similar to those presently available only up to 100 GeV/amu. This will permit us to make a definite observation on a key question in high energy astrophysics: are all cosmic ray species accelerated at their source with the same energy spectrum, and therefore by the same acceleration mechanism?

The expected measurement improvement of the iron spectrum is shown in Figure 3. The data are plotted together with an all particle cosmic ray spectrum to show the fraction of particles at a given total particle energy that are iron nuclei. The simulation assumes the spectrum of iron remains the same at higher energy. The one year mission would make it possible to observe changes in the cosmic ray composition as we approach the 'knee' at $\approx 10^{15}$eV.

Large air shower arrays presently under construction will have thresholds below $\approx 10^{14}$eV in total particle energy. They will be the only means possible to extend measurements to the highest energies known in nature, and to explore the mystery of the origin of these ultrarelativistic particles. However, a comparison and calibration with direct data in the region of overlap with the 1 year CRN flight are essential in order to analyze and understand the air shower measurements.

Conclusions

Measurements of elemental abundances in the region above 1 TeV/amu can be made with an existing, space-proven, technique if CRN is flown for a long duration. The quality of data from such an experiment can be accurately predicted from existing measurements and accelerator calibrations with this instrument. The technique of transition radiation makes possible a large area detector ($\approx 5\text{m}^2$str) at low mass (2.8 tons), which is sensitive to energies well above 10,000 GeV/amu.

The modifications to CRN which would be required for a long duration flight are minimal and not expensive. The scientific return from such a flight would be invaluable for the understanding of air shower measurements which are the only near term prospect for determining elemental composition above 10^{16}eV.

References

1. Grunsfeld, J.M., L'Heureux, J., Meyer, P., Müller, D., and Swordy, S.P., 1988, Ap. J. **327**, L31.
2. Swordy, S.P., L'Heureux, J., Müller, D., and Meyer, P., 1982, Nucl. Instr. Meth.,**193**, 591.
3. Meyer, J-P., 1985, Proc. 19th ICRC, (La Jolla), **9**, 141.
4. Cherry, M., 1978, Phys. Rev. D, **17**, 2245.

THE C SHELL, AN ACTIVE DETECTOR OF UH NUCLEI

C. Jake Waddington and Robert R. Clinton
School of Physics and Astronomy
University of Minnesota, Minneapolis, MN 55455, USA

ABSTRACT

This paper gives a brief description of the current status of our program to develop a modular array of large electronic particle detectors. These modules were designed to study the UH nuclei in the cosmic radiation with eventual deployment on the Space Station or at a lunar base. This array would determine the abundances of elements from iron to the actinides and directly measure the energies of the lower energy nuclei . If the array was deployed on the Space Station, it would use the geomagnetic threshold to place limits on the higher energy nuclei, thus studying the energy spectrum up to $\approx$ 10 GeV/n. Deployed at a lunar base it would detect nuclei with energies down to the instrumental limit. Smaller versions could be flown on balloons to test and refine the modules.

INTRODUCTION

Accurate determinations of the abundances and energy spectra of the UH nuclei in the cosmic radiation, those with $Z \geq 30$, are of great importance to reaching an understanding of the origin of the cosmic radiation and the nucleosynthesis of nuclides in the universe. Measurements of a quality to match the results now available on the lighter nuclei will require major new experimental efforts, involving very large arrays of detectors exposed in space for long periods. The two major detectors of UH-nuclei that have been flown thus far, HEAO-3, and Ariel-6, each[1,2] observed between 10 - 20 x 10^6 iron nuclei in an effective 18 to 24 months exposure. They also recorded between them a total of between 2 and 4 actinide nuclei, $Z \geq 89$. To study the nuclei of these extremely rare, but very significant, elements in detail will require an increase in collecting power of two to three orders of magnitude. This will involve either multiple exposures of recoverable passive arrays, such as the Heavy Nuclei Collector that it is currently planned to expose during the construction phase of the Space Station, or the construction and maintenance in space of a large active array with an extended operational life. It would be desirable to adopt both approaches, since they have complementary capabilities.

BASIC CONCEPT

To meet the requirements of extremely high collecting power combined with good resolution in charge and energy, we need a modular array of detectors with multiple redundancy, capable of indefinite expansion by the addition of more modules. This imposes severe constraints on the acceptable mass, power dissipation and complexity/cost of each module. Similar constraints apply to any preliminary balloon exposures of limited numbers of these modules. Given present plans it would seem best that such an array be assembled and serviced from the Space Station. In this case it could be assembled over a period of time, with individual modules being brought up by the Shuttle on a "space and mass available" basis. In the more distant future one could envisage such an array being deployed at a lunar base, where it would be able to study the lower energy nuclei not accessable from the Space Station. As initially conceived[3], the array was assumed to consist of some 2,500 modules, each with a 1.0 m diameter plastic Cherenkov radiator, assembled into a spherical shell 30 m in diameter. Hence the title "C-Shell"*. This array could have a collecting power of some 1,100 m^2/sr, three orders of magnitude greater than that of HEAO-3. A particle traversing the array would produce two coincident Cherenkov signals, which, when combined with a measurement of the time of flight (TOF) between the two modules, would serve to identify the charge and velocity. Practical considerations of geometry, weight, power and cost, have led to a somewhat smaller cylindrical array, exposed for longer, as being adequate. A cylindrical form has the advantage that it can be expanded at will, whereas a filled sphere would have to be reconstructed to be expanded.

The wide separation between opposite modules in such a large array is an advantage for the TOF measurements, while the charge measurement depends mainly on a Cherenkov signal, which is less affected by non Z^2 or high energy terms than most other signals. Cherenkov detectors also have the major advantage, in this context, of under, rather than over, estimating the charges of particles incorrectly assumed to be fully relativistic. It has been shown that by deploying an adequate number of PMT's, enough light can be collected from large Cherenkov radiators to give charge resolution without making the radiators unduly thick. These individual modules would be relatively light and place a rather small amount of material in the path of the incident particles, thus reducing the interaction correction.

DEVELOPMENT

The development of the modules for C-Shell required the optimization of several design parameters. Of prime scientific importance were those parameters that affect the collecting power, charge resolution and timing. Other important parameters were the amount of material in the particle trajectory, the rejection of background events, the mass of the modules, the power consumption, and the cost. Based on such considerations and working within the limited available funding, we have constructed an initial version of a module, Fig. 1. This prototype module was initially studied with our ultraviolet laser test facility and was then calibrated at the LBL Bevalac in May, 1989, using beams of iron nuclei of several different energies. This served to normalize our laser test results and to establish an absolute standard for the response in terms of a known particle at a known energy. The results of this calibration are still being analysed, but preliminary assesments have shown no major problems in response or resolution, and have suggested how incremental improvements could be implemented.

In this configuration the Cherenkov light is collected by four RCA-5830 photomultiplier tubes, (PMT). Three are coupled to the edge of the radiator by light pipes spaced equally around the circumference, while the fourth is mounted on a light collecting cone so that it is looking directly at the surface of the radiator. The signal from this PMT serves to normalize those from the other three, whose individual signals depend strongly on the position, whereas the summed signal from all four PMT has a weaker positional dependence. The achievable resolution will mostly depend on the number of photoelectrons per charge, $n_{pe}(Z=1)$, collected by these PMTs. The Bevalac calibration shows that the summed signals from the four tubes, averaged over several positions, assuming a Z^2 dependence, implies $n_{pe}(Z=1) \geq 5.4$ pe for a $\beta = 1.0$ particle. The formal uncertainty in the charge estimates from a Cherenkov counter, δZ, is given by $\delta Z = 1 / (2.n_{pe}{}^{1/2})$ charge units, hence this implies $\delta Z \leq 0.22$ cu. In practice, the higher the value of n_{pe} the better, and we expected to be able to increase it by ≈ 2. The light collection efficiency was not optimized in the calibrated module and post calibration tests have already demonstrated a 20% improvement. For comparison, the

Cherenkov detector in the HEAO-3 detector had an observed resolution corresponding to $n_{pe}(Z=1) = 1.6$ pe.

The TOF determination uses the signal from a thin scintillator radiator mounted next to, but optically decoupled from, the Cherenkov radiator. The light from this scintillator is collected by a single fast EMI-9822B PMT mounted on another cone. The resolution in this counter implied signals with 2.7 to 2.0 x 10^4 pe for iron nuclei with energies between 0.56 and 1.6 GeV/nucleon. The result is a well defined signal for the TOF measurement. Presumably this signal also contains charge information, but until the effects of saturation have been calibrated we can not make a realistic estimate the significance of this information.

This calibration module had a mass of 68 kgs, made up of 22 kgs of irreducible radiators and PMTs, and reducible superstructure. A final mass of 45 - 50 kgs appears attainable. The total power required of 10 watts per module can be reduced to 8 watts when multiple modules are used and to ≈ 6 watts by the use of higher efficiency power supplies and ADC converters.

The design for the electronics depended critically on the essential need to conserve power, both for long duration balloon flights, and in any space array. In addition, the system was designed so that a given module could accept a coincidence with any arbitrary other module. The time to digital convertor (TDC) is unique in that we cannot define a single start or stop time, but must depend on absolute times from each detector separately. To achieve this TOF resolution would have required an event by event analysis of the data, done after the event. Our design goal was to achieve $\delta t < 200$ ps with circuits using ≤ 2 watt. The Bevalac calibration showed that this δt was not achieved in this version by about a factor of two. Our present TOF system requires about 3 watts per module, with another 7 watts for the remaining components.

It is important that the C-shell be sensitive to nuclei having a wide range of charges. In particular, in order to normalize the data on the UH nuclei to the intensities of the lighter nuclei, it must be able to detect iron nuclei incident vertically on the radiators. This only implies a dynamic range of ≤ 60. However, the enormous flux of iron nuclei relative to that of UH-nuclei does introduce the problem of random coincidences, which will become progressively more serious for larger arrays. For a full size C-shell, with 1000+ modules and a separation of up to 30 m, any one module could be operating in coincidence with as many as 500 modules on the other side, and the coincidence gate will have to be open for up to 200 ns. There is therefore a finite probability that a random coincidence could occur between two modules and if these are due to two nuclei having large angles of incidence, with consequently large signals, that a spurious UH event could be generated. Most of these would have an apparent $\beta > 1$ and many would be tagged by another coincidence, if the remaining modules are analyzed in anti-coincidence. Most would have apparent charges not too dissimilar to that of the individual iron nuclei. Further, by requiring that the signals in the two modules be consistent the number of significant random coincidences would be greatly reduced. Similarly, due to edge effects, requiring that the signals in the Cherenkov and scintillator are consistent will reduce the numbers of events with very oblique angle trajectories, which are those required for iron nuclei to produce signals that simulate very heavy UH-nuclei. Finally, the apparent $\beta > 1$ events will always permit an evaluation of the magnitude of this effect in each charge range. It is estimated that not even the very rare actinide events, with a minimum rate of one every few days, would be swamped by random coincidences.

POSSIBLE FLIGHT PLANS

a) **Balloons:** The relatively limited payload capacity of balloons restricts the collecting power of any feasible array. An array of two horizontal layers, separated by a vertical distance of λ, is the most practical for this application. Such an array could be launched and recovered in a mode where the two layers are packed closely together, and deployed by a simple winch system in flight. The collecting power, G, in m^2sr., of such an array depends very strongly on the number of modules, n, and on λ. Fig. 2 shows G as a function of λ, for n = 6, 8, 14 and 24 modules, as typical values that could be flown. Clearly, significant values of G, ≥ 0.5 m^2sr., can only be obtained if λ is kept smaller than the lengths necessary for the TOF to give good charge resolution of nuclei with $\beta << 1$. Consequently these balloon flights have to be restricted to low latitudes, where the geomagnetic cutoff removes the low energy nuclei. The TOF capability will then serve to eliminate those few, but significant nuclei, that are detected with sub-threshold energies, such as the re-entrant albedo. There will also be an increased dependence on the accuracy of the trajectory reconstruction. This depends on the sensitivity of the mapping from the relative pulse heights. The calibration data suggests that this sensitivity will be adequate to accommodate the non-uniformity of the total signal, and that charge resolution of the lighter UH nuclei will be possible even with λ at the physical minimum of about 2 m.

After a initial proving flight from Texas it would be appropriate to assemble as large an array as possible for exposure on a long duration balloon flight flown at a low latitude, such as one launched from Australia and recovered in Brazil. Here weight and power would be the limiting factors. We estimate that such an array could probably not have more than about 12 to 14 modules, with a total mass of 540 to 630 kgs and a power requirement of 120 to 140 watts. With the minimum λ of 2m this would give a geometry factor of about 2.3 m^2.sr, and with a ten day flight an exposure factor of $\approx 2.10^6$ m^2.sr.sec. At the low latitude for these flights the mean threshold energy will be quite high, ≥ 4 GeV/n and the charge resolution dominated by the mapping corrections. After allowing for interactions we can estimate that there will be a flux of ≈ 0.2 Fe nuclei/m^2.sr.sec in the detectors. Thus this flight would expect to detect $\approx 4.10^5$ Fe nuclei and hence detect some tens of nuclei with $34 \leq Z \leq 40$ and perhaps ten with higher Z. Although these are small numbers, inadequate to materially improve our present knowledge of the relative abundances of these nuclei, they would represent a high energy sample that could be compared with the low energy samples that should be provided by passive detectors flown on polar flights . Foreseeable improvements in long range balloon capabilities could increase these numbers by factors of two or three, but not more. Really significant advances would still await exposure for long durations in space.

(b) Space: The entire strategy of how to deploy a multi-element array such as the C-Shell must depend on the available lift and assembly capabilities. In a probable scenario, with the Shuttle lifting a relatively small number of modules during each mission to the Space Station, it would seem that the most efficient mode of operation would be to begin by assembling an array with a relatively small λ, less than 10 m. Then as more modules became available, redeploying the array into progressively larger separations, until a full scale array was formed. The data obtained during these intermediate phases would serve to define how far it was necessary to continue the expansion. For example, since determining the flux of actinde nuclei represents the ultimate challenge, if it became clear that this flux is on the high side of the present wide range of uncertainty, it would be possible to meet the scientific goals and to stop with a smaller array than if the flux turns out to be on the low side.

An alternative, but more distant, scenario would be to construct such an array at a lunar base, or after some years of exposure on the Space Station to move the assembled array to a lunar base. In this location there would be no geomagnetic threshold and nuclei would be detectable down to the Cherenkov threshold. This would serve to define the energy spectra of these nuclei with greater precision.

CONCLUSIONS

Many practical questions remain concerning the development of these modules. The responses have not been optimised and considerably more refinement is possible. The TOF systems need to be improved to reach the design goals. The long term durability of the radiators and other components in the space environment have not yet been considered. None of these features appear to be critical but all need further study. Given the time frame of the Space Station program it would be possible to develop an optimized system in the time available if NASA were prepared to support such advance technology development with such a time frame.

We are indebted to A. Becker for his technical assistance.This work is partially supported by NASA under Grant No. NGR 24-005-050

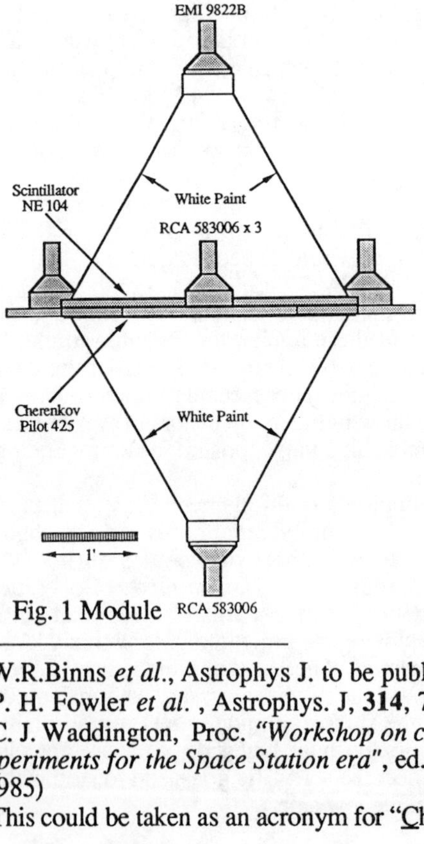

Fig. 1 Module

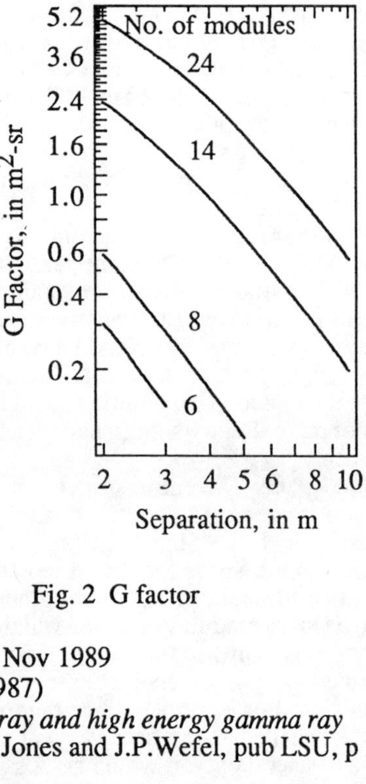

Fig. 2 G factor

[1] W.R.Binns *et al.*, Astrophys J. to be published, Nov 1989

[2] P. H. Fowler *et al.* , Astrophys. J, **314**, 739 (1987)

[3] C. J. Waddington, Proc. *"Workshop on cosmic ray and high energy gamma ray experiments for the Space Station era"*, ed. W.V. Jones and J.P.Wefel, pub LSU, p 259 (1985)

* This could be taken as an acronym for "Cherenkov Space Heavy Element Leviathan"

Measurement of the Composition and Energy Spectrum of Cosmic Rays Above 10^{15} eV

D. Berley, National Science Foundation, Washington, DC 20550
R.W. Ellsworth, George Mason University, Fairfax, VA 22030

1 INTRODUCTION

The availability of a launch vehicle with the capability of carrying a heavy payload, would make possible several definitive experiments in cosmic ray physics including: 1) the determination of the composition and energy spectrum of cosmic rays up to 10^{16} electron volts (eV) and 2) the observation of gamma rays from compact sources, up to energies of 10^{12} eV. The instrument proposed, weighing about 30 tons, is designed to address these fundamental questions.

The only known technique to achieve these measurements is calorimetry. Calorimetric energy measurement and elemental identification to extend the energy of present data above the energy of the break in the cosmic ray spectrum, 10^{15} eV, were discussed at the Workshop on Cosmic Ray and High Energy Gamma Ray Experiments for the Space Station Era held in Louisiana in 1985. Ormes and Streitmatter proposed a High Energy Space Station (HESS) [1] and Yodh [2] discussed a similar idea. Since that time more work has been done on understanding the acceptance-solid angle of a space borne calorimeter and the energy range which could be studied with the instrument.

The critical parameter in determining the cost and feasibility of the project is the mass of material needed for the requisite area (for flux gathering power) and depth (for energy resolution). In this paper we therefore concentrate on what physics can be learned with a given mass of material. The question of determining the charge of the incident cosmic ray without being plagued by backscattering from the calorimeter material [3] can be solved. The problem of sampling the development of showers in the calorimeter is also serious and challenging, but we believe it too can be solved for almost any configuration of calorimeter material. In this preliminary study we have dealt with the more fundamental questions of geometric acceptance and energy resolution.

2 SCIENTIFIC BACKGROUND

The energy achievable in the proposed experiment is above the break ($\sim 10^{15}$ eV) in the cosmic ray energy spectrum, where the otherwise smooth energy dependence of the cosmic ray flux shows structure. The region of the break could be a threshold of magnetic rigidity where the galactic magnetic fields can no longer contain the cosmic radiation or be the limiting energy of acceleration by supernova shock waves [4]. The highest energy measurement of the all particle spectrum, by direct observation, was performed in a pioneering experiment in a satellite by Grigorov [5] to about 10^{15} eV. The elemental composition has been determined from direct measurement up to about 5×10^{14} eV [6]. At higher energies, the spectrum and composition are indirectly determined by ground based experiments measuring showers of particles produced by the primary cosmic rays when they strike the atmosphere [7]. The interpretation of these experiments depends upon assumptions about the elemental composition of the primary cosmic rays and upon hadronic interactions beyond accelerator energies where they have been measured. The results are model dependent.

The measurements of gamma rays from compact sources at energies from tens of GeV to 1 TeV would complement existing data. We have no measurements at energies between a few GeV where satellite borne gamma ray detectors begin to be limited by the available flux and at 100 GeV where showers initiated in the atmosphere become large enough to be observable by ground based detectors. With the exception of radiation from the Crab, the ground based instruments have detected signals that are either episodic or of marginal statistical significance [8] [9]. One interpretation of the experiments, is that the source of these gamma rays may be sufficiently intense also to account for the flux of high energy cosmic rays observed up to energies of 10^{16} or 10^{17} eV.

3 CALORIMETER ACCEPTANCE AND ENERGY RESOLUTION

The two most significant features of a space station calorimeter for achieving the goal of extending measurements to higher energies are:

1. The counting rate for a given flux and

2. The quality of the energy measurement

The counting rate is determined by the size and shape of the calorimeter. The energy resolution, calculated elsewhere [10], depends upon the incident energy, the medium, the sampling and the total effective depth. The effective depth is particularly important; it must be large to contain enough of the cascade shower in the calorimeter for the energy resolution not to be degraded by fluctuations in the shower development. Large calorimeter mass is therefore required.

A practical limit on the mass of the calorimeter medium is imposed by the lift capabability of the launch vehicle . Accordingly, a 25 ton limit was placed on the calorimeter mass. An additional 5 tons was estimated as the mass of the ancillary equipment, to make a total of 30 tons.

We investigated the use of a low density material for the calorimeter. One possibility has been suggested, that a medium with multiple uses such as water would be a good choice. A sphere of water of weight 25 tons has a diameter of 3.5 m, only 4.1 interaction lengths. By comparison a sphere of iron weighing 25 tons would have many more interaction lengths potentially availble; its diameter would be 1.7 m corresponding to 10.4 interaction lengths. To obtain a water calorimeter with a path length necessary for good energy resolution, a cylindrical geometry might be better than a cubic geometry. The geometric factors (area times solid angle), $A\Omega$, for calorimeters of water of several geometries and of a sphere of iron were computed by Monte Carlo [10] sampling. For a payload weight of 30 Tons and a path length of more than four interaction lengths ($\lambda > 4\lambda_{int}$) the geometric factor of a water calorimeter is always less than that of an iron calorimeter. The comparison of $A\Omega$ for a sphere of iron and for a sphere of water is shown in Figure 1. At short interaction lengths, $\lambda < 3.5\lambda_{int}$, the geometric factor, $A\Omega$, is larger for water than for iron. The mass limitation imposed and the need for a deep calorimeter (as discussed below) cause iron to be a more advantageous calorimeter medium than water.

An indication of the quality of the energy measurement is its response to a change in the energy spectrum. The energy spectrum $\frac{dN}{dE}$ is represented by a power law in the energy E with spectral index γ

$$\frac{dN}{dE} \propto E^{-\gamma}$$

Indirect measurements of air showers suggest a change in the spectral index near the bend in the cosmic ray spectrum

$$\gamma \cong 2.7 \text{ for } E < 10^{15} \text{ eV}$$
$$\gamma \cong 3.2 \text{ for } E > 10^{15} \text{ eV}$$

A study [10] was made of the response of an iron calorimeter, mass 25 tons, to an energy spectrum having such a change in spectral index. The calorimeter response and resolution obtained with this simulation agreed well with measurements at accelerators. The resolution obtained as the energy varies is shown in Figure 2 for a fixed depth of 8λ. Above 10^{12} eV the resolution starts to increase because of fluctuations in the energy "leaked" from the calorimeter.

The results of the simulation of an exposure lasting one year are presented.

First, the resolution in spectral index. It is assumed that the spectral index changes by one half unit at the bend in the spectrum, i.e. a change from $\gamma = 2.7$ to $\gamma = 3.2$ at an energy $E = 10^{15}$ eV. An example of the true spectrum generated with this model is illustrated in Figure 3. The energy spectra as would be measured in the calorimeter were simulated for three values of the minimum potential path lengths for the incident cosmic ray, $3\lambda_{int}$, $4\lambda_{int}$ and $6\lambda_{int}$. To illustrate the results obtained, the simulation of the measured spectrum is illustrated in Figure 3 for the minimum potential path equal to 4λ. Preliminary results indicate that admitting

Table 1: The number of events accumulated in one year in an iron calorimeter weighing 25 tons. The minimum potential path length is $4\lambda_{int}$.

E(eV)	Total Events	H	CNO	Fe
5×10^{13}	160,000	73,000	16,000	18,000
1×10^{14}	42,000	19,000	4,200	4,600
5×10^{14}	3,700	1,702	370	410
1×10^{15}	1,300	600	130	140
5×10^{15}	120	55	12	13
1×10^{16}	28	13	3	3

path lengths of $3\lambda_{int}$ would distort the spectrum and result in poor resolution in the determination of the energy of the break and the spectral index above the break. A minimum path length of $4\lambda_{int}$ is chosen.

Second, the number of events accumulated. The number of events accumulated in one year in a calorimeter of iron weighing 25 tons, a minimum path length of $4\lambda_{int}$ required, is shown in Table 1. About 1300 events per year are obtained above the break in the cosmic ray spectrum.

The simulation was repeated for a smaller calorimeter, 12.5 tons. Reducing the weight by a factor of two, roughly reduced the event rate by a factor of 2.

4 GAMMA RAYS FROM COMPACT SOURCES

The hadronic calorimeter surrounded by a shell of anticoincidence scintillator and a shell of electromagnetic calorimeter, could extend to higher energy the presently available measurements on cosmic ray gamma rays. The electromagnetic calorimeter would have finer granularity than the hadronic calorimeter. The electromagnetic calorimeter might consist of plates, each $1rl$, separated by tracking chambers. The tracking could be drift chambers of current technology or perhaps silicon. With such a geometry, $3mrad$ angular resolution has been obtained in a large accelerator experiment and $1mrad$ resolution in test beams [11]. The anticoincidence shell would provide a neutral particle signature (for example a photon) and the calorimeter would provide a hadron signature. Estimates have been made base on the flux from Cygnus X-3 observed by Kiel [12]. The ratio of on source signal to off source background events at 1 TeV is estimated to be

$$\frac{N_\gamma}{N_p} \cong 0.1$$

We expect between more than 10 on source events over background per year with an energy above 1 TeV.

References

[1] J. Ormes, R. Streitmatter, Proceedings of the Workshop on Cosmic Ray and High Energy Gamma Ray Experiments for the Space Station Era, Louisiana State University, edited by W. V. Jones and J. P. Wefel, p. 340- 349.

[2] G. B. Yodh, ibid. pp 528 - 539

[3] R. W. Ellsworth et. al, NIM, **203**, 167 (1982)

[4] P. O. Lagage and C. J. Cesarsky, Astronomy and Astrophysics, **18**, 289 (1982); A. K. Harding, to be published in Proceedings of the Workshop on Physics and Experimental Techniques of High Energy Neutrinos and VHE and UHE Gamma Ray Astrophysics ed. G. B. Yodh and D. C. Wold, (May 11-13, 1989).

[5] N. L. Grigorov, et. al., Proceedings of the 12th International Cosmic Ray Conf., **2**, Hobart, Tasmania, (1971) p. 206.

[6] J. Grunsfeld, et. al., Ap. J., **327**, L31 (1988); M. Ryan, et. al., Phys. Rev. Lett., **28**, 985 (1972); and erratum op. cit. p. 1497; M. Simon, et. al., Ap. J., **239**, 712 (1980); T. H. Burnett, et. al., in Proceedings of International Symposium on Cosmic Rays and Particle Physics, Tokyo, Japan; edited by A. Ohsawa and T. Yuda, (Inst. for Cosmic Ray Res. publication, 1984), p. 468.

[7] See H. Freudenreich, et. al. Phys Rev (to be published) and references therein.

[8] T. C. Weekes, Physics Reports, **160**, 1 (1988)

[9] D. E. Nagle, T. K. Gaisser and R. J. Protheroe, Ann. Rev. Nucl. Part. Sci. **38**, 609 (1988); J. A Goodman, to be published in Proceedings of the Workshop on Physics and Experimental Techniques of High Energy Neutrinos and VHE and UHE Gamma Ray Astrophysics ed. G. B. Yodh and D. C. Wold, (May 11-13, 1989).

[10] R. W Ellsworth, Monte Carlo Calculation of Calorimeter Acceptance, unpublished

[11] L. Mo, Private Communication

[12] M. Samorski and W. Stamm, Astrophys. J., **268**, L17 (1983)

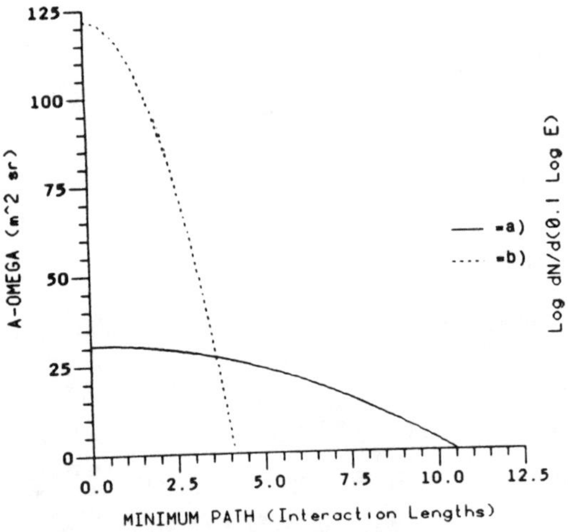

Figure 1. The geometric factor, $A\Omega$, as a function of potential path length, λ. Calorimeters of two sampling materials are compared: a) a 25 ton sphere of iron, $\lambda_{int} = 16.7cm$, diameter $= 1.76m$ and b) a 25 ton sphere of water, $\lambda_{int} = .85cm$, diameter $= 3.52m$.

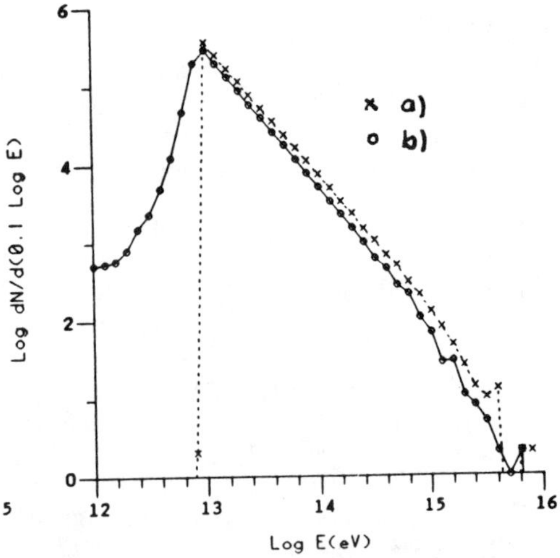

Figure 3. Monte Carlo generated spectra. a) The true spectrum starting at an energy of $E = 10^{14}eV$ and the spectral index changing from $\gamma = 2.7$ to $\gamma = 3.2$ at an energy $E = 10^{15}$eV; b) the detected spectrum for potential path length of $\lambda = 4\lambda_{int}$.

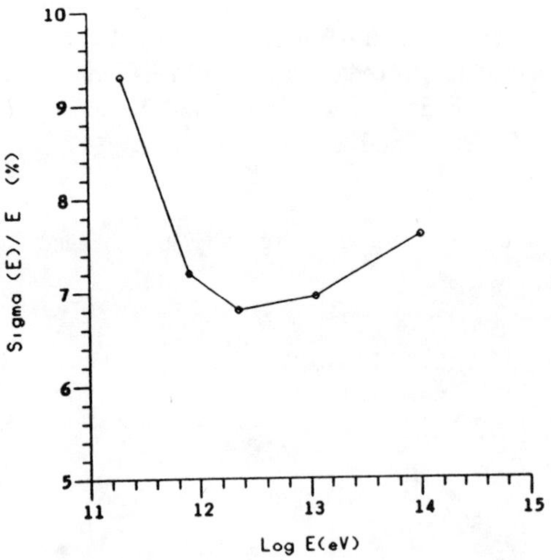

Figure 2. The energy resolution $\frac{\sigma(E)}{E}$ as a function of the logarithm of the energy, $\log_{10} E(GeV)$, for a calorimeter of fixed depth, $8\lambda_{int}$.

Theoretical Prospects

SOLAR ACCELERATED PARTICLES: COMPARISONS OF ABUNDANCES AND ENERGY SPECTRA FROM PARTICLE AND GAMMA-RAY OBSERVATIONS

R. RAMATY, R. J. MURPHY*, and J. A. MILLER**
Laboratory for High Energy Astrophysics
Goddard Space Flight Center, Greenbelt, MD 20771

ABSTRACT

We review the evidence for the existence of two distinct classes of solar accelerated particles; we present new results on abundances from gamma-ray spectroscopy and on electron energy spectra; and we discuss the implications of comparisons of particle and gamma-ray observations.

INTRODUCTION

Particle acceleration is a wide-spread phenomenon in nature, one that occurs at a variety of sites ranging from the Earth's magnetosphere to distant objects such as supernovae, active galaxies and quasars. Evidence for particle acceleration in astrophysics comes from both direct observation of the accelerated particles, such as the cosmic rays, and observations of the neutral radiations, such as gamma rays and X-rays produced by the accelerated particles at or near their acceleration sites. But only for solar flares has it been possible so far to observe simultaneously the accelerated particles and the neutral radiations.

Particles accelerated at the Sun have been observed in the Earth's atmosphere and in interplanetary space for several decades[1,2], but only recently has the existence of two distinct classes of solar energetic particle events been clearly established[3,4,5]. Particle events belonging to the first class are distinguished from events of the second class by larger ratios of relativistic electrons to protons[3,6,7], larger ^{3}He-to-^{4}He ratios[8,9], enhanced abundances[5,10,11] of elements heavier than O relative to C and O, and higher ratios of α particles to protons[12]. Additional distinguishing characteristics of first-class events are the energy spectra of the relativistic electrons, which exhibit a spectral hardening with increasing energy not seen in second-class events[13], and the mean charge of Fe ions, which is significantly higher in first-class events than in second-class events[14]. Since second-class events are characterized by large proton fluxes, they are often referred to as large proton events.

Studies[15] of soft X-ray emission from solar flares have been used to classify flares as impulsive or gradual. The duration of the soft X-ray emission in impulsive flares is on the order of tens of minutes, and, when observed on the limb of the Sun, these flares are seen to have relatively small volumes and to occur deep in the solar atmosphere. On the other hand, the soft X-ray emission in gradual flares lasts for hours, and these flares have larger volumes and occur higher in the atmosphere. The association of solar flares with magnetized flare loops is quite evident from these studies. From the correlation of the relativistic electron-to-proton ratio observed in particle events with the duration of the soft X-ray emission of the associated flares, it has been shown[3] that it is more probable for first-class events to originate in impulsive flares and for second-class events to originate in gradual flares.

Particles in first-class events are probably accelerated from hot flare plasma, a conclusion based on the temperature (10^7K) implied by the charge state of Fe ions observed[14] in these events. Because it is known from soft X-ray observations that such plasma exists in solar flare magnetic loops[15], particles in first-class events are probably accelerated in closed loops and they leak out of the loops onto open field lines connected to interplanetary space. But it is also possible that material outside the loops is heated as well, and that the acceleration occurs directly on open field lines. In addition, as an intermediate case, particles could be pre-accelerated inside loops and, after escape, further accelerated on open field lines. Particle acceleration in closed loops is known to give rise to hard X-ray and gamma-ray emission, and such acceleration is often referred to as first-phase acceleration[16]. The hard X-rays are most likely bremsstrahlung produced by nonrelativistic electrons and there is evidence for loop structure from hard X-ray imaging[17]. In the gamma-ray band, ion interactions produce line emission, which is superposed on continuum due to relativistic electron bremsstrahlung[16]. No gamma-ray imaging observations have yet been carried out, but there is strong indirect evidence for acceleration and trapping in closed loops[18] based on the comparison[16,19] of the number of protons that escape from the flare (from direct particle observations) with the number that remain and interact at the Sun (from gamma-ray line observations of the same flare). This comparisons shows that the interacting protons are generally more numerous than the escaping ones.

Particles in large proton events (i.e., second-class events) are probably accelerated from ambient coronal gas. This conclusion is based on the following: the observed elemental abundances in large proton events, after corrections for charge-dependent acceleration effects, are consistent[20] with coronal abundances; the temperature inferred[14] from the charge state of Fe ions observed in these events is consistent with the coronal temperature of $\sim 2 \times 10^6$K. Since there are large proton events[21] for which the number of escaping protons exceeds the upper limits on the number of interacting protons set by gamma-ray observations, it is very likely that in such proton events the particles are accelerated on open field lines in the corona. This acceleration, often referred to as second-phase acceleration[16,22], is probably due to shocks driven by the coronal mass ejections[23]. In addition, there are also flares (e.g., 1982 June 3) which show evidence for both first- and second-phase acceleration[24]. Evidence for first-phase acceleration in the 1982 June 3 flare is provided by direct particle observations[25], showing a very high interplanetary electron-to-proton ratio, high $^3He/^4He$, and enhanced heavy element abundances, as well as by gamma-ray observations[26], which indicate that the majority of the particles remained trapped at the Sun[24]. Evidence for second-phase acceleration is provided by the very low α particle-to-proton ratio observed[25] in the escaping particles and by the extended time profile of the gamma-ray emission[26], showing a very significant hardening of the energy spectrum of the interacting protons as a function of time, which was interpreted as second-phase acceleration[27].

It is thus quite obvious that much can be learned by comparing the properties of the accelerated particles which interact at the Sun with the properties of the particles observed in interplanetary space. We devote the next section to the presentation of new results on abundances derived from gamma-ray observations, which show that the abundances of the accelerated particles which produce the gamma rays are much more consistent with first-class particle abundances than with large proton-flare abundances. Following that section, we compare observed bremsstrahlung spectra[28] from 16 flares with calculated bremsstrahlung spectra obtained by assuming that for each flare the injection energy spectrum of the interacting electrons is the same as the spectrum of the directly observed[13] electrons from the same flare.

ABUNDANCES FROM GAMMA-RAY SPECTROSCOPY

The principal mechanisms[16,29] for the production of gamma-ray lines are nuclear deexcitation, neutron capture and positron annihilation. Nuclear deexcitation lines result from the bombardment of ambient gas with accelerated protons and α particles, and from the inverse reactions in which ambient hydrogen and helium are bombarded by accelerated ^3He, carbon and heavier nuclei. The former lead to narrow lines whose widths are determined by the recoil velocities of the heavy targets, while the latter produce broad lines whose widths are due to the velocities of the accelerated heavy projectiles. Because of their low relative abundances, interactions between accelerated and ambient heavy nuclei are not very important. Furthermore, because H and He have no bound excited states, proton-proton and proton-α particle interactions can be ignored in considerations of deexcitation-line production. However, the interaction of α particles with ambient He produces two strong lines, at 0.478 MeV from ^7Li and 0.431 MeV from ^7Be, as well as a line at 3.564 MeV from ^6Li. This line is also produced by accelerated ^3He interacting with ambient ^4He.

Nuclear cross sections, obtained from both gamma-ray and particle measurements, and the relevant nuclear kinematics were incorporated[29,30] into a computer code which produces a complete deexcitation spectrum for a given accelerated particle energy spectrum and given compositions of both the accelerated particles and the ambient medium. We have updated this code using new cross sections accurately measured in the laboratory, and will publish the details elsewhere. In addition to the ^{7}Li and ^{7}Be lines, the strongest nuclear deexcitation lines are at 6.129 MeV from ^{16}O, 4.438 MeV from ^{12}C, 1.779 MeV from ^{28}Si, 1.634 MeV from ^{20}Ne, 1.369 MeV from ^{24}Mg, and 0.847 MeV from ^{56}Fe. The strengths of the narrow lines depend primarily on the ambient gas abundances, while the strengths of the broad lines depend primarily on the accelerated particle abundances. There are also lines from neutron capture at 2.223 MeV and positron annihilation at 0.511 MeV. All of these lines are superposed on a continuum due to bremsstrahlung of accelerated electrons.

It has been shown[31] that a theoretically-calculated[32] gamma-ray spectrum, in which both the ambient gas and the accelerated particles have photospheric composition, provided an unacceptable fit to the observations of the 1981 April 27 flare, but that an improved fit could be obtained by enhancing the abundances of ambient Ne, Mg, Si and Fe relative to C and O. Subsequently, in a detailed study[33], the best fitting ambient gas abundances were derived for this flare. In that study, the accelerated particle composition was held fixed at a composition similar to that observed from large proton flares. This resulted in ambient gas abundances as follows: Mg/C, Si/C and Fe/C enhanced relative to the photosphere, but consistent with coronal values; Ne/C larger than the coronal value; and O/C consistent with both the photospheric and coronal values. However, motivated by the recent finding[34] that the derived strengths of the narrow lines are sensitive to the underlying continuum to which the broad lines make a significant contribution, we have relaxed the constraint that the accelerated particle composition is fixed. This allows the determination of both ambient medium and accelerated particle abundances. As we now show, we not only obtain a better fit to the data but also derive accelerated particle abundances which are much more consistent with compositions observed in first-class solar flare particles events than with compositions seen in large proton events.

We fit the 1981 April 27 data[34] with our theoretical spectrum folded through the response function of the SMM gamma-ray spectrometer. We use 449 channels of gamma-ray data ranging from 0.3 to 8.5 MeV. Our model has 22 free parameters: 16 abundances (ambient ^{4}He, C, N, O, Ne, Mg, Si, Fe relative to H; accelerated C, N, O, Ne, Mg, Si, Fe relative to protons and ^{3}He/^{4}He), the 2.223 and 0.511 MeV line fluxes, the amplitude and spectral index s of the bremsstrahlung power law, the spectral parameter[35] αT of the accelerated particles, and α/p , the ratio of accelerated α particles to protons. We assume that all accelerated particle species have the same spectrum as a function of energy/nucleon and we fix the ambient ^{4}He/H at 0.067. Since the calculated gamma-ray spectrum depends linearly on the first 19 parameters (but not linearly on the last 3), the best fitting abundances can be directly obtained by matrix inversion[30] for a given choice of s, αT, and α/p. We then vary these three parameters to minimize χ^2. We find a best fit at $\alpha/\mathrm{p} = 1.0 \pm 0.4$, $\alpha\mathrm{T} = 0.015 \pm 0.004$ and $\mathrm{s} = 2.90 \pm 0.03$, where the uncertainties are estimated using $\Delta\chi^2 = 1$. These values of αT and s are quite reasonable[16,28]. However, $\alpha/\mathrm{p} = 1$ is not only much larger than values of α/p observed in large proton events (where α/p is[12] typically 0.01), but also larger than those seen in ^{3}He-rich events (where α/p does not exceed[12] 0.4). We present results for $\alpha/p = 0.2$, a case for which the fit is almost as good as for $\alpha/p = 1$ (see below). We find that the derived abundances, except the ambient ^{4}He/C (see below), are fairly insensitive to the exact value of α/p.

The ambient medium abundances derived for the above best fitting values of αT and s, and $\alpha/p = 0.2$ are compared with photospheric abundances[36] in Table 1. As above, the uncertainties are estimated using $\Delta\chi^2 = 1$. Also shown in the table are the minimum χ^2, the number of degrees of freedom, and the probability $P_{\chi^2} = 0.103$ that random observations of the actual solar spectrum would produce a χ^2 as large or larger than the one obtained with the model. The results for C–Fe are similar to the previous results[33], in that the Ne, Mg, Si and Fe abundances are enhanced, while O/C is essentially photospheric. Furthermore, by fixing the ambient abundances at photospheric values[36] and allowing the accelerated abundances to vary, we find that P_{χ^2} is reduced to 0.050 (Table 2, column 3). Thus we can reject the assumption that the ambient abundances are photospheric, but only at the 95% level.

TABLE 1

Ambient Medium Abundances from the Full-Fit Gamma-Ray Analysis Compared with Photospheric Abundances[36]

	Gamma-Ray Abundances	Photospheric Abundances
^{4}He	195 ±28	269^{+23}_{-21}
C	1.00 ±0.18	$1.00^{+0.10}_{-0.09}$
N	<0.82	$0.31^{+0.03}_{-0.03}$
O	2.40 ±0.28	$2.34^{+0.20}_{-0.18}$
Ne	0.86 ±0.14	$0.34^{+0.09}_{-0.07}$
Mg	0.24 ±0.11	$0.105^{+0.013}_{-0.011}$
Si	0.47 ±0.15	$0.098^{+0.012}_{-0.011}$
Fe	0.26 ±0.11	$0.130^{+0.009}_{-0.009}$
χ^2	464.3	
ν	427	
P_{χ^2}	0.103	

The derived accelerated particle abundances are given in columns 2 and 3 of Table 2, where all abundances except that of accelerated ^{3}He are normalized to C. We compare these results with abundances observed in large proton events[37,38] (column 4) and in ^{3}He-rich events[39] (column 5). We see that, while the uncertainties are large, there is an indication that the abundances of the accelerated particles responsible for gamma-ray production resemble more closely ^{3}He-rich flare abundances rather than abundances in large proton flares. In fact, we find that if we fix the accelerated particle abundances at large proton flare abundances[37] the fit is quite poor ($P_{\chi^2} = 0.006$), and the model can be rejected at the 99.4% confidence level. We note in addition that if both the ambient and accelerated particle abundances are fixed (at photospheric and large proton flare values respectively), we find that the fit is absolutely unacceptable ($\chi^2 = 1055$ with $\nu = 441$ implying that $P_{\chi^2} < 10^{-7}$).

TABLE 2

Accelerated Particle Abundances from the Full-Fit and Fixed Ambient Abundance Gamma-Ray Analyses Compared with Large Proton Flare Abundances[37] and Abundances in ^{3}He-Rich Flares[39]

	Full Fit GR Analysis	Fixed Ambient Abundance GR Analysis	Large Proton Flare	^{3}He-Rich Flare
^{3}He/^{4}He	0.64±0.40	0.76±0.40	< 0.05[38]	> 0.1
C	1.00 ±0.50	1.00±0.65	1.00 ±0.10	1.00
N	< 0.82	5.95±2.69	0.29 ±0.02	0.2–0.8
O	1.98 ±0.85	2.44±1.08	2.30 ±0.06	1–6
Ne	<0.87	2.64±1.22	0.33 ±0.03	0.3–5
Mg	3.75 ±0.69	6.25±0.86	0.45 ±0.02	0.5–4
Si	1.15 ±0.94	3.17±0.77	0.37	0.5–3
Fe	4.23 ±0.76	8.50±0.77	0.35 ±0.04	1–10
χ^2	464.3	482.5		
ν	427	433		
P_{χ^2}	0.103	0.050		

We return now to the question of α/p and also consider ^{3}He/^{4}He. As mentioned above, the derived ambient ^{4}He/C is a strong function of α/p. This dependency, illustrated in Table 3, arises because of the fixed observed ratio of the strengths of the ^{7}Li - ^{7}Be lines (due to α particle interactions with ambient He) to the strengths of the other narrow lines (due to proton and α particle interactions with ambient heavy nuclei). Since the suggested[36] photospheric He/C is ~ 270, we conclude that α/p is probably not smaller than 0.15, since it is unlikely that the He/C ratio in the ambient gas significantly exceeds the photospheric value. Since α/p in large proton events is typically 0.01, this result provides additional evidence that the accelerated particle composition resembles that of first-class events. In addition, we see in Table 3 that P_{χ^2} is 0.149 for the best fit and only slightly smaller for $\alpha/p = 0.2$, justifying our use of this value rather than $\alpha/p = 1$. In Table 2 we provided results for the accelerated ^{3}He/^{4}He, but these are quite uncertain as they are based on a very weak feature in the gamma-ray spectrum.

TABLE 3

Ambient ^{4}He/C as a Function of Accelerated α/p from the Full-Fit Gamma-Ray Analysis

α/p	P_{χ^2}	^{4}He/C
2.0	0.140	96.7 ±20.1
1.0	0.149	84.6 ±20.2
0.2	0.103	194.5 ±44.6
0.067	0.076	485.4 ±111.9
0.01	0.065	3077 ±727

The calculated count spectrum (for α/p = 0.2; αT = 0.015; s = 2.90; and the abundances of columns 2 of Tables 1 and 2) is shown in Fig. 1, along with the background-subtracted data[34] for the 1981 April 27 flare. In this figure, the best-fitting bremsstrahlung power law has been subtracted from both the calculated spectrum and the data to show the narrow-line structure more clearly. The spectrum has been broken into four energy ranges for clarity. The nuclei responsible for the various lines are indicated (e^+ denotes the positron annihilation line and n denotes the neutron capture line).

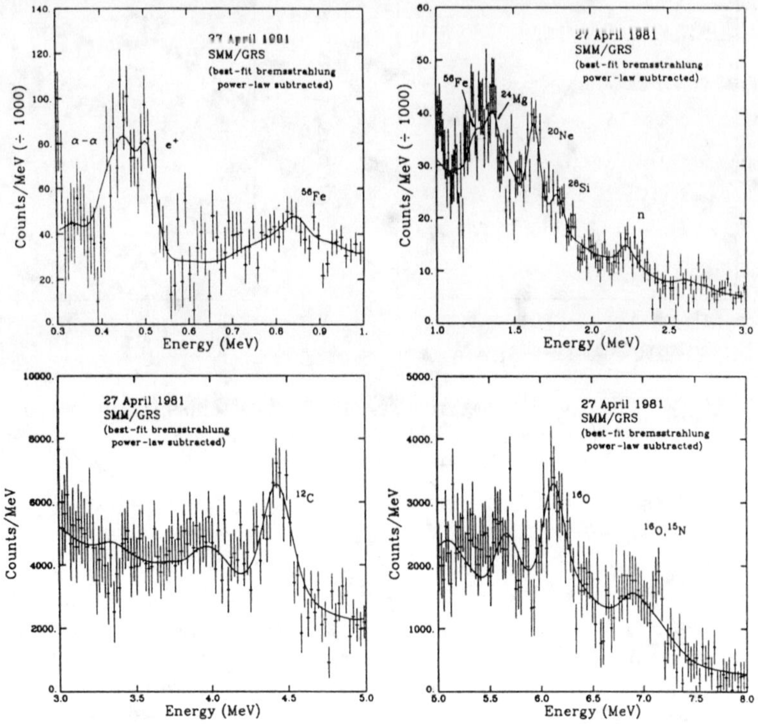

Fig. 1 Best-fitting calculated spectrum compared with the data.

RELATIVISTIC ELECTRON ENERGY SPECTRA FROM PARTICLE AND GAMMA-RAY OBSERVATIONS

The comparison of the energy spectra of electrons observed in interplanetary space with the energy spectra of electrons which interact at the Sun to produce gamma-ray continuum emission can provide important information on the relationship between the two classes of particles and hence on the two phases of particle acceleration. As mentioned in the Introduction, the energy spectra of relativistic electrons harden[13] with increasing energy in first-class events but not in second-class events. In the example shown in Fig. 2 (from ref. 13), the data are fitted with two power laws. The spectrum on the right is from an impulsive flare, while that on the left is from a gradual flare.

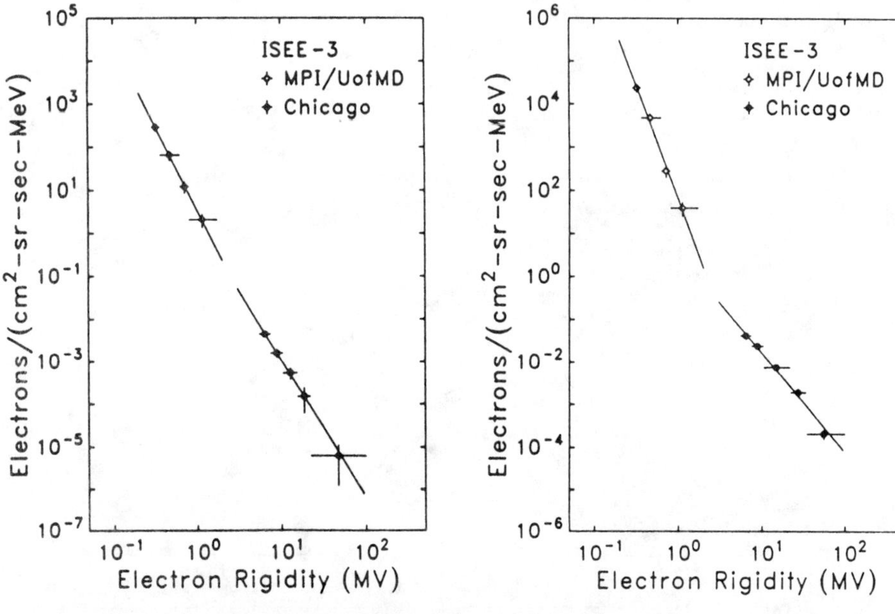

Fig. 2 Observed[13] electron spectra in first-class (right panel) and second-class (left panel) events.

Of the 55 flares for which the electron spectrum was measured[13], there are 16 flares for which the 0.3–1.0 MeV gamma-ray continuum spectrum was determined[28], and approximately 70% of these are impulsive flares. For all 16 of these flares we have calculated the bremsstrahlung spectrum in this energy range, assuming that electrons with the same spectra as those observed in interplanetary space (approximated by two power laws) impinge on a thick target[24] in which the electrons are isotropized before interacting. (See ref. 40 for the effects of an anisotropic electron distribution.) We have used the electron energy-loss rates and bremsstrahlung cross section[41] appropriate for neutral H, and we have ignored synchrotron losses. We expect that the assumption of isotropy will not introduce much error in the energy range 0.3–1.0 MeV if the distributions of interacting electrons are similar to those calculated[41] from transport in magnetic loops with MHD pitch-angle scattering. In addition, Coulomb scattering could further isotropize the electrons.

Results are given in Fig. 3, where the calculated bremsstrahlung spectral index is plotted versus the observed 0.3–1.0 MeV index for each of the 16 flares. Closed (open) symbols represent impulsive (gradual) flares. We see that the calculated spectra are systematically harder than those observed, regardless of whether the flare is impulsive or gradual. We conclude that the electrons observed in interplanetary space and the electrons responsible for gamma-ray production cannot be drawn from the same accelerated particle population without the former being affected by re-acceleration or energy-dependent escape.

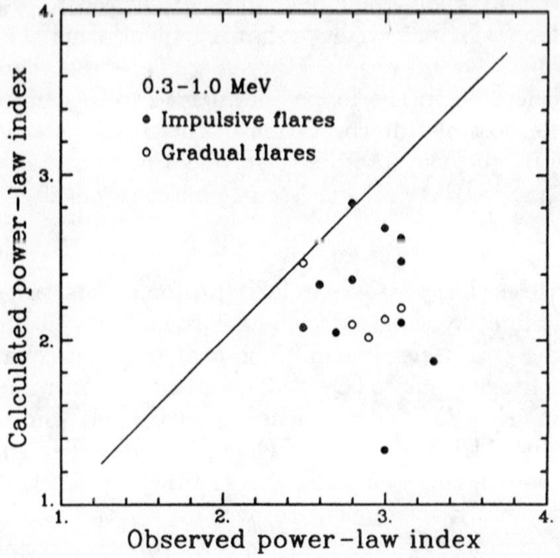

Fig. 3 Calculated vs. observed[28] bremsstrahlung spectral indexes.

CONCLUSIONS

Using gamma-ray observations of the 1981 April 27 flare, we have derived elemental abundances of both the ambient gas and the accelerated particles which interact with this gas to produce the gamma rays. For the abundances that we deduce, the calculated gamma-ray spectrum provides a fit which is acceptable at the 15% confidence level. These abundances require that Ne/C, Mg/C, Si/C and Fe/C be enhanced and O/C be consistent, where the comparison is made with the photosphere for the ambient gas and with large proton flare abundances for the accelerated particles. The model in which the accelerated particle abundances are fixed at large proton flare abundances can be rejected at a 99.4%

confidence level; the model in which the ambient gas abundances are fixed at photospheric values can also be rejected but only at a 95% level.

We have reviewed the evidence supporting the existence of two distinct classes of solar accelerated particles. We pointed out that particles of the first class, probably accelerated from hot flare plasma, are characterized by high e/p, high $^3He/^4He$, high α/p and enhanced heavy element abundances. We found that the accelerated particles which produce the gamma rays in the 1981 April 27 flare also have enhanced heavy element abundances, as mentioned above, and high α/p. Previous studies have shown that the accelerated protons which produce the gamma rays are accelerated in closed loops and are subsequently trapped and forced to interact in the loops. Thus the first-class particles should be likewise accelerated in such loops. But it is also possible that first-class particles are accelerated on open field lines or, at least, re-accelerated after escape from the loops. In fact, we have shown evidence that the energy spectrum of the electrons observed in interplanetary space is harder than the spectrum of the electrons accelerated in the loops, as inferred from gamma-ray continuum observations. One possible interpretation of this result is that the gamma-ray producing electrons and the interplanetary electrons have a common origin in the loops, but that the latter have been re-accelerated after escape from the loops.

Second-class particles (i.e., particles in large proton events) are probably accelerated on open field lines by shocks driven by coronal mass ejections. The average composition of these particles is similar to that of the corona, differing[37] only slightly due to a charge-to-mass (Q/M) dependent bias introduced by the acceleration process. Such a Q/M dependence is most likely due to the interaction of the particles with MHD turbulence which, in the case of shock acceleration, serves to confine the particles near the shock. However, the Q/M bias inferred[37] for second-class events is insufficient to account for the much larger deviation of a first-class abundance ratio such as Fe/C from both coronal or photospheric values, particularly since the difference in Q/M between Fe and C is less in the 10^7K flare plasma than in the 2×10^6K corona. This suggests that the acceleration processes of the two classes are different. It has been shown[16,35] that stochastic acceleration due to MHD turbulence can account for many of the accelerated particle properties derived from gamma-ray observations. While this has not yet been fully researched, it is reasonable to assume that much larger Q/M effects are possible for stochastic acceleration than for shock acceleration, since for the former the turbulence serves not only to trap the particles but also to accelerate them[42].

As is evident from the above, the simultaneous observation of accelerated particles and gamma rays from solar flares has outstanding scientific merits. The only unambiguous signatures of ion acceleration in solar flares are the accelerated particles themselves and gamma-ray line emission. As we have seen, among the many correlated studies, perhaps the most significant is the simultaneous determination of the elemental and isotopic abundances of the particles which interact at the Sun and those which escape from the Sun. Abundance determination is perhaps the most powerful technique in the study of cosmic evolution, and is central to all of particle astrophysics. But only for solar flares can abundances be simultaneously determined inside a source and in the accelerated particles which escape from it. In addition, the correlated gamma-ray and particle observations can also provide unique information on particle acceleration, escape and transport, which are central issues in the problem of the origin of the cosmic rays.

Acknowledgement. We would like to acknowledge D. V. Reames for useful comments on this paper.

REFERENCES

1. I. Lange and S. E. Forbush, *Terrestrial Magnetism and Atmospheric Electricity*, **47**, 185 (1942).
2. F. B. McDonald, C. E. Fichtel and L. A. Fisk, in *High Energy Particles and Quanta in Astrophysics*, ed. F. B. McDonald and C. E. Fichtel (Cambridge:MIT), p. 212 (1974).
3. H. V. Cane, R. E. McGuire and T. T. von Rosenvinge, *Astrophys. J.*, **310**, 448 (1986).
4. R. P. Lin, *Rev. Geophys.*, **25**, 676 (1987).
5. D. V. Reames, *Astrophys. J. Suppl.*, in press (1990).
6. P. Evenson, P. Meyer, S. Yanagita and D. Forrest, *Astrophys. J.*, **283**, 439 (1984).
7. E. I. Daibog, *20th Internat. Cosmic Ray Conf. Papers*, **3**, 45 (1987).
8. D. V. Reames, T. T. von Rosenvinge and R. P. Lin, *Astrophys. J.*, **292**, 716 (1985).
9. D. V. Reames and R. P. Lin, *19th Internat. Cosmic Ray Conf. Papers*, **4**, 273 (1985).
10. G. M. Mason, D. V. Reames, B. Klecker, D. Hovestadt and T. T. von Rosenvinge, *Astrophys. J.*, **303**, 849 (1986).
11. D. V. Reames, *Astrophys. J. Letters*, **330**, L71 (1988).
12. R. Ramaty *et al.*, in *Solar Flares*, ed. P. A. Sturrock, (Colorado Assoc. Univ.: Boulder), p. 117 (1980).
13. D. Moses, W. Dröge, P. Meyer and P. Evenson, *Astrophys. J.*, **346**, 523 (1989).
14. A. Luhn, B. Klecker, D. Hovestadt and E. Möbius, *Astrophys. J.*, **317**, 951 (1987).
15. R. Pallavicini, S. Serio and G. S. Vaiana, *Astrophys. J.*, **216**, 108 (1977).
16. R. Ramaty and R. J. Murphy, *Space Science Reviews*, **45**, 213 (1987).
17. B. R. Dennis, *Solar Physics*, **100**, 465 (1985).
18. T. Bai, *Astrophys. J.*, **308**, 912 (1986).
19. R. J. Murphy and R. Ramaty, *Adv. Space. Res.*, **4**, No. 7, 127 (1984).
20. J.-P. Meyer, *Astrophys. J. Suppl.*, **57**, 151 (1985).
21. E. W. Cliver *et al.*, *Astrophys. J. Suppl.*, in press (1990).
22. J.P. Wild, S. F. Smerd and A. A. Weiss, *Ann. Rev. Astron. Astrophys.*, **1**, 291 (1963).
23. S. W. Kahler *et al.*, *J. Geophys. Res.*, **12**, 209 (1984).

24. R. J. Murphy, C. D. Dermer and R. Ramaty, *Astrophys. J. Suppl.*, **63**, 721 (1987).
25. M. A. I. van Hollebeke, F. B. McDonald and J. P. Meyer, *Astrophys. J. Suppl.*, in press (1990).
26. E. L. Chupp *et al.*, *Astrophys. J.*, **318**, 913 (1987).
27. R. Ramaty, R. J. Murphy and C. D. Dermer, *Astrophys. J. Letters*, **316**, L41 (1987).
28. W. T. Vestrand *et al.*, *Astrophys. J.*, **322**, 1010 (1987).
29. R. Ramaty, B. Kozlovsky and R. E. Lingenfelter, *Astrophys. J. Suppl.*, **40**, 487 (1979).
30. R. J. Murphy, *Gamma Rays and Neutrons from Solar Flares*, Ph.D. Dissertation, University of Maryland (1985).
31. D. J. Forrest, in *Positron Electron Pairs in Astrophysics*, eds. M. L. Burns, A. K. Harding, and R. Ramaty (New York:AIP), p. 3 (1983).
32. R. Ramaty, R. J. Murphy, B. Kozlovsky and R. E. Lingenfelter, *Solar Physics*, **86**, 395 (1983).
33. R. J. Murphy *et al.*, *19th Internat. Cosmic Ray Conf. Papers*, **4**, 249, 253 (1985).
34. R. J. Murphy, G. H. Share, J. R. Letaw and D. J. Forrest, *Astrophys. J.*, in press (1990).
35. M. A. Forman, R. Ramaty and E. G. Zweibel, in *The Physics of the Sun*, eds. P. A. Sturrock *et al.*, Vol. II, p. 249 (1986).
36. N. Grevesse and E. Anders, *Cosmic Abundances of Matter*, ed. C. J. Waddington, (New York:AIP), p. 1 (1988).
37. H. H. Breneman and E. C. Stone, *Astrophys. J. Letters*, **299**, L57 (1985).
38. R. E. McGuire, T. T. von Rosenvinge and F. B. McDonald, *Astrophys. J.*, **301**, 938 (1986).
39. D. V. Reames, H. V. Cane and T. T. von Rosenvinge, *Astrophys. J.*, in press (1990).
40. C. D. Dermer and R. Ramaty, *Astrophys. J.*, **301**, 962 (1986).
41. J. A. Miller and R. Ramaty, *Astrophys. J.*, **344**, 973 (1989).
42. J. A. Miller, R. Ramaty and N. Guessoum, *21st Internat. Cosmic Ray Conf. Papers*, in press (1990).

* Universities Space Research Association

** Universities Space Research Association; also University of Maryland

COSMIC RAYS IN THE HELIOSPHERE: PRESENT STATUS AND FUTURE OPPORTUNITIES

J. R. Jokipii
University of Arizona

Abstract

The general theory of charged-particle transport and acceleration is reviewed and application to the heliosphere discussed in detail. The present status of our understanding is briefly surveyed, with emphasis on large-scale processes. Opportunities for increasing our understanding through the confrontation of theory, modelling and observations are outlined. It is concluded that the largest advances in our understanding in the next few years will come from coordinated theoretical and observations studies of large-scale phenomena requiring energetic-particle, magnetic-field and plasma data from several spacecraft. Fundamental studies of the interactions between cosmic rays and turbulent plasma to study transport coefficients, details of interactions with shocks etc, are also important in the longer term.

INTRODUCTION

Figure 1 shows, schematically the entire observed quiet-time spectrum of cosmic-rays nuclei as a function of kinetic energy T, from less than 1 MeV to $\approx 10^{20} eV$. In addition, near the sun are observed transient emissions of particles associated in solar flares. The latter are discussed separately in this symposium and will not be considered explicitly here.

The goal of cosmic-ray astrophysics is to understand the origin and transport of the energetic particles and, in the process, to further understand the astrophysical sites in which they are accelerated and through which they propagate. The cosmic rays may be conveniently divided into four distinct species, based on their energies, as shown. Our present understanding suggests that this grouping separates particles having different origins. At the highest energies ($T \gtrsim 10^{15-16}$ eV) are the ultra-high-energy particles whose acceleration site is still the subject of considerable disagreement and puzzlement. At the very lowest energies ($T \lesssim 10^6$ eV) we have the low-energy turnup, whose origin is also poorly known, although it is presumably in the heliosphere or in the sun itself. In the intermediate energy ranges, we have the anomalous cosmic rays (3×10^6 eV $\lesssim T \lesssim 10^8$ eV) and the galactic cosmic rays (10^8 eV $\lesssim T \lesssim 10^{15}$ eV) whose origin and transport are much better understood. In particular, those galactic particles having energies below about 10^{10} eV and the anomalous component have been the subject of intense confrontation between *in situ* spacecraft observations

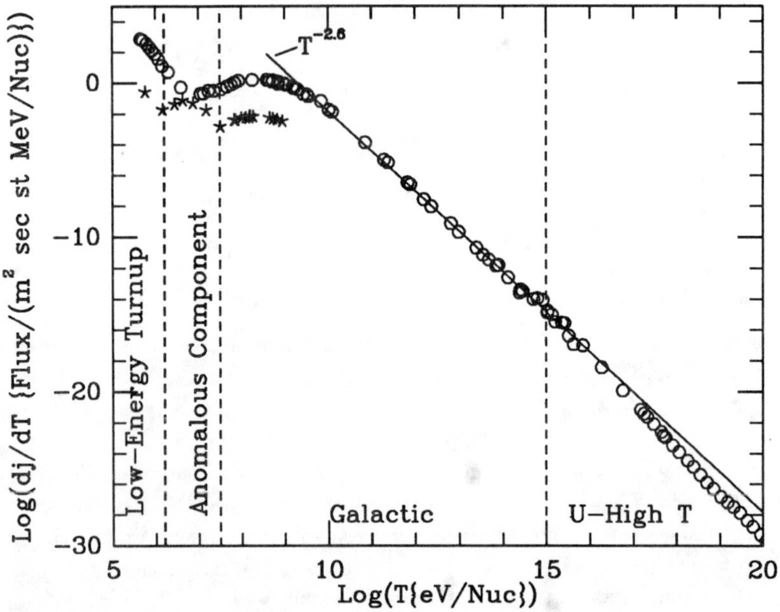

Fig. 1. The cosmic-ray differential intensity as a function of energy. The o correspond to protons for $T \lesssim 10^{15}$ eV. The composition at higher energies is not certain. The * correspond to Oxygen. The Oxygen peak at $T < 10^7$eV/Nuc is due to anomalous Oxygen. Data compiled from Meyer (1969), Linsley (1980) and Gloeckler (1979).

and detailed theoretic modelling of their interactions with the solar wind. I will show below that this has resulted in quite satisfactory agreement, suggesting that the current theory is a good first approximation.

It should be remarked here that the basic theory is quite general and applies to more-distant astrophysical sites (such as supernova blast waves) as well. The applications of this theory can be carried out with more confidence because of the detailed tests possible in the heliosphere. A major goal of particle astrophysics should be to extend the tests between this theory and observation to include the high-energy galactic particles, the ultra-high-energy particles, solar particles and the low-energy turnup. This may result in significant advances in our understanding of the astrophysics of these particles or it may result in further improvement of our theories.

GENERAL TRANSPORT THEORY

The motion of cosmic rays is generally discussed in two limiting cases – motion in smooth magnetic fields with isolated discontinuities (scatter-free transport) or in the diffusion limit which assumes that there is enough scattering by magnetic irregularities (turbulence) to keep the distribution nearly isotropic. The theory of diffusive transport is significantly more mature and describes well a much wider variety of phenomena.

Scatter-free motion has been invoked in a small number of special cases. I will discuss diffusive transport first, turning to scatter-free models in connection with small-scale interactions.

The general diffusive theory of the transport of energetic particles was developed about twenty years ago (see, e.g. Parker, 1965, Axford, 1965, Gleeson and Axford, 1967, and Jokipii and Parker, 1969; general reviews were written by Jokipii, 1971 and Völk, 1976). The equation for the pitch- angle-averaged distribution function $f(\mathbf{r}, p, t)$ as a function of position **r**, particle momentum p, and time t, may be written as

$$\begin{aligned}\frac{\partial f}{\partial t} =& \frac{\partial}{\partial x_i}\left[\kappa_{ij}^{(S)}\frac{\partial f}{\partial x_j}\right] - \mathbf{V}\cdot\nabla f - \mathbf{V}_\mathrm{d}\cdot\nabla f \\ &+\frac{1}{3}\nabla\cdot\mathbf{V}\left[\frac{\partial f}{\partial \ell np}\right] + \frac{1}{p^2}\frac{\partial}{\partial p}\left[p^2 D_{pp}\frac{\partial f}{\partial p}\right] + Q(x_i, t, p).\end{aligned} \tag{1}$$

where the successive terms on the right-hand side correspond to diffusion, convection with velocity **V**, particle drift, adiabatic cooling or heating, second-order Fermi acceleration and any local source Q. Here, for particles of velocity w, momentum p and charge q, the drift velocity is $\mathbf{V}_\mathrm{d} = pcw/(3q)\ \nabla\times\left[\mathbf{B}/B^2\right]$, c is the speed of light, and $\kappa_{ij}^{(S)}$ is the symmetric part of the diffusion tensor. We may write $\kappa_{ij}^{(S)}$ in terms of the magnetic field components B_i and $\kappa_\parallel$ and $\kappa_\perp$, the diffusion coefficients parallel and perpendicular to it as $\kappa_{ij}^{(S)} = \kappa_\perp\delta_{ij} - \left(\kappa_\perp - \kappa_\parallel\right) B_i B_j/B^2$. Recently Earl, Jokipii and Morfill (1988) derived higher- order viscous terms which must in general be added to equation (1). These appear to be small in the heliosphere. Associated with the solution to the pitch-angle-averaged distribution function f is a streaming flux **S** , which may be written in terms of f as

$$S_i = -\kappa_{ij}\frac{\partial f}{\partial x_j} - \frac{1}{3}V_i\frac{\partial f}{\partial \ell n(p)}. \tag{2}$$

with an associated anisotropy magnitude $\delta = 3\mid\mathbf{S}\mid/(wf)$.

Equations (1) and (2) are very general, requiring only that the distribution function be kept nearly isotropic ($\delta \lesssim 30\%$ should be adequate), and that the speed w of the particles be much greater than the convection speed V (again, a factor of 3 may be sufficient). Idealized step functions associated with shocks and discontinuities can be studied with this equation (e.g., Isenberg and Jokipii, 1978). Hence, this equation can be readily applied to models in which complicated transport effects such as drifts occur at the same time as shock acceleration.

If the cosmic-ray energy density is large enough, the effects of the cosmic rays on the fluid must be considered. The primary effect comes from the cosmic-ray pressure, which can become large enough to help determine the nature of the background plasma flow. A good first approximation is to include the cosmic-ray pressure $P_c = \int f(p, \mathbf{r}, t)pwd^3p/3$ in the fluid momentum equation, in addition to the usual fluid pressure. The result is a coupling between the cosmic ray transport which depends on the flow speed, and the fluid equations which now depend on the cosmic-ray pressure. These equations have been solved for acceleration and transport at shocks (see, e. g., Drury, 1983) for a review. In addition, the cosmic rays can generate magnetic irregularities which scatter the particles, resulting in yet another coupling (see, e.g. Lee, 1983). Only the latter effect has been directly tested with *in situ* observations (Kennel, et al, 1986).

The determination of the diffusion tensor in terms of the observed magnetic-field spectrum is in the area of the "microphysics' of cosmic-ray interactions. Included is the question of the validity of diffusive theory as opposed to scatter-free motion. This area is currently active, but the confrontation with observations is not now as intense. For this reason, it is not discussed in as great detail as the large-scale problems.

CONFRONTATION BETWEEN THEORY AND OBSERVATIONS

Although the equations discussed in Section 2 apply to energetic charged particles in many astrophysical contexts, the most stringent observational tests have been in the heliosphere, where *in situ* spacecraft observations are available.

A. Solar Modulation of Galactic Cosmic Rays

The phenomenon of the solar modulation of cosmic is the partial exclusion of galactic cosmic rays from the heliosphere, in which the cosmic ray intensity reaches a *maximum* at the Earth during sunspot *minimum* and is a minimum when the sun spots are at a maximum. It is the subject of the most-sophisticated and realistic application of equation (1).

As illustrated in figure 2, the heliosphere, with the sun at its center, is imbedded in the interstellar gas. The solar wind blows radially outward, carrying with it the solar magnetic field. At some radius, because of the interstellar gas, the wind undergoes a shock transition to subsonic flow. Beyond the shock is the contact surface which separates the interstellar plasma from the solar wind gas. A possible bow shock in the interstellar plasma is also shown. The outer portions of this model, beyond the inner shock are not well understood. This whole system is bathed by an isotropic, uniform distribution of galactic cosmic rays. The problem is to understand this system quantitatively in terms of equation (1), where the source Q is set equal to zero, and an external boundary condition reflecting the external bath of galactic cosmic rays is imposed. The second-order Fermi term is found to be negligible here.

The large-scale structure of the interplanetary magnetic field now seems to be well-understood (e.g., Smith and Thomas, 1986). During the years around each solar sunspot minimum, the field is organized into two hemispheres (each with an Archimedean field) separated by a thin current sheet across which the field reverses direction. At solar sunspot minimum, the current sheet is nearly equatorial. The whole field direction changes sign at each sunspot maximum. The undulation of the current sheet about the equator increases with time after sunspot minimum and for the years near sunspot maximum the structure is not simple. It should also be noted here that a recent analysis by Jokipii and Kòta (1989) suggests that the unobserved **polar** magnetic field may differ considerably from the classical model even at sunspot minimum.

With this structure, it is determined that all four effects in equation (1) – convection, diffusion, energy change and drifts, are comparable in magnitude. The large-scale

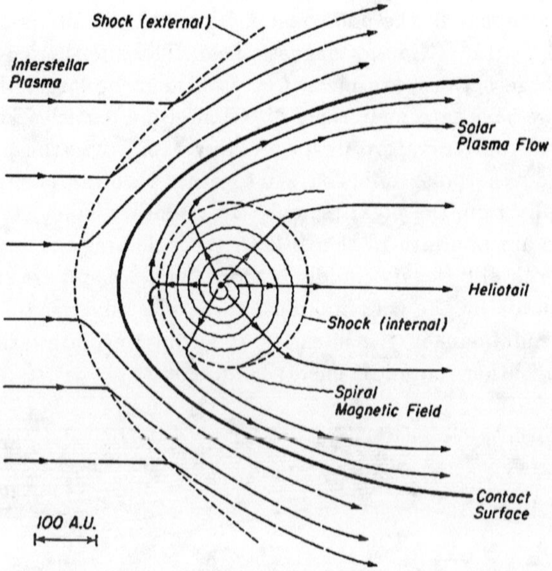

Fig. 2. Schematic illustration of a plausible structure for the heliosphere.

drifts have received much attention in the last several years. If the northern polar field is directed outward from the sun, the particles drift in over the poles and are ejected very rapidly along the current sheet which separates the northern and southern fields. This is reversed if the outward field is inward, and the particles drift rapidly in along the equatorial current sheet.

The particle drift turns out to be probably the most significant effect on the particle transport in the standard model: because they are coherent over large distances they determine the large-scale structure upon which smaller-scale and transient effects are superimposed. Forbush decreases due to shock waves and intensity changes due to merged interaction regions, etc, are definitely significant, but occur in a background distribution determined in large part by drifts. A quantitative study depends on numerical modelling.

Available numerical simulations cannot handle the full complexity of the outer heliosphere and its configuration is uncertain. For these reasons, a simpler model has been used, in which the solar-wind flow is radial and supersonic out to a spherical shock radius r_{sh}, beyond which the speed drops by a factor of 4 and begins to decrease as $1/r^2$. This all ends at a spherical outer boundary, at some 50% beyond the shock, where the cosmic-ray distribution function takes on a specified "interstellar" value, which is taken to be a power law in total energy with an index -2.65, corresponding to the observed index at high energies.

I discuss here the results from standard models using the generally accepted spiral magnetic field. The Archimedean spiral field falls off as $1/r$ near the equatorial plane beyond about 1 A.U. radius. In the polar regions the field falls off as $1/r^2$ resulting in an extremely weak ($\sim 10^{-8}$- Gauss) magnetic field. The diffusion coefficient must be correspondingly large because the mean free path must be larger than the local gyro radius. Hence, we have extremely rapid diffusion of the particles from the poles. Similarly, the gradient and curvature drifts are very rapid over the poles. Hence, in 1975, the drifts and the polar diffusion act together to bring positively charged particles in very rapidly from the polar regions. On the other hand, in 1965 or 1985, the diffusive motions are opposite to the drift in the polar regions suppressing the drift. The particles both diffuse very rapidly in from the poles and are sucked in very rapidly from the equator by the drift motions, to be ejected at some intermediate latitude. There is a minimum in the intensity at an intermediate latitude because particles coming in by either route lose energy continuously.

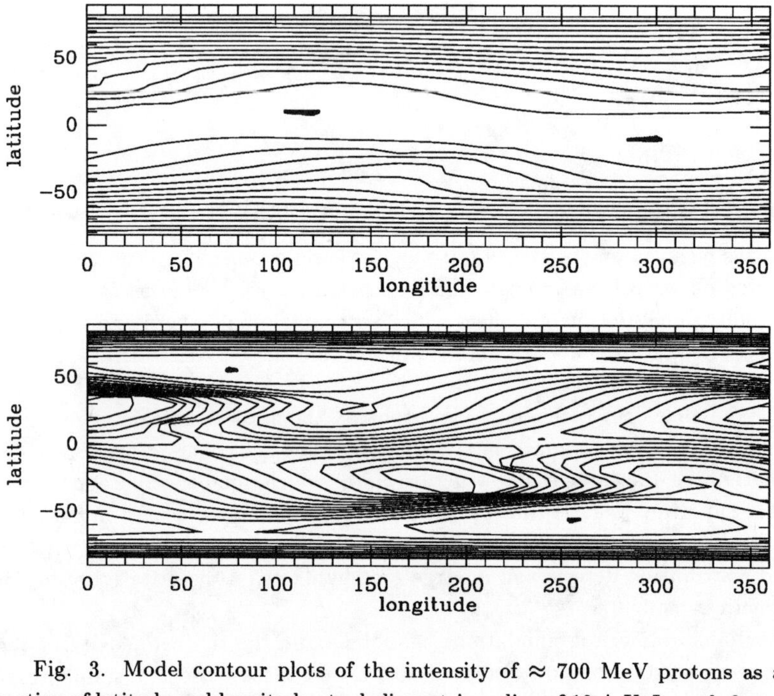

Fig. 3. Model contour plots of the intensity of ≈ 700 MeV protons as a function of latitude and longitude at a heliocentric radius of 19 A.U. In each figure the maximum occurs at the poles and the 2 minima for each plot are indicated by the small black shapes. The top curve is for $qA > 0$ (1975 minimum) and the bottom for $qA < 0$ (1965 and 1985).

These effects are illustrated very clearly in figure 3, which show contours of ≈ $1 GeV$ protons, as a function of heliolatitude and longitude, at a heliocentric radius of 19 A.U, for the two opposite signs of the classical Archimedean spiral magnetic field. A significant number of observed comic-ray effects are in very good agreement with this general picture. A number of comparisons have also been published by the group in South Africa (e.g. Webber, Potgieter and Burger, 1989)

B. The Anomalous Component and Shock Acceleration.

One other aspect of the theory of modulation and transport in the heliosphere is the study of the anomalous component of cosmic rays. It also provides a test of acceleration at shocks.

Digress first to the application of equation (1) to a shock wave. It is readily shown that the case of a planar shock yields the standard solution for diffusive shock acceleration, which illustrates the fact that the whole theory of diffusive shock acceleration is contained in our general transport equation. The solution for quasi-parallel shocks has appeared many times in the literature (see, e.g., Drury, 1983). The general case has been discussed by Jokipii (1987).

One particularly interesting situation in the heliosphere is the possible acceleration of cosmic rays at the termination shock of the solar wind, first suggested by Jokipii (1968). Pesses, Jokipii and Eichler (1981) pointed out that many features of the anomalous component in the cosmic-ray flux could be explained in terms of the acceleration of freshly-ionized interstellar particles (after Fisk, et al, 1974) at the termination shock of the solar wind, although a detailed model was not worked out. Jokipii (1986) and Potgieter and Moraal(1988) have presented numerical model calculations which are in excellent agreement with observations. In the model of Jokipii, equation (1) was solved in a two-dimensional geometry with a flat current, and the source Q set to correspond to the injection of particles at some low initial momentum p_0.

Pesses et al (1981) pointed out that, for the standard Archimedean spiral field, the termination shock is essentially a perpendicular shock everywhere but at the poles, where it is quasiparallel for a limited range of angles. The injection of low-energy particles would be preferred there. However, this gave rise to energy spectra which were not in agreement with observation. However, the new polar field suggested by Jokipii and Kòta (1989) is also transverse at the poles, suggesting isotropic injection. This new field configuration gives excellent agreement with observations (see below).

The energy density of the anomalous component is surprisingly large and increases with radius. We have investigated it within the context of our model. The procedure was as follows: the observed energy densities of anomalous hydrogen was computed from date reported in Cummings et al (1989), and the helium and oxygen energy densities were computed from numbers provided by A. C. Cummings and correspond to the values in Cummings and Stone (1988).

Figure 4 illustrates the energy density of these components and that of the solar wind as a function of radius, as obtained from a sample simulation. The solar wind value used corresponds to the slow solar wind values given by Hundhausen, 1972, scaled as $1/r^2$. In each simulation the variation of energy density with radius was computed, with the numbers normalized to give the observed energy density at 21 A. U. Very similar curves are obtained for a variety of parameters. It is clear that the solar wind energy density is much larger than that of the energetic particles within about 60 A.U., but that the anomalous hydrogen dominates beyond some 70 A.U. and

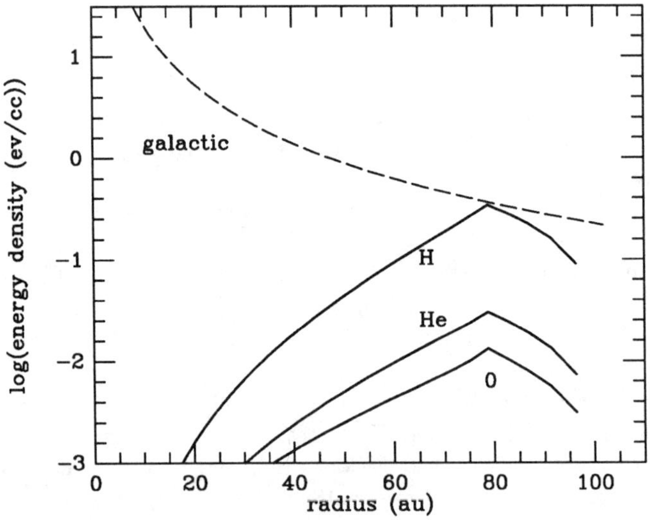

Figure 4. Plot of the energy density of anomalous protons, helium and oxygen, as a function of radius, for a numerical simulation normalized to the value observed at 21 A. U.

peaks at the radius of the termination shock, here taken to be 80 A. U. Because of the large gradient of the anomalous component, it should be dynamically more important than the galactic cosmic rays.

In any event, for these simulations, if the termination shock is significantly beyond 80 A.U., the solar-wind energy density would be substantially less than that of energetic particles, and the dynamical effects would be significant, invalidating the model or suggesting that the parameters are not correct.

C. Comparison of Model Predictions with Heliospheric Observations

Over the past several years a number of observational tests of this general picture have been devised. In particular, attention has been focussed on whether the predicted strong drift effects occur. Table 1 (adapted from one published by McKibben, 1967) is a reasonably complete survey of the situation with respect to these tests.

For the most part, this table indicates quite satisfactory agreement with data. It must be kept in mind that most of the "yes's" come as general, qualitative predictions of the overall model, without any careful tweaking of the parameters. Each one of them could, individually, be produced by a spherical model without drifts if one worked hard enough. However, in my view, this latter approach seems somewhat forced.

There are three tests which have the notation "See Text". Each one of these would have been a "No" until the past year.

Recently, J. Kòta and I (Jokipii and Kòta, 1989) have reconsidered the structure of the interplanetary magnetic field and have concluded that the standard Archimedean field is likely to be a poor representation of the actual magnetic field

Table 1 Summary (After McKibben, 1988)

$qA < 0$: 1958–1970; 1980–present
$qA > 0$: 1970–1980

PREDICTION	OBSERVED	REFERENCE
Diurnal anisotropies shift to earlier hours for $qA > 0$	YES	Ahluwahlia & Riker (1984). Swinson (1984).
22 yr. period for nucleon /electron ratio	YES	Garcia-Munoz et al. (1986).
Effect of turbulence in ecliptic greater for nucleons when $qA < 0$	YES	Shea and Smart (1981).
Radial gradients for nucleons for different signs of qA the same	See Text	Lopate et al. (1987). McDonald et al. (1986).
Correlation of modulated intensity with current sheet tilt	YES	Smith and Thomas (1986).
Duration of minimum modulation		
-Protons: longer for $qA > 0$	YES	Garcia-Munoz et al. (1987).
-Electrons: shorter for $qA > 0$	See Text	Garcia-Munoz et al. (1987).
Latitude gradients for nucleons		
- Galactic cosmic rays: negative when averaged, otherwise change sign with qA.	YES	Newkirk & Fisk (1985). McDonald, et al., (1986).
- Anomalous components : reverse with qA	YES	Cummings et al. (1987). McKibben et al. (1979).
Anomalous components for $qA < 0$		
- Intensity lower near ecliptic	YES	McKibben (1988).
- Spectral peaks at higher energy	See Text	Cummings et al. (1987)

in the distant polar regions of the heliosphere and have suggested a modified field. In the Archimedean spiral, the radial gradient for $A > 0$ is much less than that for $A < 0$, but we find that for the modified magnetic field the two gradients are very close together indicating that the observed radial gradients no longer are in contradiction to the models involving drifts. In addition, this field changes the calculated spectrum of the anomalous component to be in excellent agreement with observation.

The modulation of cosmic-ray electrons has also posed difficulties. As emphasized by Garcia-Munoz, et al. (1987), one expects the time history of electrons to be similar to that of the positively charged particles, but oppositely phased, so that the long maximum occurs in those sunspot minima when the protons have a short maximum. This asymmetry is not observed, although it is clear that the electrons behave differently from positively-charged particles. However, there are a number of effects which can change this prediction and these are currently under detailed study (Jokipii and Moraal, 1989). One important point is that most observations do not distinguish between positrons and negatrons. Available measurements are consistent with a positron fraction greater than .2 in the energy range ($\approx 600MeV$) under consideration. Such a value significantly affects the expected time history, as illustrated in figure 5. Further study is required to determine whether this alone will rectify the discrepancy (see also Burger and Potgieter, 1989).

It should also be noted that the termination shock may also accelerate electrons (anomalous electrons). Significant numbers of such electrons are not precluded by available observations and could also affect the time history. In addition, large numbers of cometesimals have been proposed by some. Secondary electrons from these could also be significant. These considerations lead to the conclusion that the fact that the electrons do not behave as protons shifted by one sunspot cycle does not at present contradict the theory. Further study of the electron component is necessary.

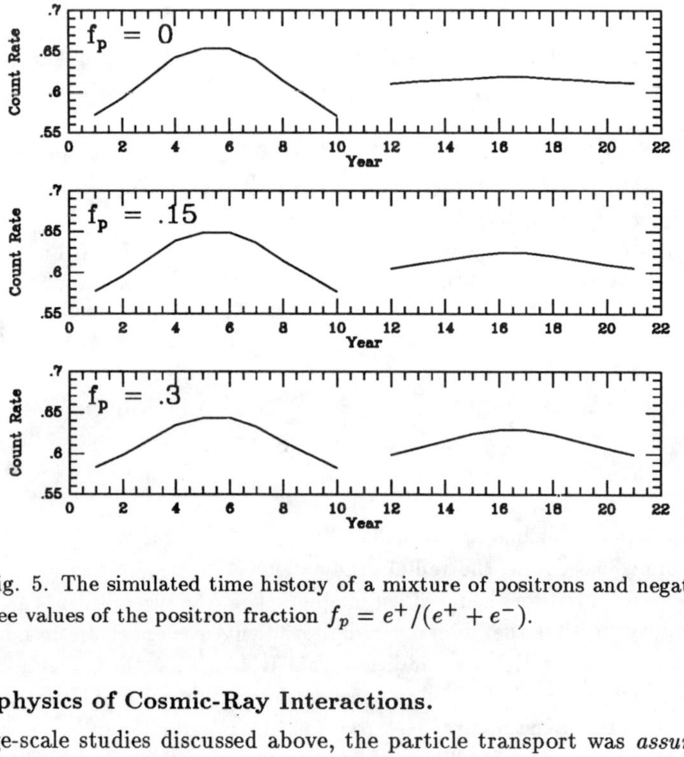

Fig. 5. The simulated time history of a mixture of positrons and negatrons for three values of the positron fraction $f_p = e^+/(e^+ + e^-)$.

D. Microphysics of Cosmic-Ray Interactions.

In the large-scale studies discussed above, the particle transport was *assumed* to be described by a diffusion equation, with a given diffusion tensor κ_{ij}. The microphysics of cosmic-ray transport concerns the details of the transport and whether or not the diffusion approximation is valid. This area of research is if considerable importance. At present, the confrontation with observations is not as intense as in the large-scale processes.

A major problem has been the difficulty in relating the diffusion mean free path to the observed magnetic field (Palmer, 1982). Simple applications of theory do not work and probably our view of the turbulence is too restrictive. Related to this is recent work on the effects of helicity on the transport coefficients (Bieber, et al, 1987). Although there is an effect in principle, it has not been established that there is in fact enough helicity in the solar wind to make a difference.

Another area of importance is the investigation of the focussing of the diverging heliospheric magnetic field on charged particles (Earl, 1981). This appears to occur in solar flare events and its study will influence our understanding of transport in general.

Very recently, Earl, Jokipii and Morfill (1988) showed that the general transport equation was incomplete in that shear in the background flow did not appear. Their corrections correspond to viscosity of the cosmic rays and have potential importance. An area of considerable controversy exists concerning the possibility that the diffusion approximation is invalid in restricted regions, such as, for example, shocks.

The term "shock drift acceleration" has been used to discuss the scatter-free acceleration at quasi-perpendicular shocks (Pesses, et al, 1982). It may be shown (Jokipii, 1988) that drift acceleration also occurs naturally in diffusive theories. Nonetheless there is a valid question as to whether scatter free acceleration occurs at shocks, especially in connection with pre- existing ambient energetic particles.

SUGGESTED FUTURE HELIOSPHERIC PROGRAM

A major goal of heliospheric cosmic-ray physics must be to continue and enhance the confrontation between theoretical and modelling on the one hand, and space-borne and Earth-based observations on the other. To accomplish this we must support both a vigorous theoretical program and ensure the continued exploration of the heliosphere by spacecraft. The fundamental importance of correlative studies of must be reflected in the institution of the COHO outer-heliosphere program. The theoretical program must include both groups and individual researchers with sufficient access to computers to permit the simulation of the very complex system we are confronting. The program of exploration must include new missions such as solar probe and the heliospheric boundary probe in addition to ensuring that the existing flotilla of spacecraft is optimally used to provide comprehensive spatial and temporal coverage.

Specific items are:

To meet long-term, general objectives -

- Establish the COHO outer-heliosphere program.
- Insure the continued coverage of existing spacecraft in deep space such as the Pioneers and Voyagers.
- Insure the presence of spacecraft near 1 A. U. to provide baseline measurements.
- Get a new start for solar probe.
- Start planning for a heliospheric boundary probe.
- Provide adequate support for a balanced effort in both fundamental theory and large-scale modelling activities.
- Provide adequate support for correlative data analysis.
- Support fundamental theoretical studies of the interaction of fast charged particles with turbulent plasmas and shocks.

To address shorter-term, specific problem areas -

- Study electrons and positrons in the heliosphere to help address questions concerning modulation.
- Address the question of the termination of the solar wind both observationally and theoretically.
- Study the energy density of the anomalous component.
- Continue to explore the microphysics of cosmic-ray interactions.
- Address observationally the response of cosmic rays to the changing heliosphere as we approach another sunspot maximum.

Acknowledgements

This work is Contribution 89-43 of the University of Arizona Theoretical Astrophysics Program and was supported in part by the National Aeronautics and Space Administration under Grant NsG-7101 and by the National Science Foundation under Grant ATM8618260. I am also grateful to J.Kòta and H. Moraal for discussions regarding many of the issues covered here.

REFERENCES

Ahluwahlia, H. S. and Riker, J. F., *Proc. Int'l Symposium on Cosmic Ray Modulation in the Heliosphere*, Morioka, Japan, 114 (1984).

Axford, W. I., *Planet. Space. Sci.* **13**, 115, (1965).

Bieber, J. W., Evenson, P. A., and Mattheaus, W. H. , *Ap. J*, **315**, 700, (1987).

Burger, R., and Potgieter, M., in press, (1989).

Cummings, A. C., Stone, E. C. and Webber, W. R., *Geophys. Res. Letters*, 14, 174, (1987).

Cummings, A. C. and Stone, E. C., *Proc. 20th Int'l Cosmic Ray Conf.*, Moscow, 3, 421, (1987).

Drury, L., *Rept. Prog. Phys*, **46**, 973, 1983.

Earl, J. A., *Ap. J.*, **251**, 739, (1981).

Earl, J.A., Jokipii, J.R. and Morfill, G.M., *Ap. J.* **331**, L91 (1988).

Fisk, L. A., Kozlovsky, B., and Ramaty, R., *Ap. J.*, **190**, L35, (1974).

Garcia-Munoz, M., Meyer, P., Pyle, K. R., Simpson, J. A. and Evenson, P., *J. Geophys. Res.*, 91, 2858, (1986).

Garcia-Munoz, M., Meyer, P., Pyle, K. R., Simpson, J. A., and Evenson, P., *Proc. 20th Int'l Cosmic Ray Conf.*, Moscow, 3, 303, (1987).

Gleeson, L. and Axford, W. I., *Ap. J.*, **149**, L115, (1967).

Gloeckler, G., *Rev. Geophys. Sp. Phys.*, **17**, 569, (1979).

Hundhausen, A. J., *Coronal Expansion and Solar Wind*, Springer-Verlag, New York, (1972).

Isenberg, P. A. and Jokipii, J. R., *Ap. J.*, **234**, 746, (1979).

Jokipii, J. R., *Ap. J.*, **152**, 799, (1968).

Jokipii, J. R. and Parker, E. N., *Ap. J.* **160**, 739, (1969).

Jokipii, J. R., *Rev. Geophys. Sp. Phys.*, **9**, 27, (1971).

Jokipii, J. R., *J. Geophys. Res.*, **91**, 2929, (1986).

Jokipii, J. R., *Proceedings of the Sixth International Solar Wind Conference*, ed by Pizzo, V. J., Holzer, T. E., and Sime, D. E., (1988).

Jokipii, J. R. and Kòta, J., *Geophys. Res. Lett.* **16**, 1, (1989).

Jokipii, J. R. and Moraal, H., in preparation (1989).

Kennel, C. F, Coroniti, F. V., Scarf, F. L., Livesey, W. A., Russel, C. T., Smith, E. J., Wenzel, K. P., and Scholer, M., *J. Geophys. Res.* **91**, 11917, 1986.

Kòta, J. and J. R. Jokipii, *Astrophys. J.*, 265, 573, (1983).

Lee, Martin A., *J. Geophys. Res.*, **88**, 6109, 1983.

Linsley, J., *IAU Symposium 94, Origin of Cosmic Rays*, ed. Setti, G., Spada, G. and R. A. Wolfendale (Dordrecht, Reidel) p. 53, (1980).

Lopate, C. R., McKibben, R. B., Pyle, K. R. and Simpson, J. A., *Proc. 20th Int'l Cosmic Ray Conf*, Moscow, 3, 409- -412, (1987).

McDonald, F. B. and Lal, N., *Geophys. Re. Letters*, 13, 781, (1986).

McDonald, F. B., von Rosenvinge, T. T., Lal, N., Trainor, J. H. and Schuster, P., *Geophys. Res. Letters*, 13, 785, (1986).

McKibben, R. B., Pyle, K. R. and Simpson, J. A., *Astrophys. J. (Letters)*, 227, L147, (1979).

McKibben, R. B., in *Proceedings of the Sixth International Solar Wind Conference*, ed by Pizzo, V. J., Holzer, T. E., and Sime, D. E., (1988).

Meyer, Peter, *Ann. Rev. Astron. Astroph.*, **7**, 1, (1969).

Newkirk, G. and Fisk, L. A., *J. Geophys. Res.*, 90, 3391, (1985).

Palmer, I. D., *Rev. Geophys. Sp. Phys.*, **20**, 335, (1982).

Parker, E. N., *Planet. Space Sci.*, **13**, 9, (1965).

Pesses, M. E., Jokipii, J. R. and Eichler, D., *Ap. J. (Letters)*, **246**, L85, (1981).

Pesses, M. E., Decker, R. B., and Armstrong, T. A., *Space Sci. Rev.*, **32**, 185, 1982.

Potgieter, M. S., and Moraal, H., *Ap. J.*, **330**, 445, (1988).

Shea M. A. and Smart, D. F., *Advances in Space Research*, Vol. 1, No. 3, 147, (1981).

Smith, E. J. and Thomas, B. T., *J. Geophys. Res.*, 91, 2933, (1986).

Swinson, D. B., *Proc. Int'l Symposium on Cosmic Ray Modulation in the Heliosphere*, Morioka, Japan 338, (1984).

Thomas, B. T. and Smith E. J., *J. Geophys. Res.* **86**, 11105, (1981).

Venkatesan, D., Decker, R. B. and Krimigis, S. M., *Proc. 20th Int'l Cosmic Ray Conf*, Moscow, 3, 385, (1987).

Volk, Heinrich J., *Rev. Geophys. and Sp. Phys.*, **13**, 547, (1976).

Webber, W. R., Potgieter, M. S., and Burger, R. A., *Ap. J.* in press, (1989).

Origin of high energy galactic cosmic rays

T.K. Gaisser

Bartol Research Institute
University of Delaware
Newark DE 19716

Abstract

In this talk I discuss two problems in high energy particle astrophysics, the flux of cosmic ray antiprotons and the chemical composition in the region of the "knee" of the cosmic ray energy spectrum ($\sim 10^{15}$ eV). Both are rather poorly known at present, yet they have great potential for informing us about the sources, acceleration and propagation of high energy cosmic radiation.

1 Introduction

It is well known that the power requirement for galactic cosmic rays is nicely matched to the energy available from supernova explosions in the galaxy,[1] which is about 3×10^{42} erg/s.* The power required to sustain the steady state cosmic ray spectrum is

$$Q_{\rm CR} = \frac{\rho_E V_S}{\tau_R} \sim 5 \times 10^{40} \frac{\rm erg}{\rm s}, \tag{1}$$

where V_S is the volume in which the sources are distributed and τ_R is the residence time in that volume. The energy density of the cosmic radiation in the local interstellar medium (LIS) is

$$\rho_E = 4\pi \int E \frac{{\rm d}N}{{\rm d}E} \frac{{\rm d}E}{\beta c} = \int \frac{4\pi E^2}{\beta c} \frac{{\rm d}N}{{\rm d}E} {\rm d}\ln E, \tag{2}$$

where E is the kinetic energy per nucleon, βc is the particle velocity and ${\rm d}N/{\rm d}E$ is the differential spectrum of nucleons. If we assume that the cosmic ray spectrum observed in the LIS is typical of that throughout the source region, then we estimate $\rho_E \sim 1$ eV/cm^3. Then if we take the source region to coincide with the disk of the galaxy ($V_S \sim 5 \times 10^{66}$ cm), and assume a residence time in the disk of 5×10^6 yrs, we arrive at the numerical estimate shown in Eq. 1.†

*This estimate is based on one supernova every 30 years with a total kinetic energy of 2×10^{51} ergs in the ejected shell.

†Note that τ_R is the residence time in the disk, not the cosmic ray "age" as determined from the measurement of the relative intensity of a radioactive isotope such as ^{10}Be.

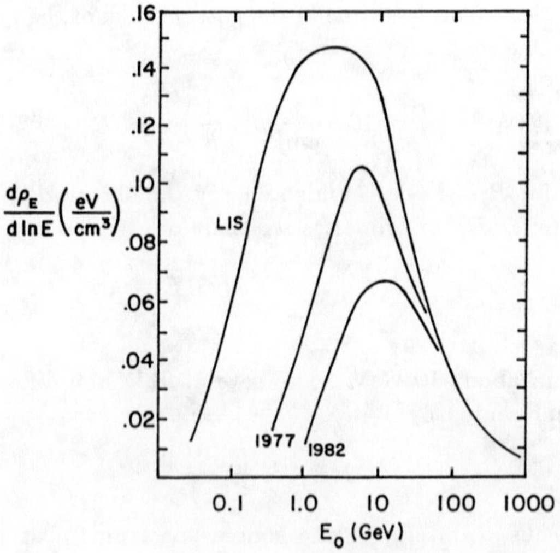

Figure 1: Energy density in cosmic rays (eV/cm^3).

It is interesting to display the contributions of different energy regions to the total cosmic ray energy density. I show this in Fig. 1 by plotting the integrand of the right hand side of Eq. 2. By plotting a quantity proportional to $d\rho_E/d\ln E$, I obtain a graph in which physical areas on the semilogarithmic plot are proportional to their contributions to the integral. The lines labelled 1977 and 1982 correspond to fluxes of protons measured at Earth at periods of minimum and maximum solar modulation respectively. The curve for the LIS is obtained from the demodulated local interstellar spectrum of protons as given by Evenson.[2] The area under the proton curve is 0.83 eV/cm^3. Helium and heavier nuclei together contribute another $\sim$ 0.27 eV/cm^3. About 99% of the energy content of the interstellar cosmic ray flux is contained in particles with energy less that 1 TeV/nucleon. Indeed, over half the energy is carried by particles with energies below several GeV and is lost to the solar wind as particles penetrate the heliosphere.

A great deal of our current picture of cosmic ray origin and propagation comes from detailed studies of elemental and isotopic composition of particles with kinetic energies of several GeV and below. The status of this research has been reviewed recently by Mewaldt[3] and Binns et al.[4] (see also Ref. 5 for a comprehensive review). One well-known result is the fact that the ratio of secondary to primary nuclei, such as boron to carbon, decreases as energy increases. This is interpreted as a rigidity-dependent decrease in the residence time (or increase in the spatial diffusion coefficient).[‡] The

‡In a model with spatial diffusion in one dimension (perpendicular to the disk) the relation between τ_R and spatial diffusion coefficient is[6] $\tau_R = hH/D$, where H is the thickness of the halo and h the thickness of the disk.

higher energy particles spend less time in the gaseous disk of the galaxy and so create fewer secondaries. One recent fit[7] gives

$$\lambda_{esc} = \beta c \rho \tau_R = 10.8 \frac{\mathrm{g}}{\mathrm{cm}^2} \beta \left(\frac{4}{R}\right)^{0.6}, \quad R > 4\,\mathrm{GV}, \tag{3}$$

and $\lambda_{esc} = 10.8\,\beta$ for $R < 4$ GV. At high energy the observed spectrum, $J(E)$, and the source spectrum, $Q(E)$, of primaries are therefore related by

$$J(E) \approx E^{-\delta} \times Q(E), \tag{4}$$

with $\delta \approx 0.6$.

At energies from about 10 GeV up to several TeV, the differential spectrum of nucleons can be represented by

$$J(E) \propto E^{-\gamma}, \tag{5}$$

with $\gamma \approx 2.7$. Thus, from Eq. 4 the source spectrum must be proportional to $E^{-(\gamma-\delta)} \sim E^{-2.1}$. The notion that cosmic rays are powered by supernova explosions was reinforced with the advent of the theory of first order Fermi acceleration by astrophysical shocks, which provides a natural explanation for this particular power law dependence of the energy spectrum.[8,9] A nice scenario for the origin of the bulk of the galactic cosmic rays, therefore, is that they are accelerated by first order Fermi acceleration at supernova blast waves expanding into the interstellar medium. The energy derives ultimately from the kinetic energy of the exploded supernova shells.

2 Composition at the "knee"

The origin of particles with energies above $\sim 10^{14}$ eV is much more problematic. Since it takes longer to accelerate high energy particles than low energy ones with diffusive shock acceleration, the maximum energy per particle is limited by the lifetime of the shock. (For non-planar shocks, escape from the acceleration region may be a more important limitation on the energy.) The acceleration rate is inversely proportional to the spatial diffusion coefficient in the region of the shock. In the approximation that diffusion is characterized by the minimum possible scattering length both upstream and downstream, $D \approx \frac{1}{3} r_L\, c$, with the gyroradius $r_L = Pc/(Z\,e\,B)$. (Here P is the total momentum per nucleus and Z the charge.) In this approximation[10]

$$E_{max} \sim \frac{3}{20} \frac{u_1}{c} Z\,e\,B\,(u_1 T_A). \tag{6}$$

Using an interstellar field of 3 μGauss, shock speed $u_1 \sim 10^9$ cm/s and lifetime $T_A \sim$ 1000 years, gives

$$E_{max}/Z \sim 10^{14}\ \mathrm{eV}. \tag{7}$$

Axford[9] uses a longer lifetime and estimates a somewhat higher value of E_{max}.

Table 1: Fractional composition of five major cosmic ray groups.

Mass group	$<A>$	Particles ($> E/A$)	Particles ($> E$/nucleus)	JACEE (Ref. 12) ($>$ 200 TeV/nucleus)
p	1	0.96	0.47	0.19±0.12
α	4	0.035	0.18	0.22±0.07
M ($Z = 6$ to 9)	14	0.0024	0.10	0.17±0.06
H ($Z = 10$ to 20)	24	0.0007	0.07	0.30±0.15
VH ($Z = 21$ to 30)	56	0.0004	0.18	0.13±0.09

The primary spectrum, as determined by air shower experiments, steepens in an energy range between 10^{15} and 10^{16} eV. This feature is referred to as the "knee" in the energy spectrum. Axford[9] suggests this feature may be a consequence of exhausting the supernova blast wave accelerator, as just discussed. Another possibility is that an increase in the escape rate from the galaxy (or from a more local trapping region) superimposed on a uniform acceleration mechanism might be responsible for the steepening of the spectrum. In both cases the behavior of the composition through this energy region would be characteristic of a rigidity dependent steepening.[11]

Calorimetric, "all particle" spectrum

The flux of cosmic rays above 10^{14} eV is less than 1 particle per square meter per hour through a detector with an angular acceptance of π steradians. The energy spectrum in this region has so far been explored almost entirely by air shower experiments because this rate is too low for detectors on balloons or spacecraft. Since an air shower experiment is calorimetric, events are classified by energy per particle, rather than by energy per nucleon. In both of the cases discussed above, the composition should become progressively rich in heavy nuclei because of their smaller rigidity at a given total energy per nucleus.

To appreciate that this signature would be readily visible if the chemical composition of the cosmic radiation could be measured in this energy range, it is helpful to see the composition on an energy per nucleus basis as directly measured at lower energy. The composition around 100 GeV/nucleon is shown in Table 1. At low energy the chemical composition of the cosmic radiation is usually presented by giving the relative fraction of various nuclei at the same energy per nucleon, as shown in the third column of the table. On an energy per nucleon basis, only 0.03% of the cosmic rays are nuclei heavier than helium, but this fraction is 36% when nuclei are compared at the same energy per nucleus (fourth column of the table). If the steepening of the spectrum is a rigidity dependent effect, either of acceleration or escape, the fraction of nuclei with $Z > 2$ above the bend would be about 53%, an increase by a factor of ~ 1.5 over its low energy value. Such a change in composition would be quite easy to detect

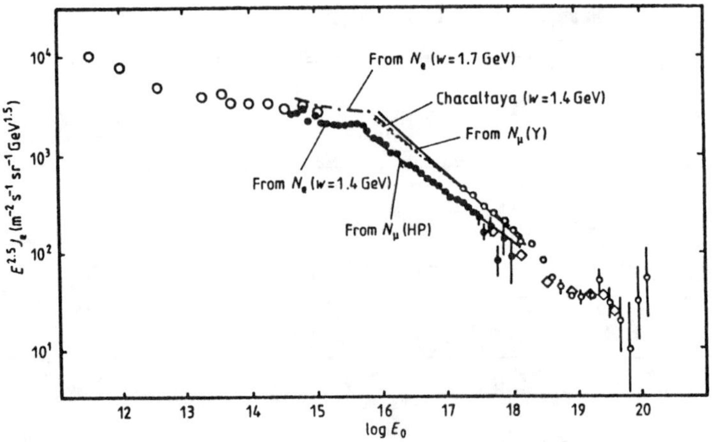

Figure 2: Primary spectrum as summarized by the Akeno group (Ref. 13).

provided only that it is possible to obtain a reasonable number of events with charge identification in the region of the "knee". (The reason for quoting $Z > 2$ relative to H and α is that air showers generated by hydrogen and helium have similar longitudinal development, while carbon and heavier nuclei reach maximum significantly more quickly and fluctuate less.)

Because of the low flux, it is not easy to obtain adequate statistics. The last column of Table 1 shows the JACEE data[12] at the highest energy for which they have events. This is the highest energy at which there is a direct measurement of the primary composition, and there are evidently a only very few events in each group. At present, going to higher energies requires use of air shower experiments, which do not see the primaries directly, but only the cascades of particles they produce in the atmosphere. Fig. 2 is a summary of calorimetric measurements of the cosmic ray spectrum - the "all particle" spectrum - from the paper of the Akeno group in Japan.[13] As is conventional, the differential spectrum has been multiplied by a factor $E^{2.5}$ to display the steepening of the spectrum which, is clearly visible in the Akeno data around 5×10^{15} eV.

Model dependence of air shower spectra

Because classic air shower experiments only sample the cascades at one depth, the conversion to primary energy and the construction of an energy spectrum is not obvious. There are at least two kinds of problems that need attention. First, the acceptance of the array is a function of energy, because bigger (i.e. more energetic and/or more deeply penetrating) showers trigger the array more efficiently. Second, the relation between observed size and primary energy depends on the properties of the hadronic interactions that make up the core of the shower, and these are not fully known at air shower energies. The latter ambiguity is illustrated by the two different representations of the Akeno data labelled N_e ($w = 1.7\ GeV$) and N_e ($w = 1.4\ GeV$) in Fig. 2. The primary energy per nucleus is given by

$$E_0 = w \times N_e(\max), \tag{8}$$

where $N_e(\max)$ is the number of particles in the shower at shower maximum. Both w and the correction from N_e at ground level to $N_e(\max)$ are model-dependent, although the correction to shower maximum can be done semi-empirically.[14] Despite the difficulties, there is a great deal of evidence from many experiments that the spectrum indeed steepens in the energy range between 10^{15} and 10^{16} eV.

Establishing the composition with air shower experiments is much more difficult than estimating the all particle energy spectrum. Very little indeed is known at present. Hillas[15] has argued that the bend in the spectrum is too sharp to be accounted for by a smooth steepening in rigidity of an underlying composition of the kind discussed above. Fichtel & Linsley[16] argue that a new source predominates above $\sim 10^{15}$ eV after the lower energy components die out, limited either by failure of the acceleration mechanism or increased escape. They further argue that this new, high energy source accelerates mostly protons. The evidence that leads to this conclusion has to do with the large fluctuations observed in various ways with air shower experiments, which would be difficult to achieve with heavy nuclei alone. The Maryland group, on the other hand, have long favored two components[17] even at relatively low energy: protons and helium with steeper spectra and heavier nuclei with flatter spectra, so that the composition is predominantly heavy already at 10^{14} eV and through the region of the break, contrary to the Fichtel & Linsley hypothesis. In my opinion, either view could be correct.

Possible mechanisms to achieve $E > 100$ TeV.

There are several ideas for new sources to provide the high energy cosmic rays. They can be divided into two categories, diffuse sources and point sources. Two ways in which the acceleration limit for supernova blast waves could be increased have been discussed in the literature. Völk & Biermann[18] point out that a supernova is likely to explode into the wind of the progenitor star rather than into the general interstellar medium, and that this can lead to somewhat higher energy than the estimate of Eq. 6. Jokipii[19] emphasizes the role of magnetic field geometry and argues that in the equatorial region of a supernova blast wave higher energy can be obtained, because there the magnetic field is nearly perpendicular to the direction of the blast wave, and oblique shock acceleration may occur more rapidly. Termination shocks of plasma winds are possible sites at which higher energy may be achieved as a consequence of higher fields or longer time scales. Cesarsky & Montmerle have discussed strong stellar winds.[20] Jokipii & Morfill[21] have suggested that the termination shock of a galactic wind may be able to accelerate cosmic rays even to 10^{19} to 10^{20} eV.

Possible point sources of cosmic rays with energies $> 10^{15}$ eV, such as accreting x-ray binaries or new supernovas, are also potential sources of TeV - PeV photons. This is a subject that has received a great deal of attention in the last decade and has been extensively reviewed.[22,23,24] The photons would come from decay of π^0's produced when some of the accelerated particles collide with matter near the source. Millisecond binary pulsars, such as 1957 + 20 with a 1.6 ms period,[25] are also possible point sources.[26]

Conclusion

It is clear that a direct determination of the energy spectrum of each major component of the cosmic radiation through the region of the knee is essential to understanding the origin, acceleration and propagation of high energy cosmic particles. At lower energies, the direct measurements of the primary composition[12,27,28] already show interesting energy dependences, which now need to be followed up to higher energy. The total flux of particles above 10^{16} eV is about 1 particle $m^{-2}sr^{-1}$ per year. Thus an exposure of several hundred square meter-years with a detector of large aperture is required. Swordy[29] has proposed a lunar calorimeter to accomplish this measurement, an idea that has been discussed further during this conference. Without an overlying atmosphere, a large calorimeter on the lunar surface could be instrumented to measure the charges of the primary nuclei directly.

3 Antiprotons

Indications[30] that the flux of cosmic ray antiprotons may be higher than expected if the propagation history of protons and helium is the same as heavier primary nuclei (as deduced from the ratio of secondary to primary nuclei) have drawn attention to the possibility that there may be more than one type of source even at energies of several GeV. One possibility[31,32] is that a fraction (25-30%) of the observed cosmic rays comes from shrouded sources, in which enough material is traversed at the source after acceleration to produce many antiprotons, while breaking up heavy nuclei so as not to spoil the conventional diffusion or leakage models for boron/carbon, etc. Such simple models have difficulty confronting the observed ^{3}He/^{4}He ratio, however. The helium cross section is small enough so that this ratio would be larger than is observed.[33]

On the other hand, more exotic ideas, such as primary antimatter[34] or WIMP annihilation[35] as the source of observable antiprotons now appear even more unlikely because of the recent low upper limits on the $\bar{p}/p$ ratio below a GeV.[36,37] These models would tend to give a relatively large flux of antiprotons at low energy. In contrast, for kinematic reasons, the flux of secondary antiprotons is supressed at kinetic energies less than about a GeV.[38] The current experimental situation, as summarized in Ref. 37 (shown here as Fig. 3), suggests a secondary origin of cosmic ray antiprotons.

The main point I wish to emphasize in this talk is that antiprotons are a unique probe of acceleration and propagation of energetic particles in the galaxy because of the high threshold for their production. The minimum process for production by a cosmic ray nucleon on interstellar hydrogen is

$$p + H \rightarrow p + p + p + \bar{p}, \tag{9}$$

which has a threshold kinetic energy 5.63 GeV. The production cross section rises from zero at threshold, increasing to about 1 mb at 100 GeV and 5 mb at 1000 GeV.

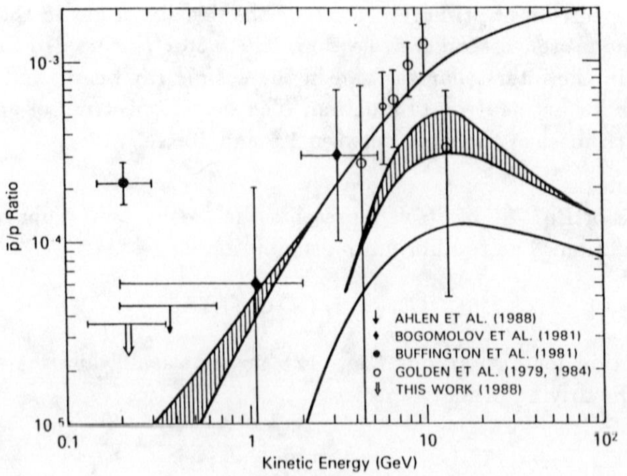

Figure 3: Summary of antiproton data and models from Ref. 37. Other data shown is from Refs. 36a, 39, 40, 41. The final results of the new low energy experiments (Ref. 36b and Streitmatter, private communication) are somewhat lower than shown here.

In contrast, spallation cross sections for processes such as

$$^{16}O + H \rightarrow {}^{10}B + \cdots \tag{10}$$

are nearly constant above a GeV and have a resonance structure below 100 MeV.[42] In addition, in nuclear spallation processes the secondary nuclei emerge with nearly the same energy per nucleon as the primary. Antiprotons, on the other hand, have considerably less energy than the projectile nucleon. (At high energy, the ratio is typically a tenth.)

To see the consequences of these features, consider the following simple approximation for the equilibrium flux of antiprotons:

$$\begin{aligned} \frac{\mathrm{d}J_{\bar{p}}(E)}{\mathrm{d}t} = 0 = \; & Q_{\bar{p}}(E) - \frac{1}{\tau_R(E)} J_{\bar{p}}(E) \\ & - \frac{J_{\bar{p}}}{\tau_i} + \frac{1}{\tau_i}\int_E^\infty \frac{\mathrm{d}n_{\bar{p}\to\bar{p}}(E,E')}{\mathrm{d}E} J_{\bar{p}}(E')\mathrm{d}E' \\ & - B\, J_{\bar{p}}(E) + B\int_0^\infty \frac{\mathrm{d}P(E,E')}{\mathrm{d}E} J_{\bar{p}}(E')\mathrm{d}E'. \end{aligned} \tag{11}$$

The first line of Eq. 11 is the balance between production of antiprotons and loss from the galaxy. If the antiprotons are secondaries, the source term is

$$Q_{\bar{p}}(E) = 2\frac{c\rho\sigma}{m_H} F \int \frac{\mathrm{d}n_{p\to\bar{p}}}{\mathrm{d}E_{\bar{p}}} J_p(E_p)\mathrm{d}E_p, \tag{12}$$

where J_p is the spectrum of primary protons, ρ the density of gas in the interstellar medium and σ the proton-proton cross section. The factor F is needed to account for effects of nuclei in the interstellar gas and in the cosmic ray beam, and the explicit factor 2 accounts for antineutron production. The source spectrum of antiprotons is quite narrow, with most of the yield between 1.5 and 15 GeV.

Leaky box model.

The first line of Eq. 11 by itself represents the "leaky box" approximation for antiproton production. The equilibrium $\bar{p}$ distribution is

$$J_{\bar{p}}^{(0)}(E) = \tau_R(E)\,Q_{\bar{p}}(E). \tag{13}$$

The source function for antiprotons (Eq. 12) asymptotically has the same energy dependence as the driving primary spectrum, i.e. as $E \to \infty$,

$$Q_{\bar{p}}(E) \to \text{constant} \times J_p(E).$$

Thus if antiprotons are produced throughout the galactic disk by a spectrum of nucleons that is the same as observed locally, then the $\bar{p}/p$ ratio will decrease at high energy because of the energy dependence of τ_R (Eq. 3). This is illustrated by the lowest solid curve[§] in Fig. 3 from the calculation of Protheroe.[43] (Note the important point that τ_R is to be evaluated at the energy of the $\bar{p}$, not at the higher energy of the particle that produced it.) If some antiprotons are produced in matter traversed by primary protons near the source, then the driving primary spectrum in Eq. 12 may have a harder spectrum than the observed spectrum, for example, $E^{-2.1}$ (recall the discussion of Eq. 5). This would give an asymptotically constant contribution to the $\bar{p}/p$ ratio.

Closed galaxy model

The second line of Eq. 11 represents attenuation of antiprotons, including annihilation as well as interactions in which an antinucleon emerges with reduced energy. (The latter dominate at high energy.) The interaction rate is $1/\tau_i$. Normally this term is unimportant (except for $E_{\bar{p}} < 1$ GeV) because $\tau_R \ll \tau_i$. In a closed galaxy model,[44] however, a fraction of the observed cosmic radiation is an "old" component, for which the confinement time is infinite. The $\bar{p}$ flux produced by this component is given by an equation like (13) with $\tau_R \to \tau_i$, which can significantly enhance the antiproton flux.[45] In this model the $\bar{p}/p$ ratio eventually approaches a constant because τ_i = constant. This occurs, however, only at very high energy, after the "young" component of the primary cosmic radiation (for which τ_R decreases according to Eq. 3) becomes negligible. A $\bar{p}/p$ ratio for a closed galaxy model is represented by the upper curve in Fig. 3.[43]

Reacceleration

The last line of Eq. 11 represents the possibility of reaceleration. B is the rate of encounters with reacceleration sites (e.g. weak interstellar shocks).[46] The function

§The escape length used for this curve is $\sim 50\%$ lower below 10 GV than Eq. 3.

$dP(E,E')/dE$ characterizes the reacceleration by giving the normalized distribution of energies E that emerges from a reacceleration event for a group of particles injected with identical energy E'. The implicit assumption in this approach[46] is that the time to redistribute the energies of the particles is much less than the characteristic time scales $1/B$ and τ_R. There is a great deal of recent work on reacceleration or distributed acceleration of cosmic rays, which is summarized in Cesarsky's rapporteur talk at Moscow.[47] Although it is apparently not possible to account for all cosmic ray acceleration as distributed acceleration simultaneous with propagation, some (weak) reacceleration is possible. Indeed, if solar cosmic rays in the heliosphere are any guide, reacceleration in the galaxy must be inevitable. For a given model of reacceleration, the energy dependence of the escape length (see Eq. 3) must be readjusted from its conventional value to reproduce the observed ratios of secondary to primary nuclei. Simon et al.[48] have discussed the effect of weak reacceleration on the production of secondary antiprotons. This is the middle curve in Fig. 3.

The general character of the Simon et al. result can be understood by considering an iterative solution of Eq. 11 valid when reacceleration is a small effect. For simplicity I neglect attenuation of antiprotons (second line of Eq. 11). Taking Eq. 13 as the unperturbed solution, the first approximation to a solution of Eq. 11 can be written as

$$J_{\bar{p}}(E) \approx J_{\bar{p}}^{(0)}(E) + \tau_R(E)\,B\int_0^{\infty}\frac{dP(E,E')}{dE}\left(J_{\bar{p}}^{(0)}(E') - J_{\bar{p}}^{(0)}(E)\right). \qquad (14)$$

If reacceleration only increases the energy (as assumed by Simon et al.[48]), then the upper limit to the integral in Eq. 14 is E. The unperturbed solution, $J_{\bar{p}}^{(0)}(E)$, has a characteristic maximum at $E = E_c \sim 2$ GeV of kinetic energy.[38] Below that critical energy $J_{\bar{p}}^{(0)}(E) > J_{\bar{p}}^{(0)}(E')$ for $E' < E$. For $E > E_c$ the slope is reversed, so reacceleration decreases (increases) the $\bar{p}$ flux at $E < E_c$ ($E > E_c$). In other words, if reacceleration always results in increased energy per particle, then the effect is simply to shift the injection spectrum, $J_{\bar{p}}^{(0)}(E)$, to higher energy. Since the observed proton flux, to which $J_{\bar{p}}(E)$ is compared, decreases with increasing energy, shifting the $\bar{p}$ spectrum to higher energy increases the $\bar{p}/p$ ratio at high energy, as shown by the middle curve[48] in Fig. 3. The increase tends to zero at very high energy because of the energy dependence of $\tau_R(E)$ that multiplies the second term on the right side of Eq. 14. At low energy the $\bar{p}/p$ ratio will actually fall below the leaky box result if all reacceleration increases particle energies. If both acceleration and deceleration occur, however, then the injected spectrum will be spread to lower as well as to higher energies. Also, for $E < E_c$ effects of solar modulation are particularly important,[49,38] and must be taken into account. In addition, Webber & Potgieter[50] have recently emphasized that modulation is likely to depend on the sign of the charge differently at different epochs of the solar cycle.

The solution (14) can also be used to show the effect of reacceleration on secondary nuclei, such as boron, by replacing $J_{\bar{p}}$ by $J_{\text{secondary}}$ and using the observed flux of primary nuclei in Eq. 12 to calculate the injection spectrum of the secondaries. Since the injection spectrum of a secondary nucleus always decreases with increasing energy (at least above $\sim$ 100 MeV) the reacceleration term on the right side of (14) always adds to $J^{(0)}_{\mathrm{secondary}}(E)$ - more at low energy and less at high energy because of the decrease with energy of $\tau_R(E)$. If the energy dependence of $\tau_R(E)$ from the leaky box model [Eq. (3)] is used as input the secondary/primary nucleus ratio becomes steeper than observed. This is corrected by adjusting δ so that $\tau_R(E)$ decreases more slowly.[51]

Conclusion

In their recent paper, Webber & Potgieter[50] argue that the antiproton flux in the 6-12 GeV range[39] may be no more than a factor of two higher than expectation in the simplest leaky box model when the correct grammage and the correct interstellar primary spectrum are used and effects of nuclei are properly accounted for. The second two factors need further investigation, but even if it turns out that there is little or no discrepancy between the antiproton flux in this energy range and a leaky box calculation, a high precision measurement of the energy spectrum of cosmic ray antiprotons has the potential to provide important information about cosmic ray acceleration and propagation. Secondary $\bar{p}$'s are injected with a relatively sharply peaked spectrum that makes them a unique probe. They give information complementary to secondary nuclei because the $\bar{p}$ production cross section is strongly energy dependent, increasing to high energy. In combination with measurements of helium and hydrogen isotopes and of positron and electron fluxes[52,53] a much more complete picture of cosmic ray origin becomes accessible.

ACKNOWLEDGMENTS. I am grateful for helpful conversations with Paul Evenson, Dick Mewaldt, Randy Jokipii, Robert Schaefer, Todor Stanev and Gary Zank. This research is supported in part by NASA grant NAGW-1644.

4 References

1. V.L. Ginzburg & S.I. Syrovatskii, *The Origin of Cosmic Rays* (Pergamon Press, 1964).

2. P.A. Evenson, "Time variation of galactic cosmic rays," *Proc. Conf. on Interplanetary Particle Environment* (JPL Pub. 88-28, ed. Joan Feynman and Stephan Gabriel) 149-161 (1988).

3. R.A. Mewaldt, "The abundances of isotopes in the cosmic radiation," in *Cosmic Abundances of Matter* (A.I.P. Conference Proceedings 183, ed. C. Jake Waddington), 124-146 (1989).

4. W.R. Binns et al., "The abundances of the heavier elements in the cosmic radiation," *Ibid.,* 147-167.

5. J.A. Simpson, "Elemental and isotopic composition of the galactic cosmic rays," *Ann. Revs. Nucl. Part. Sci.* **33** 323-381 (1983).

6. V.L. Ginzburg, Ya. M. Khazan & V.S. Ptuskin, "Origin of Cosmic Rays: Galactic Models with Halo", *Astrophys. Spa. Sci.* **68**, 295-314 (1980).

7. M. Gupta & W.R. Webber, "The Abundance of ^{14}N at the cosmic-ray source: a study using new fragmentation cross sections", *Ap. J.* **340**, 1124-1134 (1989).

8. R. Blandford & D. Eichler, "Particle acceleration at astrophysical shocks: a theory of cosmic ray origin," *Phys. Reports* **154** 1-75 (1987).

9. W.I. Axford, "Cosmic ray acceleration," *Proc. 20th Int. Cosmic Ray Conf.* (Moscow) **8**, 120-126 (1987).

10. P.O. Lagage & C.J. Cesarsky, "The maximum energy of cosmic rays accelerated by supernova shocks," *Astron. Astrophys.* **125**, 249-257 (1983).

11. B. Peters, "Origin of cosmic radiation," *Nuovo Cimento (Suppl.)* **14** 436-456 (1959). Also *Proc. 6th Int. Cosmic Ray Conf.* (Moscow) **3**, 157 (1960).

12. T.H. Burnett et al. "Energy spectra of cosmic rays above 1 TeV per AMU by JACEE," OG6.1-10, XXI ICRC, Adelaide (1990).

13. M. Nagano et al. "Energy spectrum of primary cosmic rays between $10^{14.5}$ and 10^{18} eV," *J. Phys. G* **10**, 1295-1310 (1984).

14. A.M. Hillas, "A note on various versions of the primary energy spectrum," *Proc. Cosmic Ray Workshop* (U. of Utah, ed. T.K. Gaisser) pp..16-24 (1983).

15. A.M. Hillas "The Origin of Ultra-High-Energy cosmic rays," A.M. Hillas, *Ann. Rev. Astron. Astrophys.* **22**, 425-444 (1984).

16. Carl E. Fichtel & John Linsley "High-energy and ultra-high-energy cosmic rays," *Ap. J.*, **300**, 474-487 (1986).

17. H.T. Freudenreich et al. "A study of hadrons at the cores of extensive air showers and the elemental composition of cosmic rays at 10^{15} eV," UMDCR Preprint 90-024 (1989). See also J.A. Goodman et al., "Delayed hadrons in extensive air showers: Evidence for the iron group nuclei in primary cosmic ray flux at energies $\sim 10^{13}$–10^{15} eV," *Phys. Rev.* **D26** 1043 (1982).

18. H.J. Völk & P.L. Biermann "Maximum energy of cosmic-ray particles accelerated by supernova remnant shocks in stellar wind cavities," *Ap.J.* **333**, L65-L68 (1988).

19. J.R. Jokipii, "Mechanism of Particle Energy Gain at Shocks: The Role of Magnetic Field Geometry," Proc. 6th Int. Solar Wind Conf. (Estes Park, Colorado, August 1977) NCAR TN-306, vol 2, p. 481. See also Ap. J. **313**, 842 (1987) and **255**, 716 (1982).

20. C.J. Cesarsky and T. Montmerle, "Gamma Rays from Active Regions in the Galaxy: The Possible Contribution of Stellar Winds," *Space Sci. Revs.* **36**, 173-93 (1983).

21. J.R. Jokipii & G. Morfill "Ultra-high-energy cosmic rays in a galactic wind and its termination shock," *Ap. J.* **312**, 170-177 (1987).

22. T.C. Weekes, "Very high energy gamma-ray astronomy," *Phys. Reports* **160** 1-121 (1988).

23. D.E. Nagle, T.K. Gaisser & R.J. Protheroe, "Extensive air showers associated with discrete astrophysical sources," *Ann. Rev. Nucl. Part. Sci.* **38**, 609-657 (1988).

24. J.-M. Bonnet-Bidaud & G. Chardin, "Cygnus X-3, a critical review," *Phys. Reports* **170**, 325-414 (1988).

25. A.S. Fruchter, D.R. Stinebring & J.H. Taylor, "A millisecond pulsar in an eclipsing binary," *Nature* **333**, 237-239 (1988).

26. A.K. Harding & T.K. Gaisser, "Acceleration by pulsar winds in binary systems," in preparation (1989).

27. John M. Grunsfeld et al. "Energy spectra of cosmic-ray nuclei from 50 to 2000 GeV per AMU," *Ap. J.* **327**, L31-L34 (1988).

28. J.A. Esposito et al., "Measurement of the cosmic ray iron spectrum between 60 and 200 GeV/nucleon," NASA/GSFC preprint 89-043 (1989).

29. S.P. Swordy, "The measurement of elemental abundances above 10^{15} eV at a lunar base," these proceedings (1989).

30. S.A. Stephens & R.L. Golden, "Recent observations of cosmic ray antiprotons and a critical assessment of the theories of their origin," *Astron. Astrophys.* **202** 1-7 (1988). See also S.A. Stephens & R.L. Golden "The role of antiprotons in cosmic-ray physics," *Space Science Reviews* **46**, 31-91 (1987).

31. P.O. Lagage & C.J. Cesarsky, "Antiprotons from thick cosmic-ray sources," *Astron. Astrophys.* **147**, 127-134 (1985).

32. R. Cowsik & T.K. Gaisser, "Antiprotons from galactic sources of cosmic rays and gamma rays," *Proc. 17th Int. Cosmic Ray Conf.* (Paris) **2**, 218.

33. W.R. Webber, R.L. Golden & R.A. Mewaldt, "A reexamination of the cosmic-ray helium spectrum and the ^{3}He/^{4}He ratio at high energies," *Ap. J.* **312**, 178-182 (1987).

34. F.W. Stecker & A.W. Wolfendale, "The case for antiparticles in the extragalactic cosmic radiation," *Nature* **309**, 37-38 (1984).

35. F.W. Stecker & A.J. Tylka, "The cosmic-ray antiproton spectrum from dark matter annihilation and its astrophysical implications: a new look," *Ap. J. Letters* **336**, L51-54 (1989).

36. S.P. Ahlen et al., "New limit on the low-energy antiproton/proton ratio in the galactic cosmic radiation," *Phys. Rev. Letters* **61**, 145-148 (1988). See also M.H. Salamon et al. "Limits on the antiproton/proton ratio in the cosmic radiation from 100 MeV to 1580 MeV," *Ap.J.* (to be published Jan. 1990).

37. R.E. Streitmatter et al. "Experimental limit on low energy antiprotons in the cosmic radiation," *Advances in Space Research* **9**, 65 (1989).

38. T.K. Gaisser & E.H. Levy "Astrophysical implications of cosmic-ray antiprotons," *Phys. Rev.* **D10**, 1731-1735 (1974).

39. R.L. Golden et al., "Evidence for the existence of cosmic-ray antiprotons," *Phys. Rev. Letters* **43**, 1196-1199 (1979) and "Energy dependence of the $\bar{p}/p$ ratio in cosmic rays," *Astrophysical Letters* **24** 75-83 (1984).

40. E.A. Bogomolov et al., "Galactic antiproton spectrum in the 0.2-5 GeV range," *Proc. 20th Int. Cosmic Ray Conf.* (Moscow) **2**, 72-75 (1987).

41. A. Buffington, S.M. Schindler & C.R. Pennypacker, "A measurement of the cosmic-ray antiproton flux and a search for antihelium," *Ap. J.* **248**, 1179 (1981).

42. M. Garcia-Munoz et al., "Cosmic-ray Propagation in the galaxy and in the heliosphere: the path-length distribution at low energy," *Ap. J. (Supplements)* **64**, 269-304 (1987).

43. R.J. Protheroe, "Cosmic ray antiprotons in the closed galaxy model," *Ap. J.* **251**, 387 (1981).

44. B. Peters & N.J. Westergaard,"Cosmic ray propagation in a closed galaxy," *Astrophys. Space Sci.* **48**, 21 (1977).

45. G. Steigman, "Secondary antiprotons: A valuable cosmic ray probe," *Ap. J. (Letters)* **217**, L131-133 (1977).

46. Amri Wandel et al. "Distributed reacceleration of cosmic rays," *Ap. J.* **316**, 676-690 (1987).

47. C.J. Cesarsky, "Cosmic ray propagation in the interstellar medium," *Proc. 20th Int. Cosmic Ray Conf.* (Moscow) **8** 87-119 (1987).

48. M. Simon, U. Heinbach & Ch. Koch, "Antiproton production of propagating cosmic rays under distributed reacceleration," *Ap. J.* **320** 699-702 (1987).

49. J.S. Perko, "Solar modulation of galactic antiprotons," *Astron. Astrophys.* **184**, 119-121 (1987).

50. W.R. Webber & M.S. Potgieter, "A new calculation of the cosmic-ray antiproton spectrum in the galaxy and heliospheric modulation effects on this spectrum using a drift plus wavy current sheet model," *Ap.J.* **344**, 779-785 (1989).

51. M. Simon, W. Heinrich & K.D. Mathis, "Propagation of injected cosmic rays under distributed reacceleration," *Ap.J.* **300**, 32-40 (1986).

52. Dietrich Müller & Kwok-Kwong Tang, "Cosmic-ray positrons from 10 to 20 GeV: A balloon-borne measurement using the geomagnetic east-west asymmetry," *Ap. J.* **312** 183-194 (1986).

53. R.L. Golden et al. "Observation of cosmic ray positrons in the region from 5 to 50 GeV," *Astron. Astrophys.* **188**, 145-154 (1987).

Related Observations

COSMIC RAY STUDIES WITH THE GAMMA-RAY OBSERVATORY

D. A. Kniffen
NASA Goddard Space Flight Center, Greenbelt, MD 20771

ABSTRACT

Even prior to the first undisputed observations of the diffuse galactic gamma-ray emission, it was well understood that the dominant contribution to the observed galactic gamma-ray emission above 100 MeV would be that due to the decay of neutral pi-mesons resulting from the interaction of the galactic cosmic rays with the interstellar gas. The first unambiguous detection of this radiation was made in 1968 by the OSO-3 counter telescope[1], and details sufficient to allow preliminary studies of the implications of the observations were provided by the spark chamber experiments aboard NASA SAS-2 satellite[2] and by the ESA COS-B mission[3]. Many theoretical and observational studies have addressed this question since these initial explorations, but interpretation has been hampered by limited angular resolution in the observations, lack of broad band spectral coverage and insufficient information on the interstellar gas distribution, and contributions from discrete gamma-ray sources. This paper will give details on how the instruments aboard the Gamma-Ray Observatory will provide the details necessary to resolve these problems and aid significantly in determining the distribution of both cosmic-ray nucleons and electrons throughout the galaxy.

INTRODUCTION

One of the major motivations leading to the development of an observational gamma-ray capability in the 1960's was the recognition that this was the most direct way of studying the galactic cosmic ray distribution outside of the solar vicinity. Pollack and Fazio[4] and Stecker[5] were among many who pointed this out. So it was no surprise when Kraushaar et al.[1] unambiguously observed this gamma-ray emission, even though the angular resolution of the measurements was very poor. (Figure 1). Interpretation of the observed emission was hampered by the lack of statistics in the gamma-ray data, and lack of information on the interstellar gas distribution, but seemed to indicate that a major enhancement in the cosmic-ray density toward the central region of the galaxy was required to explain the observations.

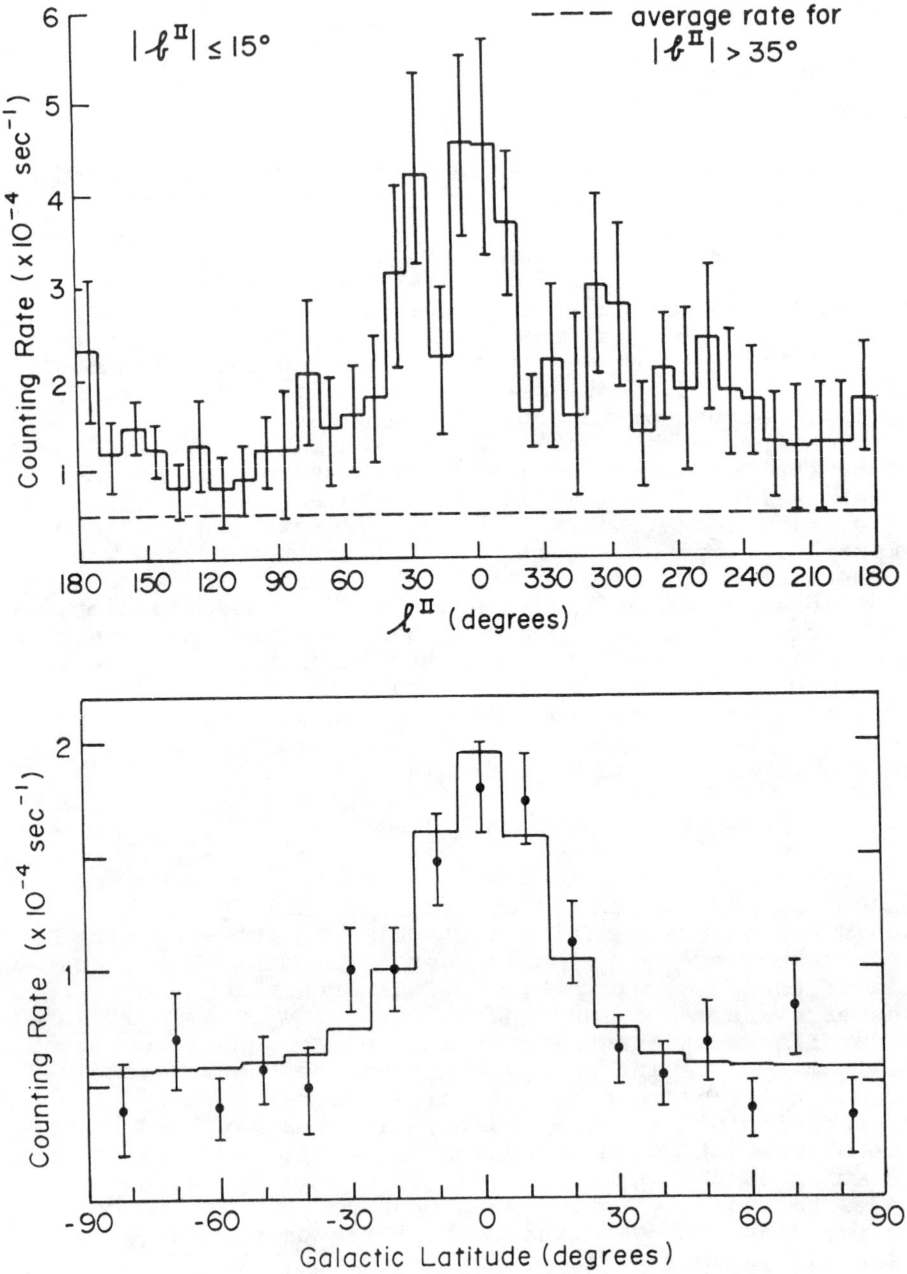

Fig. 1. The OSO-3 observations of the diffuse galactic gamma-ray emission[1] above 100 MeV.

SAS-2[2] confirmed this basic conclusion, but enhancements in the count rate coming from the directions of known spiral arms led this group[6,7,8] to conclude that

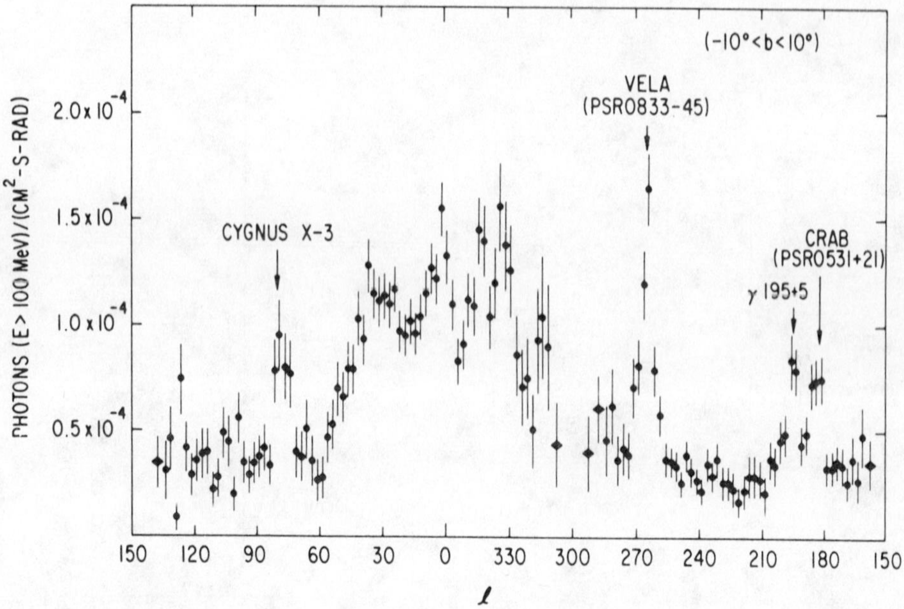

(a)

Fig. 2. SAS-2 observations of the galactic gamma-ray emission above 100 MeV[2]. (a) Galactic longitude distribution. (b) Galactic latitude distribution.

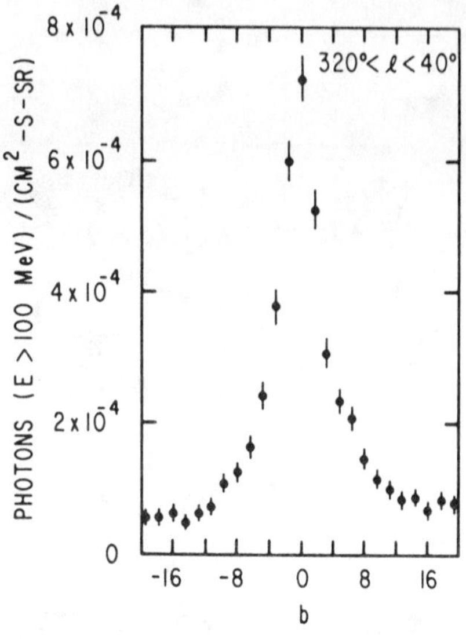

(b)

not only the gas, but also the cosmic rays must be enhanced in the spiral arms. This assumption also led to a model in which the overall galactic diffuse emission from the central region of the galaxy was enhanced relative the anticenter direction, as shown in the data[8] (Figure 2). COS-B data show neither the arm to interarm nor the center to anticenter contrast observed

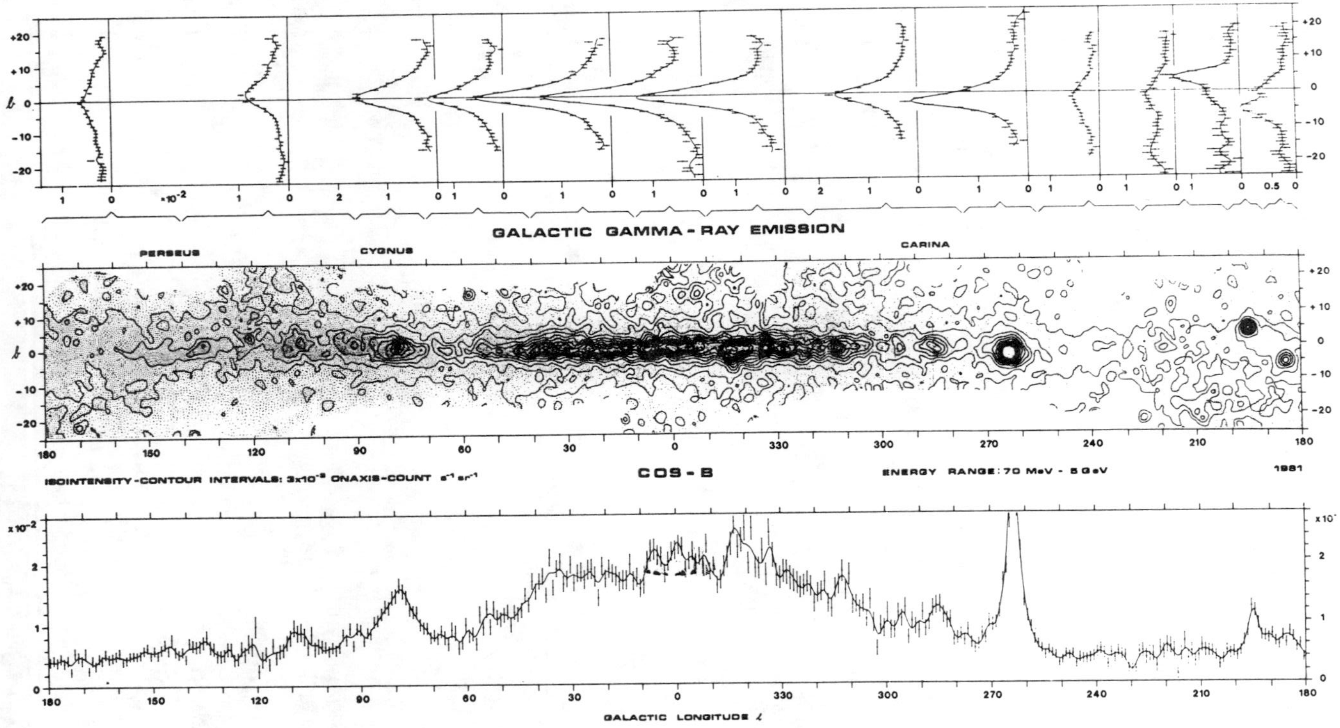

Fig. 3. COS-B observations of the galactic gamma-ray emission above 70 MeV[3].

by SAS-2 (Figure 3), and so, not surprisingly, that group reaches somewhat different conclusions about the galactic distribution of the cosmic radiation[9,10,11].

Theoretical studies related to the observations of SAS-2 and COS-B[12,13] have framed the question well, and indicate that, with better observations, it should be possible to study the distribution of not only the nucleonic component of the cosmic-rays, but also the electronic component as well (Figure 4). Improved observations at many wavelengths and a broad range of gamma-ray energies are needed. In the next section, prospects for such improved data are considered.

OBSERVATIONAL OUTLOOK

The expected intensity of gamma-rays for a telescope observing the emission from an observatory in the vicinity of our Sun is given by

$$j(E,l,b) = \frac{1}{4\pi}\int_0^\infty dr\ q(E,r,l,b) \quad (1)$$

where E is the gamma-ray energy, l and b are the galactic coordinates r is the r is the radial distance from the Sun and q is the source function in gamma-rays per unit volume, time and energy. The source function results predominantly from neutral pions and electrons produced in the interactions of cosmic ray nucleons with the intestellar gas, and may be expressed as:

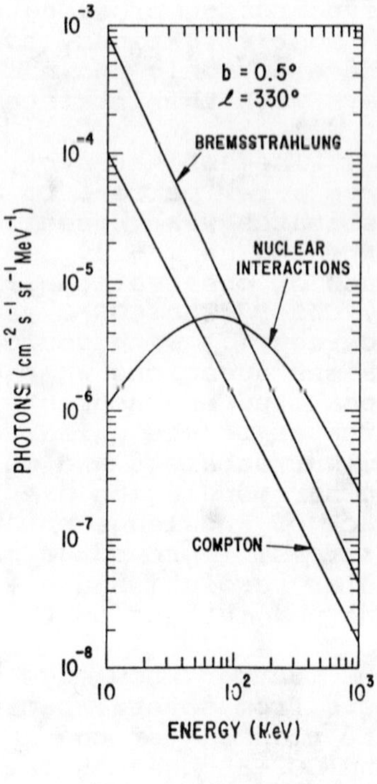

Fig. 4 Calculated γ-ray energy spectrum from the central region of the galaxy[12].

$$q = q_{\gamma_n}(E,r=0)\,g_n(r,l,b)\,f_m(r,l,b) + q_{\gamma_{e_p}}(E,r=0)\,g_e f_m + q_{\gamma_{e_s}}(E,r=0)\,g_e f_m^2 .$$

Here the subscript n refering to g(r,l,b) is the ratio of the cosmic ray density at the emission site to that locally (solar vicinity), and f(r,l,b) is the similar ratio for the target mass. The functional dependence of g and f on r,l, and b is omitted after the first term. The subscript n refers to the nucleonic component and e to the primary (p) and secondary (s) sources of electrons.

In order to be able to determine the cosmic-ray distribution, it is necessary to remove the source contributions and to know the densities of the target materials throughout the galaxy. This includes the matter (pion decay and bremsstrahlung produced gamma-rays), photons (Compton scattered gammas), and magnetic fields (synchrotron produced gammas). The dominant contributor, in most situations, comes from the interaction of both the electronic and nucleonic components of the cosmic rays with the interstellar matter.

The estimates of the source contribution have varied from a few percent to a third of the total[14-19]. These estimates were based mostly on the consideration of the expected gamma-ray emission from pulsars. However, the lack of observational identification of the sources seen by COS-B (The COS-B catalog contains almost thirty sources[20].) with sources observed at other wavelengths raises questions whether these are actually sources, or localized enhancements of diffuse emission. Better estimates of the pulsar component based on more recent pulsar catalogs and more current models of the galactic pulsar population are in progress[21], but because of the lack of knowledge on other classes of gamma-ray emitting sources, improved gamma-ray observations with better angular resolution are a more fruitful approach for obtaining new insight on this problem.

Better knowledge of the matter distribution now exists from several recent sets of observations, and more are planned for the near future. At the time of the SAS-2 and COS-B observations the only wide scale galactic surveys addressing this question were those of the neutral atomic hydrogen. Subsequent surveys of the interstellar CO distribution[22,23,24], a tracer of the galactic molecular hydrogen distribution, indicate that the molecular component probably dominates in the inner galaxy, although interpretation of the results is very difficult. A recent study by Stecker[25], has used the 100μm far infrared IRAS galactic survey as another indicator of the interstellar molecular hydrogen distribution and finds a good correlation with the gamma-ray distribution.

A major improvement in the gamma-ray observations will be provided by the Gamma-Ray Observatory. This second of NASA's "Great Observatories", currently manifested for a June 4, 1990 launch on the Space Shuttle, contains a complement of four instruments to cover almost six orders of magnitude in energy. In the following section the mission and the four instruments will be described, with emphasis on the two wide-field instruments whose galactic survey is expected to add significantly to our knowledge of the structure and dynamics of our galaxy.

THE GAMMA-RAY OBSERVATORY

The Gamma-Ray Observatory (Figure 5) is a free-flying satellite currently scheduled for a June 4, 1990 launch as STS-37 aboard the Space Shuttle Atlantis. The Observatory contains four instruments which collectively span almost six orders of magnitude in energy with a sensitivity over ten times that of any previous instrument covering any portion of the spectrum, with improved angular resolution, and good energy resolution across the spectral range. The Observatory is three-axis stabilized and may be pointed to any direction of the sky at any time. Pointing will be maintained accurate to 0.5 degrees and pointing knowledge will be good to 2 arcminutes. Timing will be maintained to an absolute accuracy of 100 microseconds to allow studies of pulsars spanning weeks of observing time, and to provide accurate timing of gamma-ray burst detections for use in inter-satellite timing. The GRO was designed as a two year mission, but analysis of the subsystems indicates that with careful use of propellant, used principally for maintenance of the orbit against atmospheric drag, should allow the Observatory to operate for six to ten years.

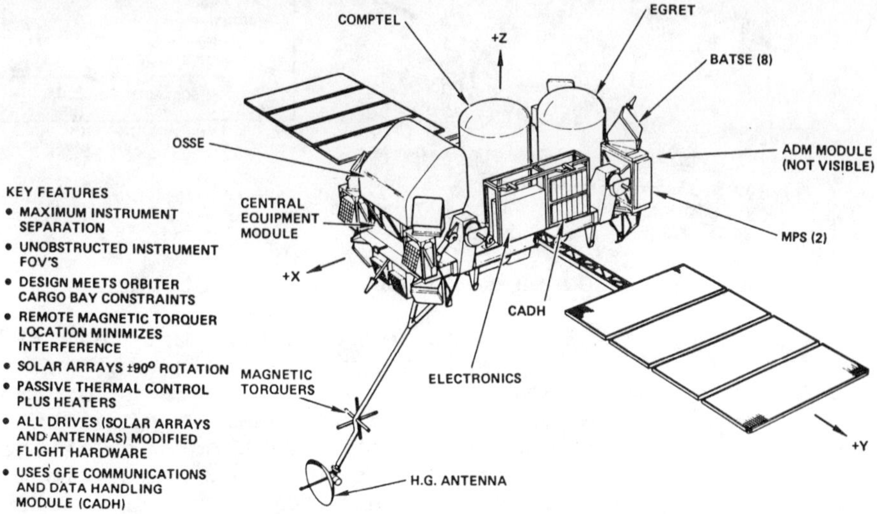

Fig. 5 The Gamma-Ray Observatory

The plan is for the Shuttle to take GRO to its 550 km altitude circular orbit where it will use its propulsion system to maintain the orbit between 440 and 450 km. Once in operations the data are packetized on board the spacecraft with all ancillary information, such as accurate timing and knowledge of attitude and position, included in the 32 kilobits per second recorded on board and transmitted to ground via the Tracking and Data Relay Satellite System once every second orbit. The observing plan provides for a full sky survey of 31 observations of two weeks each with the two wide field instruments (COMPTEL and EGRET). The narrow field spectrometer (OSSE) will spend most of this period studying known or potential discrete sources. The burst instrument (BATSE) observes the entire unocculted sky at all times.

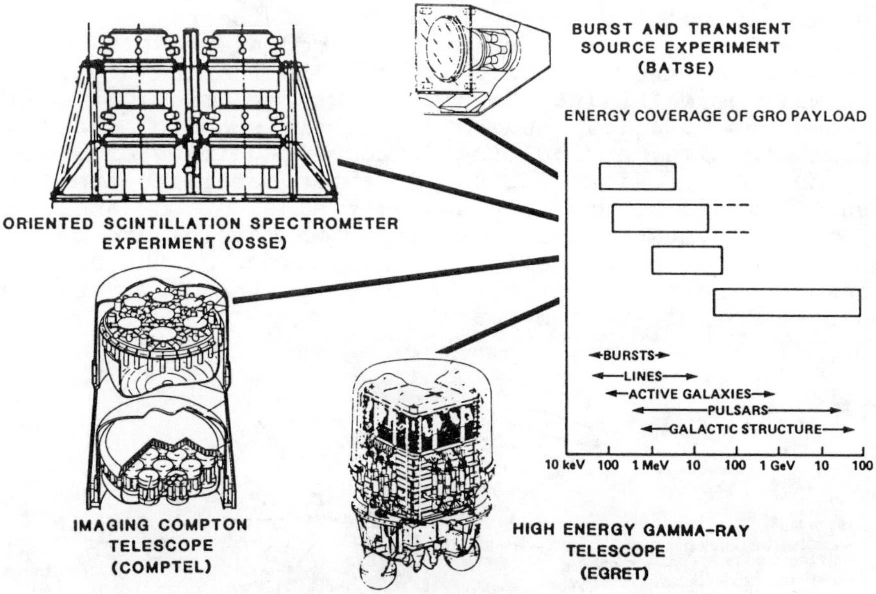

Fig. 6 The Gamma-Ray Observatory Instruments

The Burst and Transient Source Experiment (BATSE) for the GRO (Figure 6) is designed to continuously monitor a large fraction of the sky for a wide range of transient gamma-ray events. The monitor consists of eight wide field detector modules at the eight corners of the spacecraft platform. This arrangement provides maximum continuous exposure to the unobstructed sky. The instrument has a 0.1 msec time resolution, a burst location accuracy of about one degree, and sensitivity of 6×10^{-8} erg/cm^2 for a 10 second burst. The energy range covered with BATSE is 30 keV-1.9 MeV. A separate spectrometer measures the spectra of stronger bursts over the 20 KeV to 110 MeV range.

The Imaging Compton Telescope (COMPTEL) (Figure 6) is based on the concept of gamma-ray detection by Compton scattering in the 1-30 MeV range, an important range for the study of galactic gamma radiation produced by electromagnetic processes. This instrument employs a two step process in two detector layers. A Compton collision in the upper detector is followed by the total absorption of the scattered gamma-ray in the second detector element. This method, in combination with effective charged particle shield detectors, results in the suppression of the otherwise inherent instrumental background. Spatial resolution in the two detectors together with the well defined geometry of the Compton interaction permits the reconstruction of the sky image over a wide field of view (~ 1 steradian) with a resolution of a few degrees. In addition, the instrument has a capability of searching for polarization of the radiation. The instrument has good capabilities for the search for weak sources, weak galactic features, and for the search for spectral features in the extragalactic diffuse radiation. COMPTEL has a 5σ minimum source detectability of $3 \times 10^{-5}/cm^2$sec above 1 MeV for a 10^6 second observing time. The sensitivity for detection of gamma-ray lines ranges from 3×10^{-5} to $3 \times 10^{-6}/cm^2$sec. A source can be detected to an accuracy of 5-30 arcminutes.

The Energetic Gamma-Ray Experiment Telescope (EGRET) is designed to cover the energy range from 20 MeV to 30×10^3 MeV. This energy range includes the region around the 70 MeV pion decay peak characteristic of cosmic ray nucleon interactions with the interstellar material. The instrument (Figure 6) uses a multiple layer spark chamber and thin metal plates to detect gamma rays by the electron-positron pair-production process. A total energy counter using NaI(Tl) is placed beneath the instrument to provide good energy resolution over a wide dynamic range. The instrument is covered by a plastic scintillator anticoincidence dome to prevent readout on events not associated with gamma-rays. The combination of high energies and good spatial resolution in this instrument provides good source positions of as good as 5 arcminutes (2D RMS) for strong sources with hard spectra. The wide field-of-view (~ 1/2 steriadian) makes this and the COMPTEL instrument ideal for performing the all sky survey. EGRET has a 3σ point source sensitivity for a 10^6 second observation of $(5\text{-}10) \times 10^{-8}/cm^2$sec for a source not in the galactic plane. Most importantly for the study of the galactic diffuse emission, EGRET has angular resolution and point source location cabalilities an order of magnitude better than any previous experiment and its sensitivity is almost a factor of twenty greater than previous observations. A full survey of the galactic plane will be completed in the first year of observations.

The Orbiting Scintillation Spectrometer Experiment (OSSE) utilizes four large actively-shielded and passively-colliminated sodium iodide (NaI) scintillation detectors, with a 3.8° x 11.4° FWHM filed-of-view (Figure 6). The large area detectors provide excellent sensitivity for both gamma-ray line and continuum emissions. An offset pointing system modulates the celestial source contributions to allow background subtraction. It also permits observations of off-axis sources such as transient phenomena and solar flares without impacting the planned Observatory viewing program.

SUMMARY

These instruments collectively will provide the most sensitive observation of the galactic diffuse gamma-ray emission to date. Given the knowledge of the interstellar gas and dust densities which are obtained from radio and infrared observations, and improved angular resolution to allow a better estimate of unresolved discrete sources to this emission, the GRO instruments will provide the best information yet available on the galactic distribution of comic-ray nucleons and electrons. This will not only allow a better understanding of the dynamics of the galaxy, but also will shed light the question of the origin of the cosmic-rays.

REFERENCES

1. W.L.Kraushaar, G.W.Clark, G.P.Garmire, R.Borken, P.Higbie, V.Leong, and T.Thoros, Astrophys. J. 177, 341, (1972).
2. R.C.Hartman, D.A.Kniffen, D.J.Thompson, C.E.Fichtel, H.B.Ögelman T.Tümer and M.E.Özel, Astrophys. J. 230, 597 (1979).
3. H.A.Mayer-Hasselwander, K.Bennett, G.F.Bignami, R.Buccheri, P.A.Caraveo, W.Hermsen, G.Kanbach, F.Lebrun, G.G.Lichti, J.L. Masnou, J.L.Paul, K.Pinkau, B.Sacco, L.Scarsi, B.N.Swannenburg, and R.D.Wills, Astron. Astrophys. 105, 164 (1982).
4. J.B.Pollack, and G.G.Fazio, Astrophys. J. 141, 730 (1965).
5. F.W.Stecker, Cosmic Gamma Rays (Mono Book Corp., 1971)
6. G.F.Bignami, C.E.Fichtel, D.A.Kniffen, and D.J.Thompson, Astrophys. J. 199, 54 (1975).

7. D.A. Kniffen, C.E.Fichtel, and D.J.Thompson, Astrophys. J. 215, 765 (1977).
8. R.C.Hartman, D.A.Kniffen, D.J.Thompson, C.E.Fichtel H.B. Ögelman, T.Tümer, and M.E.Özel, Atrrophys. J. 230, 597 (1979).
9. F. Lebrun, K. Bennett, G.F.Bignami, J.B.G.M.Bloemen, R.Buccheri, P.A.Caraveo, M.Gottwald, W,Hermsen, G,Kanbach, H.A.Mayer-Hasselwander, T.Montmerle, J.A.Paul, B.Sacco, A.W.Strong, and R.D.Wills, Astrophys. J. 274, 231 (1983).
10. J.B.G.M.Bloemen, L.Blitz, and W.Hermsen, Astrophys. J. 279, 136 (1984).
11. J.B.G.M.Bloemen, K.Bennett, G.F.Bignami, L.Blitz, P. A.Caraveo, M.Gottwald, W.Hermsen, F.Lebrun, H.A.Mayer-Hasselwander, and A.W.Strong, Astron. and Astrophys. 135, 12 (1984).
12. D.A.Kniffen and C.E.Fichtel, Astrophys. J. 250, 389 (1981).
13. C.E.Fichtel and D.A.Kniffen, Astron. and Astrophys. 134, 13 (1984).
14. D.A.Kniffen, C.E.Fichtel, R.C.Hartman, R.C.Lamb and D.J.Thompson, Proceedings of the 12th ESLAB Symposium, Frascati, ESA SP-124, 24 (1977).
15. A.W.Strong, A.W.Wolfendale, and C.Dahanayake, Mon. Not. R. Astr. Soc. 179, 69P (1977).
16. G.F.Bignami, P.Caraveo, L.Maraschi, Astron. and Astrophys. 67, 149 (1978).
17. R.J.Protheroe, A.W.Strong, A.W.Wolfendael, P.Kiraly, Nature 277, 542 (1979).
18. A.K.Harding, Astrophys. 247, 639 (1981).19. F.W.Stecker, Cosmic Gamma-rays, Neutrinos and Related Astrophysics (eds. M.M.Shapiro and J.P.Wefel, Kluwer Academic Publishers) 85 (1989).
20. G.F.Bignami and W.Hermsen, Ann. Rev. Astron. Astrophys. 21, 67 (1983).
21. M.Bailes and D.A.Kniffen, in Progress.
22. M.A.Gordon and W.B.Burton, Astrophys. J. 208, 346 (1976).
23. N.Z.Scoville and P.M.Solomon, Astrophys. J. 199, L105 (1975).
24. T.M.Dame, H.Ungerechts, R.S.Cohen, E.J.deGeus, I.A.Grenier, J.May, D.C.Murphy, L.-A.Nyman, and P. Thaddeus, Astrophys. J. 322, 706 (1987).
25. F.W.Stecker, To appear in a Festschrift in honor of Maurice Shapiro, 1990.

GAMMA-RAY ASTRONOMY AT 1 TeV

G. Vacanti
Iowa State University, Ames, IA, 50011

ABSTRACT

An outline of some problems and ideas of TeV ground based γ-ray astronomy is given, with particular attention to the Imaging Technique. The current observational status of Cygnus X-3 and Hercules X-1 is then reviewed.

INTRODUCTION

The long interval since the last observations of γ rays with satellite born detectors has increased the interest in ground-based observations. The 80's have seen a growth in the number of observatories around the world operating more sensitive telescopes. This trend is expected to continue in the next decade.

Ground-based observations can complement observations in the 100 MeV region and, thank to the large collection areas, discover new phenomena on short time scales. Also, the detection of TeV γ rays from celestial objects points at the locations of powerful particle accelerators and can furnish part of the answer to the problem of the origin of cosmic rays.

The following sections aim to give the reader an idea of the problems of the Atmospheric Cherenkov Technique (ACT) and review the status of two interesting and controversial sources.

THE ATMOSPHERIC CHERENKOV TECHNIQUE

Above 10-20 GeV the γ-ray flux is so low and the feasible collection areas of the detectors are so small as not to make possible the development of any reasonable astronomy from space. For instance, a telescope with an effective collection area of 1,000 cm^2 would detect less than one photon per year when observing the brightest source in the Galaxy.

Fortunately, above 100 GeV the atmosphere becomes a giant (and economical!) photon detector. Entering the atmosphere, a photon with E_γ > 100 GeV generates an air shower whose secondary component survives at least until mountain altitudes where it can be detected with relatively simple and inexpensive techniques (Figure 1).

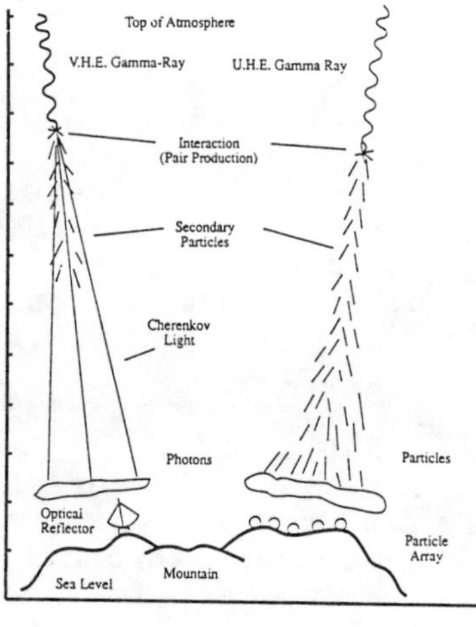

Fig. 1. Schematic development of air showers.

At energies in excess of 0.1 PeV the secondary component (electrons and muons) can be detected through an array of particle detectors. The direction of arrival of the primary can be reconstructed by fast differential timing between the different detectors. Such an array has a collection area of about 1,000 m^2 and a 1 degree angular resolution. Below 100 TeV not many secondary particles survive the journey through the atmosphere, even at the highest mountain altitudes. In this energy range (.1 TeV to 100 TeV) the ACT must be used.

Particles in the air shower are produced with ultra relativistic velocities and travelling in the atmosphere give away Cherenkov radiation. For a primary of energy 1 TeV the Cherenkov photon density at 2,000 m a.s.l. is about 100 photons/m^2. These light flashes can easily be detected with a simple telescope made coupling a medium quality mirror to a photomultiplier with some pulse counting electronics. Such a simple detector enjoys a big collection area (of the order of 10,000 m^2) with a field of view of a few degrees.

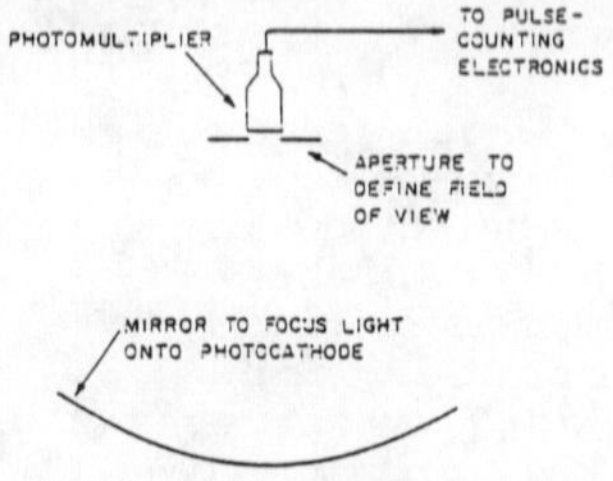

Fig. 2. Basic Cherenkov detector.

Detectors like the one shown in Figure 2 have been in use for 30 years and are still in use today.Simplicity has an important drawback: simple detectors cannot differentiate genuine photon-induced flashes from the ones induced by the interaction of cosmic rays with the atmosphere. Above 100 GeV the cosmic-ray flux is about 10,000 times the γ-ray flux and ground-based detectors have to rely on long exposures to achieve a reasonable signal/noise. At times, if the object of interest is know to have a well defined periodicity, the signal/noise can be improved by the use of phase analysis techniques. This problem was recognized since the early days of the discipline, and over the years different avenues have been explored in the attempt to reduce the influence of the background without significantly reducing the sensitivity of the detectors to the γ-ray signal[1,2]. The idea behind each approach is that different physics is involved in the development of hadronic and photonic showers and that even after many interaction lengths this should translate into differences in the Cherenkov light pools on the ground.

THE IMAGING TECHNIQUE

Figure 3 shows two simulated air showers, one (a) generated by a 250 GeV photon, the other (b) by a 250 GeV proton. Only the particle components of the showers are shown: about 8,000 Cherenkov photons per radiation length are emitted by each particle. Differences between the two showers include: (i) lateral distribution of particles; (ii) angular distribution of the Cherenkov light; (iii) temporal distribution of the light front; (iv) spectral composition. From (i) and (ii) it follows that the photon-induced showers generate a more compact Cherenkov image. The Imaging Technique (IT) exploits this property to identify likely γ-ray candidates via the use of image parameters introduced by Hillas[3] (Figure 4).

The IT has been employed by the Whipple Observatory Collaboration since 1982. The Collaboration's 10 meter optical reflector was fitted with a 37-pmt/pixel camera and transformed into an imaging system. Each Cherenkov flash detected by the camera is recorded and γ-ray candidates are selected off line. Of the 5 parameters originally proposed[3] one, Azwidth, has proven to be particularly effective. Its use in the selection process gave a gain in sensitivity of a factor of 4[4].

The 37-pixel camera had quite a coarse resolution (0.5 deg) which did not match the characteristic size of Cherenkov flashes ($\approx$0.2 deg). In Spring, 1988 a 109-pixel High Resolution Camera (resolution 0.25 deg) was installed in its place. The HRC gave a further factor of 2.5 in increased sensitivity[5].

Sensitivity gains of the two cameras with respect to an non-imaging system are summarized in Table I. The Crab Nebula is used as a standard candle.

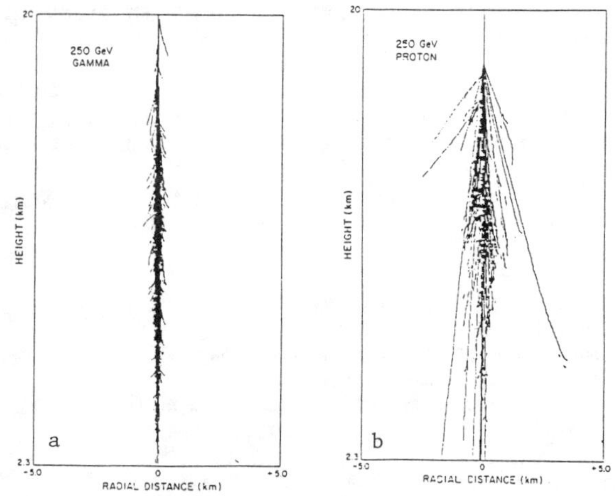

Fig. 3. Simulated air showers: (a) 250 GeV proton; (b) 250 GeV photon.

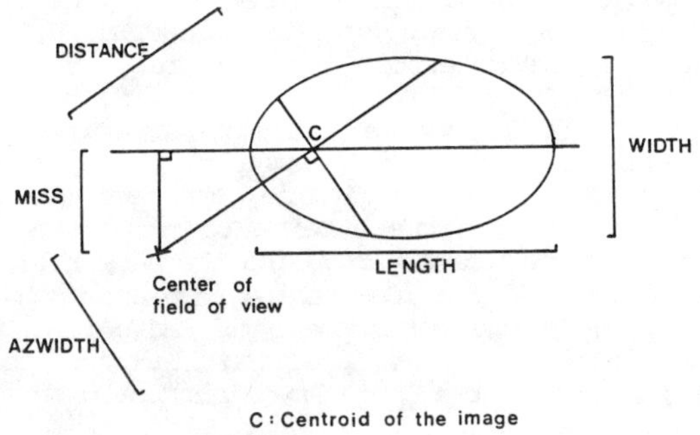

Fig. 4. Image parameters.

Table I
Sensitivities of different detectors
for observations of the Crab Nebula

Date	Imaging	Time on (h)	σ	Gain factor over non-imaging system	Ref.
1968-72	no	150	3	1	6
1986-88	yes (37)	80	9	4	5
1988-89	yes (109)	30	15	10	4

So far for the current status: what are the developments to be expected in the next decade? For one thing, there will be more imaging telescopes. The IT has gained support in the community and it is now being pursued at a number of centers, e.g. the Crimean Astrophysical Observatory, the JANZOS collaboration in Australia, the ANI project in Armenia. We can expect that the number of sources detected at a level of high statistical confidence will increase.

Recently, Hillas[7] pointed out that the Cherenkov light front contains information all the way down to a 2 arcminute scale. In principle, this is the ultimate angular resolution that an imaging detector should have. To the author's best knowledge there is one detector, still at its planning stage, that aims at such a fine resolution.

Further gain in sensitivity can be achieved also by the use of two imaging telescopes to stereoscopically view the Cherenkov showers. This will be accomplished by the GRANITE project[8], recently approved for funding. With GRANITE the imaging telescope currently in use at the Whipple Observatory will be paired to a similar detector located 120 m away. At the beginning the second detector will be a replica of the 37 pixel camera to be later upgraded to a higher resolution. Preliminary simulations show that the GRANITE system will be a factor 2-3 more sensitive than the current 109 pixel camera alone for showers seen by both detectors: the gain comes from the combined effect of a lower energy threshold and an increased ability to discriminate against background showers. Work to install the second reflector will begin in Spring, 1990.

TWO HOT SOURCES

TeV γ rays have been claimed from a variety of objects: X-ray binaries, pulsars, cataclysmic variables, extragalactic objects. Although not all the claims may be on solid grounds, they have certainly spurred a lot of interest, forcing theoreticians to re-think the physics of these systems in order to allow for the production of very energetic particles in efficient ways.

The two sources reviewed here, although far from being representative of the entire γ-ray zoo, give an idea of the problems and the potentials of γ-ray astronomy at 1 TeV.

Cygnus X-3. Subject of much attention from the astrophysical community, 20 years after its discovery as a strong X-ray source Cygnus X-3 is still a mystery. Known to emit in the radio, IR and X-ray bands, this source is probably the most controversial topic of γ-ray astronomy.

Over the past 15 years γ-ray emission from 100 MeV through 0.5 EeV has been reported by several groups. Yet, it is safe to say that for each claimed detection there has been a refutation, so that to date the picture is not clear-cut. This led some authors[9] to doubt of its very existence above 100 MeV. The picture is made more confusing by the (unconfirmed) detections of TeV signals pulsed at 9.22 ms[10] and 12.6 ms[11].

The energy spectrum of Cygnus X-3 as measured above 100 MeV is shown in Figure 5; it is very flat and can be fitted by a power law of index ≈1. It has been argued that such a spectrum is what might be expected from random fluctuation at all energies[12], but also that it is what the current VHE and UHE sensitivities would allow for a small but significant signal[1]. The spectrum, taken with different instruments at different times and averaged over different time intervals, suggests that the Cygnus X-3 system is accelerating particles up to an energy of 10-100 PeV. This has a very important consequence for the cosmic-ray energy budget in the Galaxy. For instance, it can be shown that only one source like Cygnus X-3 could account for all the galactic cosmic rays observed in the 1-10 PeV range[13].

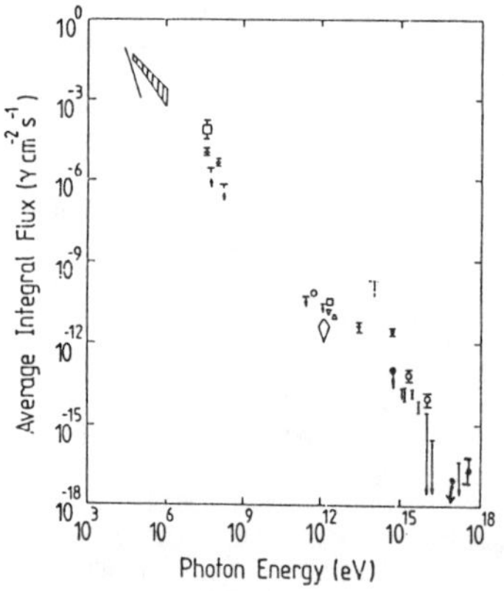

Fig. 5. Energy spectrum of Cygnus X-3[1].

Hercules X-1. Hercules X-1 is a typical X-ray binary whose characteristics, in contrast to Cygnus X-3, are largely known. The source, identified in the X-ray and visible bands, shows three distinct periodicities: 1.24s (X-ray and visible) associated with the spinning neutron star; 1.7d (X-ray and visible) associated to the orbital motion of the system; 35d (X-ray) whose nature is not clear but that is probably due to the precession of the accretion disk or the neutron star.

In the TeV range the source wa discovered by the Durham group[14] and since then confirmed in several occasions by independent groups. In each case the signal has been pulsed at a frequency close to the X-ray one. Recently the Tata group[15] reported a 42 sigma burst from the source. Unfortunately a malfunction in the recording system did not allow the identification of the source from its characteristic periodicity.

In general these detections are not statistically compelling per se, thus the case for TeV emission from Hercules X-1 is strengthened by the fact that on one occasion when two telescopes were tracking the source during the same active period they both detected the same pulsed signal[16,17].

One of the features that any theoretical model should explain is that the frequency of the TeV signal is usually slightly blue shifted with respect to the X-ray emission. This effect was particularly evident in 1986 when three groups[18,19,20] detected a signal blue shifted by the same amount (0.16% of the X-ray frequency; Figure 6). For different reasons amount (0.16% of the X-ray frequency; Figure 6). For different reasons (intensity of the episodes, longitude of the observatories, weather) none of the episodes was visible by more that one detector at the time. This set of observations taken over a three month period constitutes strong evidence for the detection of TeV γ rays from Hercules X-1. However, the two groups that have the capability to identify the nature of the primary radiation, Los Alamos and Whipple, were not able to confirm the photonic nature of the signal. In the case of the Whipple detection it is worth noticing that the effect is significant at the 1% level.

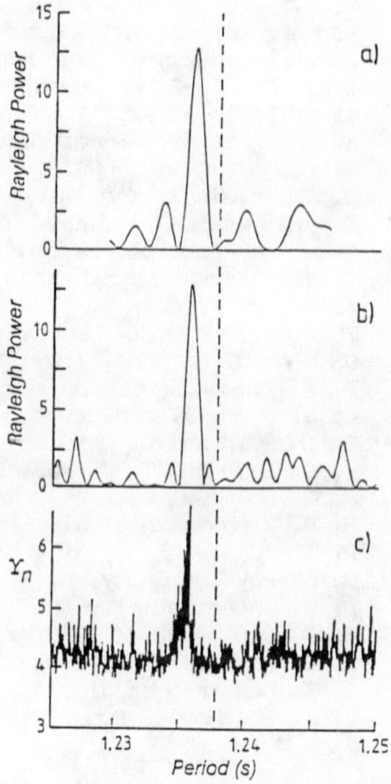

Fig. 6. Her X-1[21]: power versus period for (a) Haleakala[20] (May, 1986), (b) Whipple[19] (June, 1986), (c) Los Alamos[18] (July, 1986)

What are the conclusions that can be drawn from these findings? If all the detections are correct and the photonic nature of the signal can really be ruled out, the implication is that some neutral particle generates the observed signals. The problem is that no known particle would be able to produce the effects reported.

Another explanation could be that the nature of the interaction of photons and matter at 1 TeV and above is drastically different from what it is believed today, so that the photon induced air showers would look more hadron-like than currently expected. Both explanations required the introduction of some new kind of physics, and more observations of stronger statistical significance and a better understanding of the physics involved are necessary before we can give our blessings to any revolutionary idea.

The 90's will see the coming on line of more sensitive instruments (GAMMA-1, GRO, GRANITE, CASA and others) covering the MeV - PeV range: the next decade is going to be very exciting for γ-ray astronomy.

This work is supported by the U.S. Department of Energy.

REFERENCES

1. T. C. Weekes, Physics Reports, **160**, 1 (1988)
2. N. A. Porter and T. C. Weekes, SAO Special Report 381 (1977)
3. A. M. Hillas, Proc. 19th ICRC, 3, 445 (1985)

4. T. C. Weekes et al., Ap. J., **342**, 379 (1989)
5. M. J. Lang et al., Proc. Workshop on the Physics and Experimental Techniques of High Energy Neutrino and VHE and UHE Gamma-Ray Particle Astrophysics, Nuclear Physics B (1989), and Proc. 21st ICRC (OG 4.3-2, in press)
6. G. G. Fazio et al., Ap. J. Letters, **175**, L117 (1972)
7. A. M. Hillas, Proc. International Workshop on VHE Gamma-ray Astronomy, Crimea, USSR, pag. 134 (1989)
8. C. W. Akerlof et al., Proc. Workshop on the Physics and Experimental Techniques of High Energy Neutrino and VHE and UHE Gamma-Ray Particle Astrophysics, Nuclear Physics B (1989)
9. J. M. Bonnet-Bidaud and G. Chardin, Physics Reports, **170**, 326 (1989)
10. YU Zynski et al., Astronomical Circular Bureau of Astron. Inf. of USSR Acad. Sci. no. 1508 (1989)
11. P. M. Chadwick et al., Nature, **318**, 642 (1985), and K. T. Brazier et al., Ap. J. (in press)
12. G. Chardin and G. Gerbier, Astron. Astrophys., **210**, 52 (1989)
13. R. C. Lamb and T. C. Weekes, Science, **238**, 1483 (1987)
14. J. C. Dowthwaite et al., Nature, **309**, 691 (1984)
15. P. R. Vishwanath et al., Ap. J. (in press)
16. P. M. Chadwick et al., Proc. NATO Workshop on Very High Energy Gamma-Ray Astronomy, K. E. Turver Ed. (Reidel, Dordrecht, 1987)
17. P. W. Gorham et al., Ap. J., **309**, 114 (1986)
18. B. L. Dingus et al., Phys. Rev. Letters, **61**, 1906 (1988)
19. R. C. Lamb et al., Proc. 20th ICRC, **1**, 244 (1987)
20. L. K. Resvanis et al., Ap. J. Letters, **328**, L9
21. R. J. Protheroe, Proc. 20th ICRC, **8**, 21 (1987)

The Experimental Search for Point Sources Above 1 TeV

J. A. Goodman
Department of Physics and Astronomy
University of Maryland
College Park, Maryland 20742-4111
Presented for
The GRANDE Collaboration
A. Adams,(1) R. Bond,(1) C. B. Bratton,(2) T. Burnett,(3) V. Chaloupka,(3)
M. Cherry,(4) L. Coleman,(1) S.B. Ellison,(4) J. Gaidos,(5) J.A. Goodman,(6)
H. Gurr,(2) T. G. Guzik,(4) T. J. Haines,(6) D. Kiełczewska,(9) W. Kropp,(2)
C. Lane,(10) M. Lieber,(11) F. Loeffler,(5) D. Nagle,(12) M. Nelson,(2) M. Potter,(12)
L. R. Price,(2) F. Reines,(2) A. Rollefson,(1) J. Schultz,(2) G. Sembroski,(5)
H. Sobel,(2) R. Steinberg,(10) R. Svoboda,(4) R. Tripp,(13) J. Wefel,(4)
R. J. Wilkes,(3) C. Wilson,(5) D. Wold,(1) G.B. Yodh(2)
and
The CYGNUS Collaboration
D. E. Alexandreas,(2) D. Berley,(6) S. Biller,(2) R.L. Burman,(12) C.Y. Chang,(6)
B.L. Dingus, (6) G. Dion,(2) R.W. Ellsworth,(14) S. Freedman,(7)
J.A. Goodman,(6) T.J. Haines,(6) C.M. Hoffman,(12) D. Krakauer,(7) R. Lamb,(15)
X-Q Lu,(12) D.E. Nagle,(12) M.E. Potter,(12) V.D. Sandberg,(12) M. Stark,(6)
T. Tumer,(8) P.R. Vishwanath,(2) S. White,(8) G.B. Yodh,(2) A. Zych,(8)

(1) *University of Arkansas, Little Rock*
(2) *University of California, Irvine*
(3) *University of Washington*
(4) *Louisiana State University*
(5) *Purdue University*
(6) *University of Maryland*
(7) *Argon National Laboratory*
(8) *University of California, Riverside*
(9) *Warsaw University*
(10) *Drexel University*
(11) *University of Arkansas, Fayetteville*
(12) *Los Alamos National Laboratory*
(13) *Lawrence Berkeley Laboratory*
(14) *George Mason University*
(15) *Iowa State Univeristy*

Abstract

The search for point sources of VHE and UHE radiation is an exciting new field which bridges Astrophysics and High Energy physics. This paper describes three experiments to study this radiation: The first is the CYGNUS experiment in Los Alamos which has been operational since 1986 and is sensitive to the electromagnetic and muonic component of extensive air showers above 5×10^{13}eV. The second is the proposed GRANDE detector which will be capable of detecting all components of extensive air showers above 10^{13}eV and detecting upward coming muons from neutrino induced events in the rock below it. And third a new experiment, MILAGRO, to study the electromagnetic and muonic component of extensive air showers above 10^{12}eV is described.

CYGNUS Experiment at Los Alamos

Since the first reports of directional radiation from point sources, more questions about the nature and origin of this radiation have been raised than answered[1]. The CYGNUS experiment was one of the first experiments specifically designed to look for sources of Ultra High Energy (UHE) radiation above 5×10^{13}eV. This experiment was designed to have good angular resolution when reconstructing the direction of Extensive Air Showers (EAS) and have unambiguous detection of muons accompanying these showers.

The CYGNUS experiment consists of an air shower array and a detector to sample the muon content of these showers[3, 4]. It is located at the end station of the LAMPF accelerator at Los Alamos National Laboratory (107.6° W and 35.9° N) at an altitude of 2134m, corresponding to an atmospheric depth of 800 gm/cm^2. An air shower array (CYG I) covering $\sim 2\times10^4$ m^2 area is operating with 108 scintillation counters, each of 0.83m^2 and 10 cm thick. The average array spacing is 14m. Ninety-six additional counters (CYG II), spaced on a wider grid, are currently being deployed on the east side of the existing array. The E225 [5] detector, which is used as the muon detector, is located at roughly the center of the original 108 counters.

Our typical trigger requirement is: 20 counters with signal greater than 0.1 minimum ionizing particle. This trigger allows showers generated by primaries of E>50 TeV to be detected with good efficiency. The hardware rate for the CYG I is about 4 hz. At this time, no cuts are made and all data is written to tape. The CYG I array has operated with over 90% on-time. The CYGII array has been on-line since summer 1989 with 48 counters operational. The two arrays will each have separate and overlapping triggers. The data for both arrays will be recorded in the CYG I aquisition system with data from CYG II being sent over a fiber optic data link. The graded spacing of CYG II provides for the combined arrays to have a more nearly uniform coverage of a wide energy interval.

The ratio of signal from a point source to that from isotropic background of cosmic ray hadrons is increased by improving angular resolution. The angular resolution of our array was determined to be about 1° for small showers and about 0.5° for large showers. In order to achieve this resolution shower front curvature corrections must be made[4]. Recently thin lead sheets have been placed on the counters in CYG I. This has been reported [6] to improve the signal and timing response of the counters by converting photons in the air shower. Preliminary results from the detector confirm the improved response as measured by χ^2 of the fit and the difference between the reconstructed angle of the subdivided array. Lead will also be placed on counters in CYG II. New studies have shown that the effective shower curvature is decreased with lead on the counters. Also recently, a new satellite clock has been added to the array which gives absolute timing accuracy of $< \pm1\mu$sec.

The E225 neutrino detector originally consisted of flash chambers and scintillators surrounded by an active shield of multiwire proportional chambers (MWPC).

All sides are covered with steel and concrete of sufficient thickness to give a minimum threshold of 2 GeV for penetrating muons. The flash chambers were used initially to study the muon counting algorithm in the shield. This study showed that the algorithm was accurate to better than ± 1 muon track for 7 or less muons. In addition, a study of flash chamber response to showers whose cores are near the E225 detector show that effects of shower ' punch through' are absent.

A new set of counters, called Anasazi for the cliff dwellers who lived near Los Alamos, are being installed in tunnels drilled in a hillside near the center of the CYG II array. These counters will add an additional $75m^2$ to the muon coverage.

In summary, the CYGNUS array, which is the first array specifically designed to look for UHE radiation from point sources, has been shown to have a threshold of about 50 TeV and to have an angular resolution better than 1° for most showers. The upgraded array will continue to provide information on UHE showers well into the next decade.

GRANDE

The next generation of air shower detectors need to be more than merely enlarged and improved versions of existing devices. They must have, in addition to large size and improved energy and angular resolutions, the capacity to study, in detail, other properties of the gamma-induced air showers. Further, the new experiments should bridge the differing energy domains of the current generation of air Čerenkov and extensive air shower experiments.

The study of the neutrino radiations from these cosmic accelerators is still in its infancy. Existing detectors such as IMB, Kamiokande, and Baksan have seen no indication of an extraterrestrial source of high-energy neutrinos. Yet there are sound theoretical reasons to believe that a significant improvement in sensitivity will enable the sources to be detected.

GRANDE (Gamma Ray and Neutrino DEtector) is an imaging, water Čerenkov detector, which combines in one facility an extensive air shower array and a high-energy neutrino detector.

GRANDE utilizes the proven technology of water Čerenkov detectors. A feasibility study, completed last year, has shown that the powerful background discrimination which is inherent in the directional property of the Čerenkov light and in the large size of the detector, will allow successful surface operation with an acceptable trigger rate. A site has been selected (a water-filled quarry near Little Rock, Arkansas) and an engineering firm has completed the preliminary design of the detector structure. Preliminary studies of the site show that it is well suited for the facility.

The Neutrino Telescope The primary goals of the neutrino telescope are the observation and study of point sources of high-energy extra-terrestrial neutrinos, and the study of very-high energy neutrino interactions. These will be accomplished by detecting upward-going muons, i.e., muons generated in the rock and water below the detector by neutrinos which entered from the opposite side of the earth. This detector takes a very different approach than previous detectors, such as DUMAND or IMB. Instead of going to great depths, either underground or undersea, to reduce the atmospheric muon background, it uses the idea of a *surface* detector in a shallow body of water (about 70 meters deep); the up-down muon discrimination is obtained from the directionality of the Čerenkov light, and from the time of flight through the detector's three 17 meter layers. In addition, GRANDE's Gamma-Ray Telescope provides an efficient cosmic-ray anticoincidence. The neutrino detector's aperture will be from a nadir angle of 0° to 70°. As indicated in the detector schematic (Fig. 1), GRANDE consists of several optically-isolated layers. The three lower layers, the Neutrino Telescope, are instrumented with planes of downward-looking PMTs (open circles in the figure), while two of the upper three layers, the Gamma-Ray Telescope, have planes of upward-looking PMTs (closed circles). Light from the upward-going particle is seen, with high efficiency, only by the downward-facing PMTs; the upward-facing tubes see little or none.

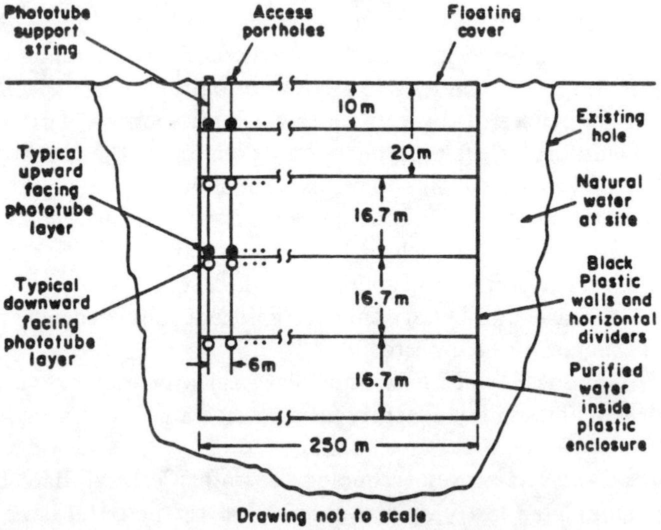

Figure 1: A schematic of the GRANDE detector.

The upward-going, neutrino-induced muon signal from a typical source is expected to be tens of events per year, easily observed above the essentially isotropic atmospheric neutrino background. This background can in turn be used to study neutrino interactions at energies above 1 TeV, and to search for evidence of neutrino oscillations.

The Gamma-Ray Telescope The gamma-ray telescope has the same surface area (60,000 m^2) as the neutrino detector, and employs much of the same electronics, ancillary equipment, and facilities. It uses primarily the two planes of *upward-looking* photomultipliers.

A 10 meter layer of surface water constitutes a *fully active* total absorption calorimeter for the electromagnetic and hadronic components of the showers. This layer will be viewed by one plane of upward-facing PMTs. The second plane of PMTs, deeper and optically isolated from the calorimeter layer, will observe the muon component.

The detector's salient features include:

- Largest air shower telescope which possesses full sensitivity over its entire area
- Sufficient size (60,000 m^2) to observe known sources with excellent statistics, and to search for new sources with improved sensitivity
- A trigger threshold of about 10 TeV, low enough to overlap measurements made with the air Čerenkov technique.
- The ability to measure over the whole detector the three major components of air showers: electromagnetic, muonic, and hadronic
- Unmatched angular resolution (~0.3°) for superior source location and background rejection
- A hadronic-shower rejection factor of at least 100 for conventional gamma ray showers
- Excellent energy resolution
- Good coverage of many known or potential cosmic sources
- Better cosmic-ray primary-mass resolution than does any other air shower experiment.

The gamma telescope's size was chosen so as to be able to detect (i) steady-state emissions at flux levels as small as 10^{-15} $cm^{-2}sec^{-1}$ and (ii) bursts at flux levels of 10^{-11} $cm^{-2}sec^{-1}$at 100 TeV of less than an hour's duration from Cygnus X–3-like sources, both at a confidence level of $> 5\sigma$. In addition, its unique coverage for muons and hadrons will make it possible to examine the properties of the primary particle and its interaction with atmospheric nuclei and thereby investigate and establish the possible existence of new interactions and particles.

MILAGRO

The Cygnus collaboration is proposing a new gamma ray observatory to cover the energy range above 1 TeV thus covering a largely unexplored region of the spectrum. Its purpose is to study of the energy dependence of the anomalous gamma-ray interactions and a search for more VHE sources. This detector would be located near Los Alamos (2650m) and is similar in design to the GRANDE *Gamma Ray Telescope*, however an order of magnitude smaller. The unique feature of this facility is ability to push the threshold of EAS measurement to the

VHE energy range. The combination of the high altitude and the ability of the detector to respond to the large number of photons in low energy showers, make it possible to obtain an energy threshold much lower than otherwise possible (< 1 TeV). The water Čherenkov technique appears to be optimum for a large sensitive-area detector that is needed to observe the lower energy gamma rays.

The site of this detector would be in the Jemez mountains where there is a water pond owned by Los Alamos National Laboratory at an elevation of 2650m with a plastic liner The air shower would be detected by 500 photomultiplier tubes spaced three meters apart, located one meter below the surface; 150 tubes near the bottom of the pond and optically isolated from the top layer would measure the muon content of the showers. The pond, which measures 60-m by 80-m at the water surface and is about 8-m deep, is currently used as a reservoir for the hot-dry rock geothermal program at Los Alamos, but would be available to be used as a new detector.

Figure 2 shows an artist's conception of the detector. In Figure 3, the expected energy spectrum of all detected events is shown; note that there are a large number of events even below 1 TeV. Figure 4 shows the expected angular resolution as a function of the energy of the primary; angular measurements of this precision are unprecedented in this field and imply greatly increased sensitivity. Monte Carlo studies show that the detector will be able to search for anomalous muon content in gamma ray showers down to 5 TeV; this includes an energy region for which the nature of photon interactions will soon be measured in accelerator experiments.

Conclusions

The CYGNUS experiment represents a significant step toward understanding the source and nature of UHE cosmic radiation. The next generation of water Čherenkov detectors will expand the energy coverage by almost two decades and will provide substantially more information about the nature of the interactions and sources of this radiation. In addition, GRANDE will open a new window into the galaxy - neutrino astronomy.

References

[1] J. A. Goodman, Univ. of Maryland preprint: PP-90-030, June 1989 to appear in Nuclear Physics B procedings suppliment.

[2] B.L. Dingus et. al., Phys.Rev.Letters,**61**,1906 ,1988.

[3] B.L. Dingus, et. al., Phys. Rev. Letters, **60**,1785, 1988.

[4] B. L. Dingus, University of Maryland Ph. D. thesis, Univ. of Maryland preprint: PP-88-253, June 1988 (unpublished).

[5] R. C. Allen, et. al., Phys. Rev. Lett.,**55**, 2401, 1985.

[6] J Cronin, et al., Nuclear Instruments and Methods (1988).

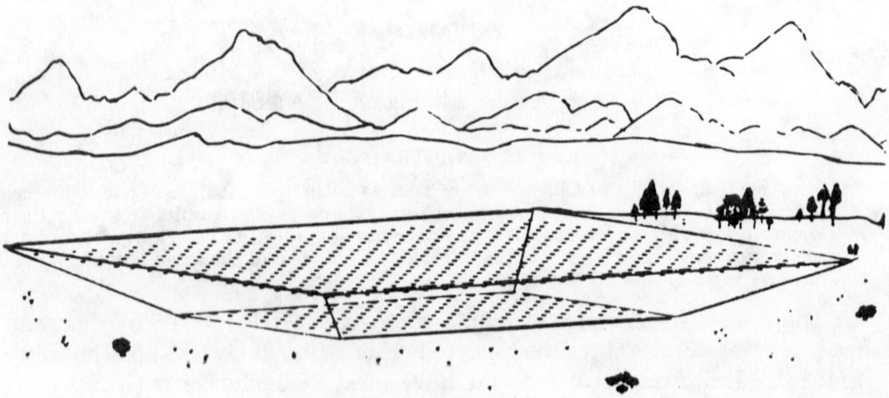

Figure 2: Artists view of the MILAGRO detector

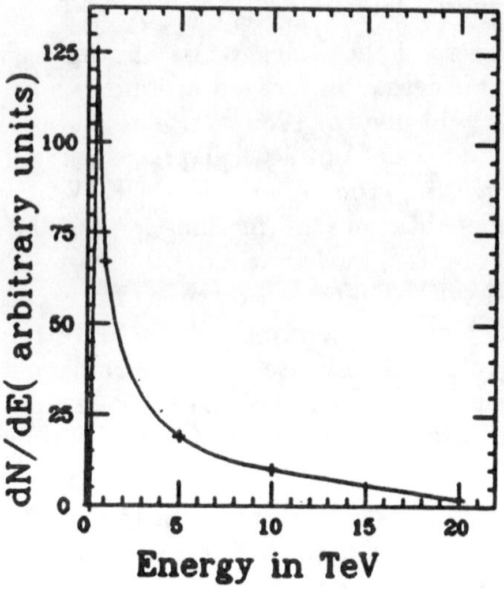

Figure 3: Detected energry spectrum in MILAGRO

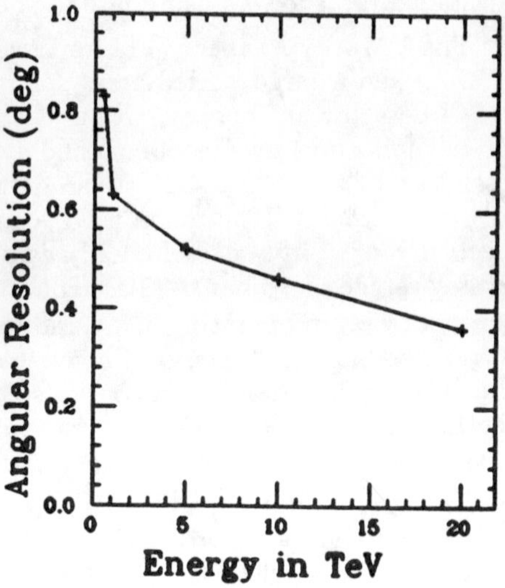

Figure 4: Angular resoltuion in MILAGRO

The UMC Extensive Air Shower Array: Results and Prospects

J. Matthews
Department of Physics, University of Michigan
Ann Arbor, Michigan, USA 48109

ABSTRACT

The Utah-Michigan-Chicago array addresses the need for greater measurement sensitivity in searches for ultra-high energy gamma rays. Results from the Utah-Michigan experiment demonstrate the power of muon rejection. Even if photons interact in unconventional ways, the combination of CASA and the large muon array will measure these effects.

There have been many attempts to observe UHE ($> 10^{14}$eV) gamma rays from celestial sources since the reported observation of Cyg X-3 by the Kiel group in 1983[1]. However, results so far have been inconclusive with most positive reports having weak statistical significance[2]. Those experiments which have attempted to distinguish gamma-ray induced showers from the plentiful hadronic cosmic ray background have reported the surprising observation that the gamma showers have a muon content similar to ordinary cosmic rays.

The Utah-Michigan-Chicago (UMC) Air Shower Array is designed to provide a definitive answer to the question of whether UHE sources exist. The technique increases the size and improves the resolution of the ground array method and rejects hadron showers using large-scale muon detection.

The UMC experiment (Figure 1) consists of the Chicago Air Shower Array (CASA) and a large buried array for muon detection located around the Fly's Eye II detector at Dugway, Utah. CASA will involve 1089 scintillator stations spread uniformly over a 500m$\times$500m grid. Fast timing will provide angular resolution of $\approx 0.5°$; a sophisticated triggering scheme will record EAS above 10^{14} eV at 20 Hz. The muon array components[3] are 2.5 m^2 plastic scintillator sheets arranged in banks of 64 adjacent counters, buried at a depth of 3m. The total coverage of 16 banks (1024 counters with 2560 m^2 total area) far exceeds that of any other operating array and provides a hadron rejection factor of $10^2 - 10^3$, depending upon energy. Shower calculations indicate that electromagnetic punch-through to the muon counters is negligible when they are buried to this depth. We have made measurements with a test arrangement of buried counters at two depths to confirm the simulations.

The size, resolution, and hadron rejection abilities will result in 15 γ-showers day^{-1} from Cyg X-3 (with a signal to noise factor of 33) if the source intensity is like that reported by the Kiel experiment. Put another way, a 5σ excess over background would emerge if the source intensity was only 10^{-2} the Kiel report.

The current Utah-Michigan array (Figure 1) is the first stage of the UMC experiment and has been in operation since late 1987. There are 33 surface stations, each with four plastic scintillators, arranged over an area of radius 100m. The present configuration of 8 banks (512 counters totalling 1280 m^2) is the largest muon detector of any air shower array now operating. Triggering and resolution are more fully described elsewhere[4].

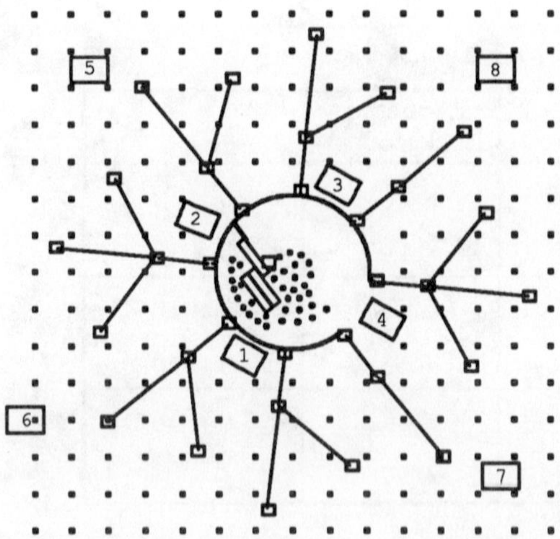

Figure 1. Layout of the central region of the UMC experiment showing the currently operating Utah-Michigan array. Grid marks are 15 m spacings and represent the location of CASA detectors under installation. The numbered rectangles indicate 64-counter muon banks. The smaller connected rectangles indicate positions of the present 33 unit surface array. The Fly's Eye II and electronics trailers are at the center. The full UMC array extends beyond that shown in the diagram and will include a total of sixteen muon banks and 1089 CASA stations.

The location and direction of the shower axis are found by fitting the pulse heights and arrival times of the surface counter hits. The electron and muon sizes N_e and N_μ are computed from maximum likelihood fits of surface data to an NKG function[5] and muon counter hits to a Greisen muon density function.[6]

The directional resolution $\delta\theta$ is defined such that 72% of events from a point source will reconstruct within $\delta\theta$ of the source direction. This definition maximizes $S/\sqrt{B}$ for a signal S in the presence of a uniform background B. The resolution is estimated by dividing the array into two parts, fitting each half separately, and comparing the results. For cores within 100m of the center of the array and $N_e > 10^4$, $\delta\theta = 3°$. Systematic pointing error is negligible, determined by comparison to data obtained by a tracking air-Cerenkov telescope operated in coincidence with the arrays.

Calculations predict 98% of gamma ray induced showers will have less than one-tenth the mean number of muons in hadron showers with similar N_e and zenith angle. We accept showers meeting this μ-poor criterion as γ-ray candidates.

Table 1 displays results of a search for excess showers from within 3° of Cyg X-3. The data is divided into subsets of increasing energy based on cuts on electron size N_e. The γ-ray energy threshold E_0 is defined as the energy at which our acceptance of γ showers has attained 25% of its maximum value. The background is determined by using the measured rate of all off-source events in local coordinates to predict the rate of ordinary events from the source direction as it moves across the sky. The data are shown with and without cuts on the muon content.

We find no excess activity from the source, with or without a μ-poor cut. The power of muon rejection is evident in Table 1. The sample is reduced by factors of $10^{-2} - 10^{-3}$ and flux limits are improved by a factor of ten. The flux limits obtained for μ-poor showers are approximately ten times lower than previously reported observations[2] at these energies (see Figure 2).

Table 1. Cygnus X-3 Observations from 4 April 1988 to 3 August 1989

$\log_{10}N_e$	E_0(eV)	Observed Events	Expected Background	Flux(90%CL) ($cm^{-2}sec^{-1}$)
4.5–6.0	2.1×10^{14}	25890	26277	$< 1.3 \times 10^{-13}$
	(μ-poor)	129	141	$< 1.3 \times 10^{-14}$
5.0–6.0	4.7×10^{14}	5214	5378	$< 6.9 \times 10^{-14}$
	(μ-poor)	6	6.38	$< 4.9 \times 10^{-15}$
5.5–6.0	1.1×10^{15}	662	715	$< 3.6 \times 10^{-14}$
	(μ-poor)	0	0.22	$< 3.5 \times 10^{-15}$

We have searched in a similar manner data taken during the large radio outbursts from Cyg X-3 in June and July 1989[7]. We find no evidence for any enhanced emission from the source on timescales from 15 minutes to 1 day during this period. Flux limits are a factor of ten below those reported during similar radio flares in October 1985 by the Baksan collaboration[8].

Similarly negative results have been obtained in searches for emission from Her X-1[9] and a whole sky survey for sources[10]. We have also analyzed data from the Crab Nebula[11]. This object is perhaps emerging as the leading candidate as a standard candle in VHE gamma rays. The persistent emission reported by two groups[12] has an apparently softer spectrum than that presumed from Cyg X-3. Consequently, our flux limits are about twenty times higher than the VHE flux levels extrapolated to 200 TeV.

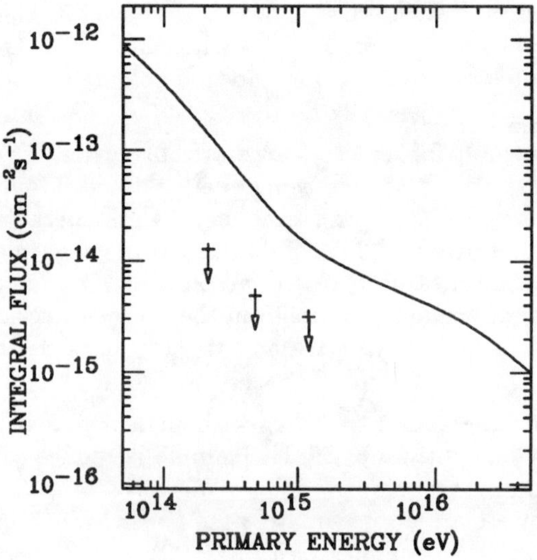

Figure 2. Upper limits (90% C.L.) on the γ-ray flux of Cygnus X-3 from the Utah-Michigan array using muon-poor showers (from Table 1). The solid line is the γ flux expected from the source based an extrapolation of x-ray and VHE results. This line also approximates results and limits obtained above 10^{14} eV (see reference 2). The dip in the spectrum corresponds to the expected loss of UHE γ rays which interact with the 3° cosmic background radiation.

The Chicago Air Shower Array (CASA) will permit flux sensitivity substantially below previous reports of Cyg X-3, Her X-1, and similar to that expected from the Crab Nebula. CASA offers eight times the target area and improves the angular resolution by a factor of six over the existing apparatus. The energy threshold will be somewhat lower due to the higher density of counters. Using $S/\sqrt{B}$ as a figure of merit, we anticipate more than a factor of ten improvement.

The first 500 units of CASA are now in place with data taking to begin in November, 1989. The remaining elements of this array are scheduled for installation in 1990 along with the final eight muon patches.

In summary, large scale muon detection and surface measurements will offer a definitive resolution of the current puzzle regarding UHE gamma ray emission from astrophysical objects. Flux limits for Cyg X-3 using Utah-Michigan array data are at levels comparable to previous experiments; use of μ-rejection improves these tenfold. Thus, if photons interact conventionally, muon rejection provides demonstrated sensitivity to fluxes well below those previously reported. Should there be new physics and/or anomolous interactions, the CASA array will be more sensitive to such activity than prior attempts and the muon structure will be examined in detail.

REFERENCES

[1] M. Samorski and W. Stamm, Astrophys. J.**268**, L17-21,(1983); M. Samorski and W. Stamm, Proc. 18th Int. Cosmic Ray Conf. **11**, 244 (Bangalore, 1983).

[2]For reviews see R.J. Protheroe, Proc. 20th Int. Cosmic Ray Conf.**8**, 21 (Moscow, 1987); W. Hermsen et al.,Astron. Astrophys. **175**, 141 (1987).

[3]D. Sinclair, Nucl. Instrum. Meth. **A278**,583 (1989).

[4]G.L. Cassiday et al., Proc. 21st Int. Cosmic Ray Conf., paper HE 3.4-1 (Adelaide, 1990).

[5]K. Greisen, Prog. Cosmic Ray Phys. **3**, 1 (1956); E.J. Fenyves et al., Phys. Rev. **D37**,649 (1988).

[6]K. Greisen, Ann. Rev. Nucl. Sci. **10**, 63 (1960).

[7]D. Ciampa et al., Proc. 21st Int. Cosmic Ray Conf., paper OG 4.1-12 (Adelaide, 1990); G.L. Cassiday et al., to appear in Phys. Rev. Lett. (1989).

[8]V.V. Alekseenko et al., Proc. 20th Int. Cosmic Ray Conf., **1**, 229 (Moscow, 1987); V.S. Berezinsky, Nature (London) **334**, 506 (1988).

[9]D. Ciampa et al., Proc. 21st Int. Cosmic Ray Conf., paper OG 4.2-9 (Adelaide, 1990).

[10]D. Ciampa et al., Proc. 21st Int. Cosmic Ray Conf., papers OG 4.6-19, OG 4.7-3 (Adelaide, 1990).

[11]S.C. Corbato et al., Proc. 21st Int. Cosmic Ray Conf., paper OG 4.3-11 (Adelaide, 1990).

[12]T.C. Weekes et al., Astrophys. J. **342**, 379 (1989); C. Akerlof et al.,University of Michigan preprint UM-HE-89-11, to appear in Proc. GRO Science Workshop, NASA/Goddard SFC (1989).

STATUS OF THE MACRO EXPERIMENT AT GRAN SASSO

Presented by Steve Ahlen
Workshop on "The NASA Cosmic Ray Program for the 1990's and Beyond,"
Goddard Space Flight Center, November 6, 1989

The MACRO (Monopole Astrophysics and Cosmic Ray Observatory) detector is being installed in Hall B of the Gran Sasso Laboratory near L'Aquila, Italy. When completed the detector will have dimensions 72mx12mx10m, providing an acceptance of 10,000m^2sr. The experiment is located under the Gran Sasso Mountains at an average depth of 3800 meters water equivalent (mwe) and a minimum depth of 3000 mwe. A single supermodule of MACRO (12mx12mx5m) has been operational since February 1989. An additional 5 supermodules will be operational by September 1990. It is expected that the entire MACRO detector will be running by 1991.

Each supermodule of MACRO consists of a sandwhich array of two layers of liquid scintillation counters (top and bottom), 10 layers of streamer tubes separated by rock absorbers and several layers of plastic track-etch detectors across the central plane of the supermodule. The sides and extreme ends of the detector are covered with a single scintillator layer and 6 layers of streamer tubes.

The streamer tube system consists of 5000 wires per supermodule. Eight tubes, each having dimensions 3cmx3cmx12m are combined in a single chamber. The tubes utilize 100 micron anode wires and a graphite cathode. The tubes operate in the limited streamer regime with a gas mixture of Ar, CO_2 and n-pentane. In the future, helium and neon will be used in the tubes to provide redundant detection efficiency for low velocity monopoles via the Drell effect. A two dimensional readout is performed using signals from the anode wires (X-view) and 26° stereo pickup strips (D-view). The overall tube efficiency is 97% (due to wall geometry), while the pickup strips have 98% efficiency relative to the wires. Spatial rms resolution for the X-view is 1.1cm while that for the D-view is 1.2cm, implying an intrinsic angular resolution of 0.1°. Multiple scattering in the rock dominates the actual resolution however (angular differences between pairs of muons within muon bundles has an rms dispersion of 0.6°).

The liquid scintillation counters are devoted mainly

to an accurate measure of the energy loss and timing of crossing particles. The horizontal layers are PVC tanks (12mx75cmx25cm) lined with FEP teflon and equipped with two 20cm diameter hemispherical photomultipliers (PMTs) at each end. The vertical layers which close the sides and ends of the detector have dimensions of 12mx50cmx24cm and are viewed by one PMT at each end. The liquid scintillator is based on an ultra-pure low paraffin mineral oil with a mixture of pseudocumene, wavelength shifters, and anti-oxidants. The measured attenuation length of the mix falls in the range 12 to 14m. A muon traversing the counter at a distance of 12m from a PMT yields 200 photoelectrons at the PMT. This number will increase in the future when the PMT end chambers are filled with plain mineral oil. The PMT readout system consists of: ADCs on the anode and one dynode, covering a dynamic range up to 10,000 photoelectrons; TDCs used to distinguish upward from downward going muons; analog and digital systems to reconstruct energy deposited in a counter independent of impact point down to a threshold of 5MeV; slow (20MHz) and fast (50MHz) waveform digitizers to record precisely the light yields and, hence, the energy loss and velocity of slow and fast particles crossing the counters. The standard deviation of the time measurement of a single end of a counter is 1.7ns and the energy resolution is $30\%E^{\frac{1}{2}}$ where E is in MeV. The spatial standard deviation for muons is 15 cm.

The physics results achieved with the running of the first supermodule from February 1989 to May 1989 are summarized below.

MAGNETIC MONOPOLE SEARCH

No magnetic monopoles were observed. At the 90% confidence level (CL) this corresponds to an upper limit on monopole flux of $4\text{x}10^{-14}/(\text{cm}^2\text{-sr-s})$ for a velocity range from 0.00025c to 0.015c. These results represent more stringent monopole flux limits at lowere monopole velocities than reported in previous experiments.

SEARCH FOR POINT SOURCES OF DOWNWARD MUONS

The energy threshold for a downward moving muon to penetrate to MACRO is 1.4TeV at the mountain surface. No point sources of such muons were found. The limit on the flux from point sources at the 90% CL is $7.6\text{x}10^{-12}/(\text{cm}^2\text{-s})$. The 90% CL upper limit for unmodulated muons from Cyg X-3 in a 3° by 3° bin is $4.1\text{x}10^{-12}/\text{cm}^2$-

s). The 90% CL upper limit for modulated muons in a 10° by 10° bin around Cyg X-3 is $4.3x10^{-12}/cm^2$-s).

UPWARD MOVING MUONS

For the Spring 1989 run only one upward moving muon was seen. We expected only several such muons due to atmospheric secondaries during this period for one supermodule.

MULTIPLE MUONS AND COSMIC RAY COMPOSITION AT HIGH ENERGIES

Abundant data on muon bundles were obtained during the Spring run. During this time 245,000 muon triggers were collected, coresponding to 1890 hours of live time. About 10,000 double muons were observed as well as numerous occurrences of multiplicities up to 22. There were 5 events with 19 or more muons per bundle observed. It is expected that in the near future (when Monte Carlos of acceptance efficiencies are completed) we will be able to claim a cosmic ray composition at 10^{14} eV (top of atmosphere) based on these observations.

SEARCH FOR STELLAR COLLAPSE

Two complementary trigger and readout systems were used to monitor the horizontal scintillator counters, during the Spring run for stellar collapse. A mass of 42 tons of scintillator was used for this search. No events were seenwith more than 4 neutrino induced events at greater than 10MeV for 2 second intervals. This corresponds to a gravitational collapse equivalent to SN1987A at the center of the Galaxy.

Splinter Group Reports

COSMIC RAY STUDIES OF SOLAR AND HELIOSPHERIC PHYSICS: GOALS FOR THE 1990'S AND BEYOND*

Mark E. Wiedenbeck
Enrico Fermi Institute and Department of Physics
University of Chicago, Chicago, Illinois 60637 USA

ABSTRACT

A summary is presented of discussions by the splinter group on solar and heliospheric physics, which met as part of the workshop on "The NASA Cosmic Ray Program for the 1990's and Beyond", held at the Goddard Space Flight Center on November 6–8, 1989. This group identified major science themes and discussed a complement of new missions to address these themes.

SCIENCE THEMES

High priority objectives identified for solar and heliospheric physics fall into the following broad categories:

- Exploration of the heliospheric boundary and the local interstellar medium.
- Investigations of particle acceleration and transport in solar flares through combined measurements of solar energetic particles, solar gamma rays, and solar neutrons.
- Extension of solar energetic particle composition investigations to include "ultraheavy" elements and isotopes.
- Studies of the large scale structure of the heliosphere and the modulation of energetic charged particle fluxes.
- Studies of the "microphysics" of the interaction of the interplanetary magnetic field and solar wind plasma with energetic charged particles.

RECOMMENDED PROGRAMS

A set of missions were identified for meeting the above scientific objectives. Of highest priority is the prompt execution of the missions and experiments now under development or selected for this area of science: Ulysses, CRRES, NOAA-I/EHIC, SAMPEX, ISTP/WIND, EOS/POEMS, and ACE. In addition, high priority should be given to the proposed Global Heliosphere Initiative, which would facilitate in-depth studies of heliospheric phenomena through the use of coordinated observations made with existing and planned spacecraft. New initiatives include the following:

- An "Interstellar Probe" mission to study the heliospheric boundary and the very local interstellar medium.
- Inclusion of instrumentation for the study of impulsive solar flares on the proposed Mercury Orbiter mission.

---

*This work was supported, in part, by NASA grant NGL 14–001-005.

- A mission to measure the abundances of ultraheavy nuclei in solar energetic particles, using a lunar base, an Explorer outside the magnetosphere, or a platform in polar orbit.
- A small Explorer mission to measure solar gamma rays, to be flown at the same time as ACE, in order to enable correlative studies of gamma ray and charged particle observations in solar flares.
- An Explorer mission, possibly inside 1 AU, devoted to the study of particle acceleration and transport processes in solar flares.
- Full utilization of future "missions of opportunity" to carry instruments for measuring particles and fields in the inner and outer heliosphere, enabling studies of the large scale structure of the heliosphere and of solar modulation.
- Development and utilization of small satellite arrays for investigations of the microphysics of interplanetary particle acceleration and propagation.

PRESENT STATUS

Important progress has been made in solar and heliospheric physics over the past 10 to 15 years through a variety of energetic charged particle investigations coupled with significant theoretical developments. Figure 1 illustrates in a schematic way the current status (solid bars) of our knowledge in five broad categories of scientific objectives in this field. Contributions to our present base of knowledge have come from charged particle experiments on the deep space missions Pioneers 10 & 11 and Voyagers 1 & 2, together with the IMP and ISEE spacecraft near Earth and the Helios 1 & 2 missions inside 1 AU. Balloon-borne measurements of electron fluxes throughout the past solar cycle have contributed to our understanding of the solar modulation process. Solar gamma ray and neutron investigations, particularly on SMM, have addressed the question of particle acceleration in solar flares in a way which complements the solar energetic particle studies.

A number of new cosmic ray experiments which will address problems in solar and heliospheric physics will soon be launched. These include investigations on the Ulysses spacecraft which will make the first studies at high heliographic latitude, and experiments on the CRRES and NOAA-I spacecraft in Earth orbit. Also presently under development are important experiments for the WIND, Geotail, and SOHO spacecraft in the International Solar–Terrestrial Program (ISTP), and for the SAMPEX Small Explorer mission. The next major steps beyond these upcoming experiments will be made in the mid- to late-1990's by the recently selected Advanced Composition Explorer (ACE) mission and the POEMS experiment for the EOS program. The goals and capabilities of each of these upcoming experiments were discussed in invited talks at this workshop, and will not be repeated here. In Figure 1 the advances in our knowledge expected from these investigations are illustrated by the shaded bars.

STATUS OF INVESTIGATIONS

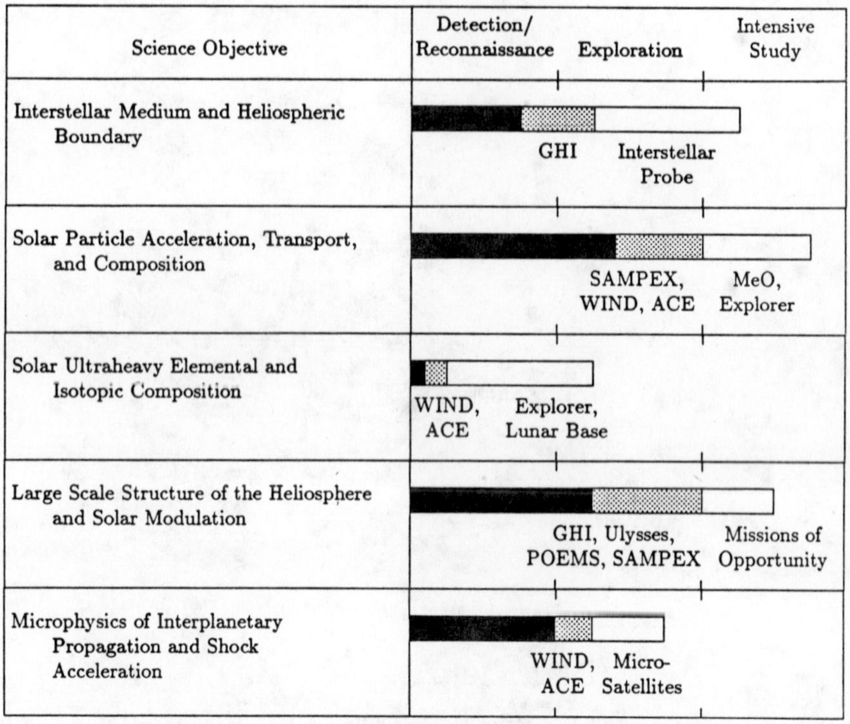

Figure 1.
Schematic diagram indicates the present and projected status of studies of key science objectives. Solid bars represent the current status, shaded bars indicate the advances expected with missions and experiments in the currently approved program ($\sim$ 1990–2000), and open bars show the additional progress which can be made with the recommended future initiatives ($\sim$ 2000–2010). Missions expected to contribute to future advances are indicated. "GHI" denotes the Global Heliosphere Initiative program.

Continued progress in heliospheric physics can be expected not only from new missions, but from the coordinated studies using existing missions. The Pioneer and Voyager spacecraft (Figure 2) represent an important resource for studying the outer heliosphere, which will not be duplicated in the foreseeable future. The recently formulated Global Heliosphere Initiative[1] will facilitate the use of this impressive array of spacecraft for studies of the large scale structure of the heliosphere and solar modulation.

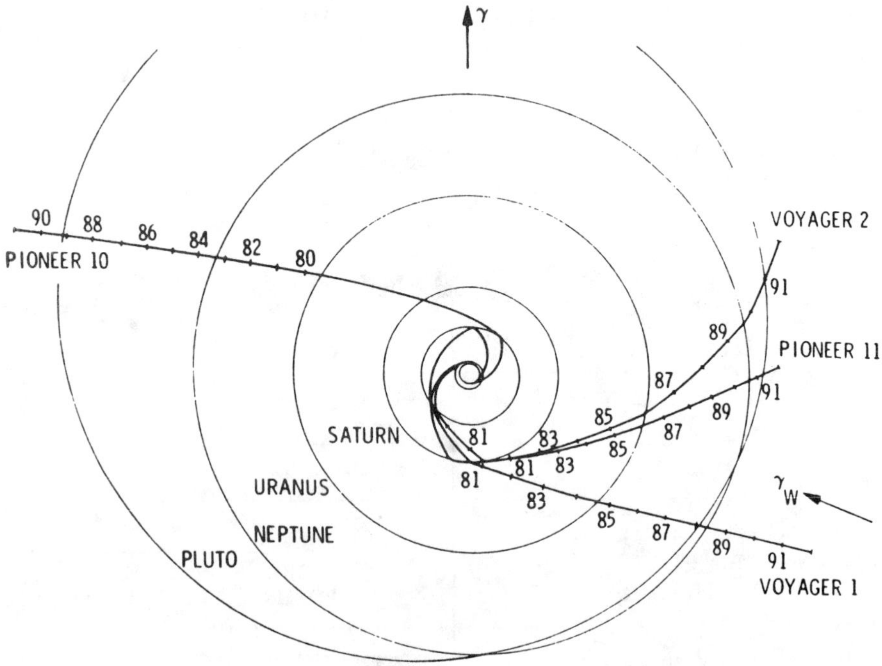

Figure 2.
Trajectories of Pioneers 10 and 11 and Voyagers 1 and 2.

PHYSICS OF THE INTERSTELLAR MEDIUM AND THE HELIOSPHERIC BOUNDARY

Cosmic ray experiments conducted in the inner solar system are limited in their capabilities for studying low energy particles (below a few hundred MeV per nucleon) accelerated in the outer solar system or the galaxy, since solar modulation effects prevent these particles from penetrating to 1 AU. Advances have been made with deep space missions, which have measured radial and latitudinal gradients of galactic and anomalous cosmic rays, leading to tentative estimates of the location of the modulation boundary. It is hoped that these missions will, in the coming years, reach the solar wind termination shock and obtain improved estimates of the unmodulated galactic cosmic ray spectrum.

A new mission will be required to make definitive measurements of the solar wind termination shock and the heliopause region, and to characterize in detail the particles and fields environment in nearby interstellar space. An "Interstellar Probe"[2] designed to leave the heliosphere at a rate $\sim$ 20 AU/yr should reach the termination shock in $\sim$ 3 to 10 years and then carry out *in situ* measurements of the interstellar medium (ISM), reaching a distance of 1000 AU in 50 years.

With such a mission one could make direct measurements of low energy cosmic ray nuclei and electrons, determining the unmodulated spectra of all major species. Such measurements would establish the contribution of cosmic rays to the energy density and to the heating of the ISM. It could carry out *in situ* studies of the interstellar plasma, magnetic field, and neutral gas, determining fundamental parameters including temperature, density, composition, ionized fraction, and magnetic field strength. It could also look for inhomogeneities in these quantities, and for anisotropies in the energetic particle fluxes.

An Interstellar Probe mission could study the solar wind termination shock, and plasma conditions in the pre- and post-shock regions. Furthermore, it could measure the dynamical response of the termination shock to varying solar wind conditions. The solar modulation boundary should be identifiable as the galactic cosmic ray flux reaches its steady interstellar value.

Particle acceleration in the outer heliosphere can be studied as the Probe penetrates that region. In particular, it should be possible to identify the site and mechanism responsible for acceleration of the anomalous component, and possibly to discover anomalous cosmic ray contributions for additional elements. The termination shock, having a scale larger then those of shocks previously studied, may accelerate particles from other populations of galactic or solar origin. Thus measurements of the variation of low energy particle spectra and anisotropies in the vicinity of the shock could result in important extensions of our understanding of particle acceleration on a variety of astrophysical scales.

SOLAR PARTICLE ACCELERATION, TRANSPORT AND COMPOSITION

Measurements of solar flare associated energetic particle events provide a means for studying the acceleration mechanism, conditions at the flare site, and particle propagation in the solar corona and in the heliosphere. Most intensively studied thus far have been those flares which produce large particle events at Earth. However, it is now thought that the more typical solar flare produces a relatively small particle increase at Earth. These small events are also characterized by impulsive acceleration, enhancement of the $^3He/^4He$ ratio, and a pattern of element enhancements which increase with increasing Z.

The study of impulsive flare events at 1 AU is complicated because the increases of particle fluxes over quiescent values are relatively modest, commonly less than an order of magnitude even at the peak of the flare, and because scattering during propagation from the Sun degrades the rise time and broadens the pitch angle distribution. The Helios spacecraft, in elliptic orbits about the Sun with aphelion at 1 AU and perihelion at 0.31 AU, succeeded in measuring a few SEP events from impulsive flares while near perihelion. As shown in Figure 3, the particle signatures at this radius were much more dramatic than when subsequently observed by ISEE-3 near 1 AU, with flux increases several orders of magnitude greater and with sharper rise times and pitch angle distributions. Thus the capability for making SEP observations over an extended period of time at a radius significantly inside 1 AU should enable a significant improvement in our understanding of impulsive flares.

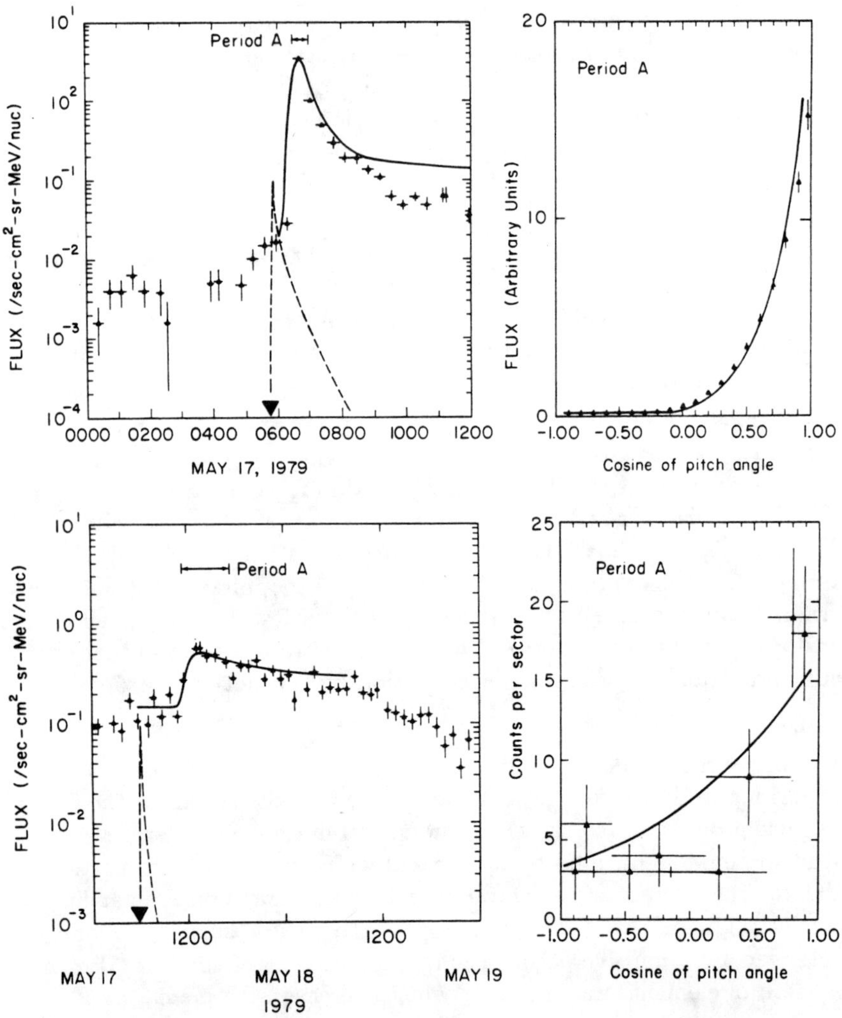

Figure 3.

Observations[3] of helium nuclei with energies $\sim 1\,\mathrm{MeV/nucleon}$ in the SEP event from an impulsive solar flare on 17 May 1979. The upper panels show measurements made on Helios-I which was at 0.32 AU, while the lower panels show measurements made on ISEE-3 near 1 AU. Comparisons of flux increases, rise times, and pitch angle distributions between the two data sets illustrate the enormous improvement which can be made in the study of impulsive flares through the use of measurements made well inside 1 AU. The different times scales used in the top and bottom panels should be noted.

The Mercury Orbiter mission, now under study, would consist of one or two spacecraft in orbit around the planet Mercury (~ 0.4 AU), emphasizing studies of the planet's magnetosphere. Instrumentation for measuring energetic particle spectra, composition, and atomic charge states, as well as solar gamma rays and neutrons, should be included as part of this mission in order to take advantage of this unique opportunity for solar flare studies.

Advances in solar gamma ray and neutron spectroscopy[4], particularly those made on the Solar Maximum Mission, have demonstrated that such observations can be an important complement to solar energetic particle measurements for determining solar abundances and for studying particle acceleration in solar flares. For example, comparison of abundances seen in a SEP event with those deduced from gamma ray observations of the flare in which these particles were accelerated can be used in attempting to identify the pool of matter from which the SEP particles were extracted. Among the objectives of the ACE mission are precise measurements of elemental, isotopic, and charge state composition over a broad energy range in SEP events. Important additional information on solar flares could be obtained with a Small Explorer mission, devoted to solar gamma ray and neutron measurements, which would fly during the same time as ACE.

In the period following ACE, a program focusing on the goal of understanding energetic particle acceleration and transport on the Sun would be a logical next step. For this purpose, a Delta-class Explorer mission combining measurements of solar gamma-rays, x-rays, and neutrons, electrons and positrons, and nuclei could provide the correlated observations needed to address these questions. For example, solar flare electron spectra exhibit a number of well defined slope changes which are indicative of features of the acceleration process. Gamma rays carry information both about the ambient matter that serves as input to the acceleration process and about the motions of energetic particles in the solar atmosphere. Secondary products of high energy collisions, including ^{2}H, ^{3}H, ^{3}He, Li, Be, B, and positrons, probe the amount of matter traversed by the accelerated particles. The charge states of the accelerated nuclei can be used as a thermometer to determine the temperature at the flare site. These and other topics have been addressed with a variety of observations of separate flares, made on different missions, and such studies will be continuing on upcoming missions, including SAMPEX, WIND, and ACE. An important future goal is to bring all of these sources of information to bear on the study of individual flares, so that interrelationships can be established and a comprehensive model of the flare acceleration process developed.

ULTRAHEAVY COMPOSITION IN SOLAR ENERGETIC PARTICLES

Measurements of energetic particles accelerated in solar flares yield both information concerning the physical conditions and processes responsible for the acceleration, and data relevant to the composition of the solar surface layers. Such composition data are an important supplement to those obtained from more traditional sources, including mass spectroscopy of terrestrial, lunar, and meteoritic samples, and spectroscopic measurements of the solar photosphere and corona. Abundances obtained from solar energetic particle (SEP) measurements are subject to a different set of systematic uncertainties than abundances obtained from these other sources. Thus comparison of abundances resulting from these different types of studies provides a powerful method for identifying systematic errors and extracting valid solar abundances. Furthermore, the SEP studies have the advantage that they yield measurements of essentially all stable nuclides, whereas there are significant gaps in the data from meteorite studies due to the depletion of volatile elements, and in the spectroscopic data due to thermal conditions which produce ionization states for some elements which do not lead to lines in the measurable part of the electromagnetic spectrum. The SEP measurements can be used to obtain *isotopic* as well as elemental composition of solar material, a capability previously available for only a few elements using other techniques. Investigations[5] using data from Voyager have shown that the deviations of SEP abundances from spectroscopic abundances measured in the corona are well correlated with the SEP charge states, enabling extraction of accurate SEP-derived coronal abundances when charge state measurements are available. Photospheric abundances have then been estimated by use of an empirically derived correction for first ionization potential dependent biasing.

Upcoming experiments on NOAA-I, SAMPEX, WIND, and Geotail should exploit these new capabilities for obtaining solar composition from SEP measurements. The ACE mission will provide a factor > 10 increase in collecting power for SEP studies, making possible precise measurements of SEP composition of even rare nuclides with $Z \leq 28$, as well as exploratory measurements in the range $28 < Z \lesssim 40$.

The next step in SEP studies of solar composition will require measuring abundances of ultraheavy nuclides (UH, $Z \gtrsim 30$). The low fluxes of these species will require instruments with geometrical factors large enough to collect $\gtrsim 10^6$ iron nuclei in a moderate sized SEP event. In addition, the fractional mass resolution requirements will become more stringent as one attempts to separate nuclides of higher mass. In one possible approach[6], using time-of-flight vs. total energy, flight paths up to 10 m would be beneficial. Thus large instruments are needed both for geometrical acceptance and for resolution. A lunar base, with its negligible atmosphere and negligible magnetic field and with facilities for deploying large instruments, would provide an excellent site for studies of UH composition in SEP events. Alternatively, significant progress in this area could be made with a Delta-class Explorer mission outside the Earth's magnetosphere. A third possibility is use of a polar platform in low Earth orbit, although in this case low energy particles can be collected only during $\sim 30\%$ of each 90 minute orbit, which complicates the analysis and increases the required geometrical factor by a factor ~ 3.

Impulsive solar flares tend to produce relatively small particle events near Earth, characterized by large enhancements of ^{3}He and heavy elements. These enhancements, which increase with increasing Z, can be more than a factor of 10 for the Fe/O ratio. Extrapolating this trend to the ultraheavy elements one might expect enhancements of factors of 100 to 1000 for the heaviest elements. Measurements of this enhancement pattern should provide a useful probe of particle acceleration and transport in these flares. Instrumentation requirements for such measurements are similar to those for SEP composition studies in larger flares, discussed above, since the small flare sizes may be balanced by the large UH enhancements.

LARGE SCALE STRUCTURE OF THE HELIOSPHERE AND SOLAR MODULATION

Our knowledge of the large scale structure of the heliosphere has been advanced greatly as it has become possible to make measurements at different heliocentric radii and heliographic latitudes, and at different times. For example, the sector structure observed in the interplanetary magnetic field in the 1970's has been shown to be due to the wavy structure of the current sheet separating two hemispheres in which the magnetic field carried outward from the Sun has opposite polarities. The studies of latitudinal effects will be greatly extended by measurements to be made on the Ulysses mission, scheduled for launch in 1990, which will make the first studies of particles and fields in the heliosphere's polar regions.

Studies of the spatial and temporal variations of cosmic ray intensity have provided the fundamental data which led to our present knowledge of solar modulation. In addition, theoretical investigations have identified the role of gradient and curvature drifts in the modulation process, providing a possible explanation of observed differences of modulation effects in alternate 11 year cycles.

The multipoint measurements which are essential for investigation of solar modulation and heliospheric structure will benefit greatly over the coming decade from coordinated measurements using Pioneer 10 and 11, Voyager 1 and 2, Ulysses, and various near-Earth spacecraft. This unique array of spacecraft, which is unlikely to be duplicated in the foreseeable future, must be fully exploited to advance these studies. In the future, it will be important to assure that modest particle and fields instruments be carried on planetary and other spacecraft travelling beyond the immediate vicinity of the Earth. To take advantage of such instrumentation, mission plans must provide for cruise-phase observations and, where appropriate, for extended missions once primary objectives have been met. A charged particles experiment on the Mercury Orbiter mission would provide important data from inside 1 AU. Further, if a major Mars exploration initiative is undertaken, particles and fields instruments should be placed in Martian orbit to establish a long duration Earth–Mars baseline for modulation and heliospheric dynamics studies.

While at present the requirement for multipoint measurements at widely separated sites makes this class of investigations dependent on secondary-science opportunities on missions in other fields, opportunities for placing a dedicated array of instruments at appropriate sites could become possible through the development of small, low cost spacecraft. Studies[7] of such "microspacecraft", focusing on payloads of no more than a few tens of kilograms, have been carried out at JPL. Cooperative work between the microspacecraft study group and the Space Physics Division could prove beneficial for optimizing the spacecraft definition for heliospheric objectives.

An important prediction of gradient and curvature drift models of solar modulation is that the modulation of particles with opposite charge sign but the same velocity and rigidity will differ, with the sense of this difference changing when the polarity of the solar magnetic field reverses. Preliminary work comparing the modulation of electrons and helium at comparable rigidities suggests the existence of such effects, but the interpretation of these results is complicated by the differing velocities of these particles and by the fact that the electron measurements do not distinguish between e^- and e^+. Precise measurements of low energy electron and positron spectra throughout a solar cycle will provide vital data for understanding this aspect of solar modulation. The POEMS experiment selected for the Earth Observing System (EOS) will provide the necessary measurements. Launch on the first EOS platform and long duration operation are important for such modulation studies.

MICROPHYSICS OF INTERPLANETARY PROPAGATION AND SHOCK ACCELERATION

In addition to studies of global characteristics of the heliosphere and its interaction with energetic charged particles, investigations are needed of the "microphysics" which leads to the large scale properties. For example, the diffusion tensor needed to account for modulation over scales of many AU must arise from particle scattering by magnetic field turbulence on scales of a particle's gyroradius. An important goal in heliospheric physics is to characterize the spectrum of turbulence and to quantitatively establish the relationship between this spectrum and the large scale motions of energetic charged particles. Such studies require the determination of multipoint autocorrelation functions of the interplanetary magnetic field and of energetic particle fluxes, and cross correlations between these two. Such determinations can be obtained from simultaneous multipoint measurements over distance scales comparable to particle gyroradii. A possible approach involves the use of a "cluster" of spacecraft, similar to successful measurements in magnetospheric physics but carried out in interplanetary space rather than in a planetary magnetosphere. A possible alternative approach, utilizing a single spacecraft, would measure single point correlations between the vector fluctuations in the magnetic field and particle flows and take advantage of the fact that the particle motions are sensitive to the field over the scale of a gyroradius. Further theoretical work is needed to determine the utility of this approach and establish the statistics and sampling rates required.

Particle acceleration in interplanetary shocks is another subject involving the microphysics of particle–field interactions. It has been found that interplanetary shock acceleration is best studied in the outer heliosphere where backgrounds due to energetic particles of solar or Jovian origin are much less significant than near Earth, and where shock structures are better formed. Shock acceleration investigations are continuing using data from the Pioneers and Voyagers. The acceleration by shocks is likely to lead to situations where the cosmic ray energy density (or pressure) is comparable to that of the ambient plasma. This would result in significant modification of the flow by the cosmic rays. Appropriate observations will help elucidate the important physics occurring in this new, unexplored regime. Further advances in the study of particle acceleration in interplanetary shocks will require multispacecraft measurements of particles and fields in the neighborhood of the shock.

Objectives related to the microphysics of interplanetary propagation and acceleration could benefit from development of microsatellites, mentioned above in connection with studies of large scale structure. In the present context, however, several such spacecraft would fly in close formation, possibly simplifying communications between spacecraft (if needed) and with the Earth.

INFRASTRUCTURE ENHANCEMENTS / NEW CAPABILITIES

In addition to major initiatives for particle studies in solar and heliospheric physics, the importance of continuing and enhancing the ongoing program should be emphasized. Specifically:

- Support for vigorous theoretical work. Theoretical input is needed both in defining mission objectives and in interpreting experimental findings. The organization of several series of topical workshops bringing theorists and observers to work on specific problems of common interest could be particularly valuable.
- Support for analysis utilizing data from multiple missions. The Global Heliosphere Initiative addresses this need in the case of the Pioneers and Voyagers. Similar provisions should be made for carrying out this type of analysis taking advantage of data from future missions
- Improvements in tracking capability. The data return from spacecraft which have completed their prime missions is often significantly reduced due to the limited availability of tracking coverage, particularly when the DSN is required. Efforts to upgrade tracking capabilities and to plan the necessary tracking support for deep space missions could significantly enhance the value of present and future missions.
- Support for developing advanced instrumentation concepts for solar and heliospheric physics. Although instruments in this field are typically not appropriate for suborbital experiments, enhanced laboratory development efforts will be needed to advance the state of the art, particularly in this era of infrequent space flight opportunities.

A new capability which may find application in the coming years involves the creation and study of artificial magnetospheres. By flying relatively large magnets (e.g, Astromag, or possibly a somewhat larger single-coil magnet) one can investigate the interaction of magnetized bodies with the solar wind, and make *in situ* measurements of plasma processes which may be of interest for scaling astrophysical observations.

CONCLUSIONS

Significant advances in the study of solar and heliospheric phenomena are expected in the coming decade as a result of cosmic ray experiments now under construction or selected. The complement of future missions identified by the solar and heliospheric splinter group and discussed above should make possible important further progress in addressing the major scientific questions in this field.

ACKNOWLEDGEMENTS

This summary benefited from the contributions of all of the participants in the solar and heliospheric splinter group. I particularly wish to thank those who read and commented on this manuscript: A. Cummings, M. Garcia-Munoz, J. R. Jokipii, and R. A Mewaldt, B. Milliken, and R. Ramaty.

REFERENCES

1. Heliospheric Science Working Group (E. Smith, chairman), "A Program of Coordinated Heliospheric Observations", Preliminary Report, April 1988.
2. R. A. Mewaldt and E. C. Stone, "Cosmic Ray Studies with an Interstellar Probe", this volume.
3. G. M. Mason, C. K. Ng, B. Klecker, and G. Green, *Astrophys. J.*, **339**, 529 (1989).
4. R. Ramaty, "Solar Abundances from Gamma-Ray Spectroscopy", in *Cosmic Abundances of Matter*, C. J. Waddington editor (Am. Instit. of Phys., New York), 1989, pp. 91–99.
5. E. C. Stone, "Solar Abundances as Derived from Solar Energetic Particles", in *Cosmic Abundances of Matter*, C. J. Waddington editor (Am. Instit. of Phys., New York), 1989, pp. 72–90.
6. M. E. Wiedenbeck, "The Abundances of Ultraheavy Nuclei in Solar Energetic Particles", this volume.
7. R. Anderson, P. Coleman, F. Dyson, J. French, R. Forward, J. Hornbuckle, M. Huebschman, A. Meinel, M. Meinel, and N. Sorensen, "Microspacecraft for Space Science Workshop", October 6, 1988.

RECOMMENDED PROGRAMS FOR GALACTIC COSMIC RAYS (< 1 TeV/nucleon) FOR THE YEARS 2000-2010

W. R. Binns
Department of Physics and McDonnell Center for the Space Sciences, Washington University, St. Louis, Mo. 63130 USA

ABSTRACT

This paper describes the discussions and recommendations from the splinter group meeting on Galactic Cosmic Rays (<1 TeV/nucleon), held as a part of the workshop on "The NASA Cosmic Ray Program for the 1990's and Beyond". The science themes which were developed are presented. These are then followed by a discussion of the recommended programs which address these science themes. In addition, other potential programs which were presented at the workshop are discussed. The importance of continued laboratory investigations, in particular cross section measurements at the Bevalac, and continued support for a strong ballooning program are emphasized.

SCIENCE THEMES

The scientific studies which are recommended for Galactic Cosmic Rays (<1 TeV/nucleon) for the decade beginning in the year 2000 are expected to result in fundamental new knowledge of the origin and history of galactic cosmic rays. The broad science goals which they address are 1) to study the origin and evolution of galactic matter and 2) to investigate acceleration, transport, and time scales of galactic cosmic rays.

The specific science themes recommended for investigation are:

1) Study the interstellar galactic cosmic rays directly on an interstellar probe mission.
2) Conduct the first isotope study of neutron capture process nucleosynthesis by studying the ultra heavy (U.H.) cosmic ray nuclei ($Z \geq 30$);
3) Investigate nucleosynthesis and special sources of cosmic rays;
4) Determine whether U.H. nuclei have a separate origin and history from that of $Z < 30$ nuclei;
5) Make measurements of previously unmeasured cosmic ray "clocks" to explore time scales for nucleosynthesis of source nuclei;
6) Study acceleration mechanisms for cosmic rays.

RECOMMENDED PROGRAMS

There are three new mission thrusts which are recommended to address these science themes, and a fourth program which is recommended as a second generation Astromag investigation. They are:

1) An "Interstellar Probe" mission to the heliopause and beyond into the interstellar medium to study the unmodulated interstellar galactic cosmic ray abundances and spectra, and to measure, "in situ", the acceleration effects of the solar wind termination shock on the anomalous cosmic ray component and on galactic cosmic ray nuclei and electrons.

2) An Explorer class isotope mission to make high resolution measurements of the isotopic composition of U.H. galactic cosmic rays.
3) A mission to make detailed measurements of the U.H. cosmic ray elemental composition for 30<Z<100 with improved collecting power over previous or planned programs. This mission could be conducted on a co-orbiting platform, an attached payload, or from a lunar base.
4) A second generation Astromag investigation to measure the spectra and composition of very high energy cosmic ray isotopes.

Fig. 1 gives the status of knowledge for each of these thrusts at the present time (shown by the black bars), the knowledge expected by the year 2000 from approved programs (cross-hatched bars), and that expected from the recommended thrusts from 2000-2010 (open bars).

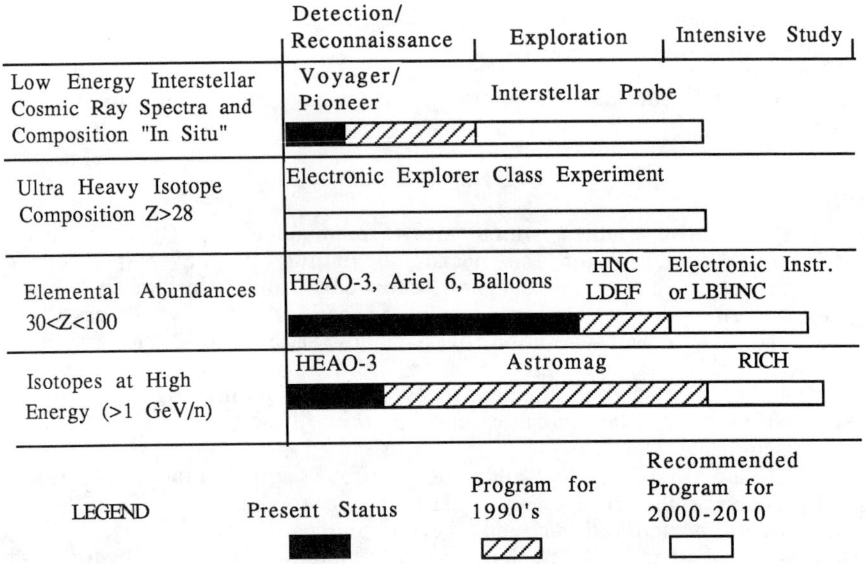

Fig. 1--Status of research for recommended investigations.

The Interstellar Probe--As shown in Fig. 1, our present knowledge of the effect of modulation in the heliosphere on the cosmic ray abundances and energy spectra can be characterized as being in the detection/reconnaissance stage. Our most direct information as to the nature and extent of the modulation region comes from instruments on the Voyager and Pioneer spacecraft. The elemental composition of nuclei with charge up through iron for energies less than about 250 MeV/n has been measured at 20-25 A.U. on the Voyager 2[1]. By the year 2000 Voyager 1 will be about 75 A.U. from the Sun with its distance increasing at a rate of about 3.7 A.U./year. Voyager 2 will be at about 60 A.U. traveling at a slower velocity[2]. However their limited lifetime due to the on board power supplies will restrict their maximum distance from the Sun to somewhat over 100

A.U., and the instruments are likely to be operating at only partial efficiency as the available power decreases.

The "Interstellar Probe" mission will build upon the foundations established by the Voyagers and Pioneer, and undertake "exploration" and the beginning of "intensive study" for measurements of the interstellar cosmic rays. This mission is capable of attaining distances of 100-200 A.U. in 5-10 years, and 1000 A.U. in ~50 years[2]. A recent estimate of the distance to the Heliopause is ~100 A.U., with an approximate factor of two uncertainty[3,2]. Thus the measurements would be expected to extend from within the solar cavity at 1 A.U., through the termination shock and heliopause, and then into the interstellar medium. The cosmic ray modulation "boundary" would be determined by direct measurement, with the measurements of abundances and energy spectra outside this boundary being free from modulation by the solar wind, thus giving a direct, "in situ", measurement of interstellar cosmic rays. One of the major components of this mission would be the study of element abundances and energy spectra for interstellar cosmic ray nuclei with E<1 GeV/n, and of the interstellar electron spectrum below ~200 MeV/n. These measurements will address a broad range of issues including 1) the determination of acceleration effects of the solar wind termination shock on galactic cosmic ray nuclei and electrons; 2) the correlation of unmodulated galactic cosmic ray abundances with solar energetic particle (SEP), solar system (SS) , solar wind (SW), and anomalous cosmic ray (ACR) abundances; 3) the measurement of interstellar parameters to normalize cosmic ray propagation models; 4) the determination of the "primary" and "secondary" contributions to cosmic ray electrons; and 5) the determination of the cosmic ray contribution to the heating, pressure, and energy balance of the local interstellar medium[2].

Other important and closely related studies, e.g. the measurement of the anomalous cosmic rays (ACR) in their acceleration region, are described in Ref. 2.

The Isotopic Composition of Ultra Heavy Cosmic Ray Nuclei--The study of the isotopic composition of the rare nuclei in the U.H. cosmic radiation is the next important step in building on our present measurements of the light isotopes, and on the anticipated measurements of isotopes up through Nickel from the Cosmic Ray Isotope Experiment (CRIS) on ACE[4], and from the Large Isotope Spectrometer for Astromag (LISA) Experiment[5]. As shown in Fig. 1, there are no isotopic abundance measurements at present in the U.H. cosmic rays. This region of the cosmic ray spectrum has special significance in that detailed measurements will yield the first *isotopic* study of neutron capture nucleosynthesis, one of the central issues in understanding cosmic ray origin. Isotopic abundances are less subject to fractionation than elements for usual fractionation processes, and thus provide a preferred probe for nucleosynthesis studies. Fig. 2 is a plot of the relative abundance of U.H. nuclides in the solar system for the charge range from Zinc to Zirconium[6]. Each of the isotopic abundances plotted is labeled with the dominant nucleosynthesis process for that isotope: R for r-process, S for s-process, E for nuclear statistical equilibrium, and x for explosive nucleosynthesis. The most abundant isotope for each element is labeled with its mass number. For the lighter U.H. nuclei, Zn through Ge, the measured abundances will address the competition between the

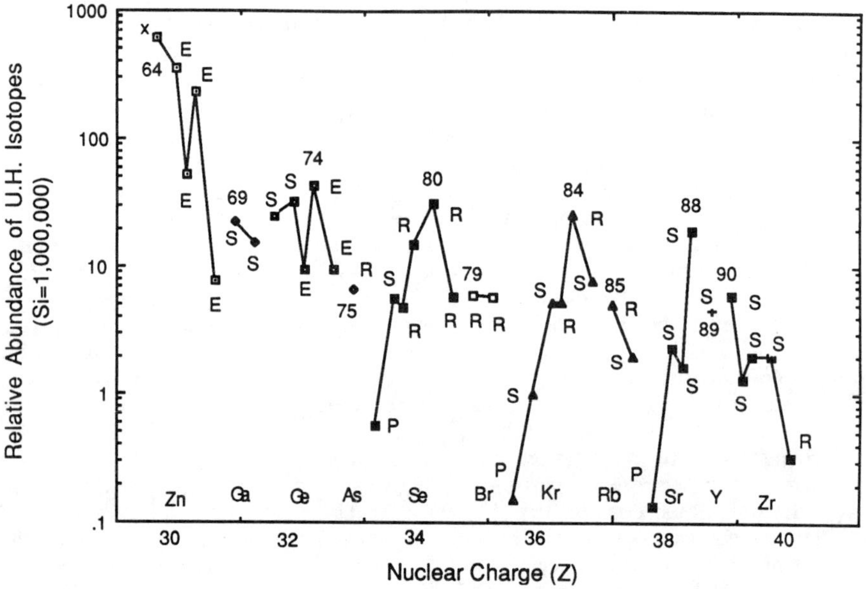

Fig. 2--Plot of the relative abundances of U.H. nuclides in the solar system[6].

equilibrium process, the s-process, and the explosive process. A study of the heavier elements, As through Zr, will give numerous measurements which should delineate the r- and s-process contributions to the cosmic ray nuclei. For example, if the abundances are similar to the solar system, r-process nucleosynthesis would be expected to dominate the production of ^{75}As, ^{77}Se, ^{78}Se, ^{80}Se, ^{82}Se, ^{79}Br, ^{81}Br, ^{83}Kr, ^{84}Kr, and ^{85}Rb, while s-process production would dominate in ^{69}Ga, ^{71}Ga, ^{70}Ge, ^{72}Ge, ^{76}Se, ^{80}Kr, ^{82}Kr, ^{86}Kr, ^{87}Rb, ^{86}Sr, ^{87}Sr, ^{88}Sr, ^{89}Y, ^{90}Zr, ^{91}Zr, ^{92}Zr, and ^{94}Zr (Refs. 7 and 8).

In addition, the Wolf-Rayet model[9] predicts that ^{67}Zn, ^{69}Ga, ^{71}Ga, ^{70}Ge, ^{80}Kr, ^{82}Kr, and ^{86}Sr could be enhanced if WR stars are significant contributors to the U.H. cosmic rays[7]. These abundances combined with those for lighter isotopes should provide a stringent test of this model.

These measurements will enable us to investigate the origin and history of the U.H. nuclei and whether it differs from that of the Z<30 nuclei. In addition, questions of acceleration and propagation of cosmic rays will be addressed from these measurements.

The best location for performing these measurements is outside of the Earths magnetosphere to maximize the number of particles detected per unit time. Thus, either an explorer class mission to the L1 libration point, or a lunar based mission would be preferred. A polar orbiter would also be possible but would only be 30-40% as efficient for the collection of low energy nuclei. In order to collect sufficient statistics for meaningful measurements a reasonable goal would be to detect 10^7 iron nuclei. Table 1 below gives the corresponding number of U.H. nuclei. The abundances were scaled from the HEAO-3 measurements of element abundances[10,11]. This represents an increase of more than two orders of magnitude in collecting power over that expected by ACE-CRIS for three years.

Element	No. of UH nuclei
Zn	5800
Ge	720
Se	490
Kr	290
Sr	300
Zr	180

TABLE 1--Approximate number of U.H. nuclei corresponding to 10^7 iron nuclei.

Two papers suggesting experimental approaches to achieve these objectives were presented.

The first of these uses a combination of multi-detector arrays of large-area silicon solid state detectors combined in a module with a scintillating optical-fiber trajectory (SOFT) detector hodoscope and is described in Ref. 7. Modules using this detector combination would be grouped to achieve the required collecting power. Silicon detectors have demonstrated the capability to resolve isotopes of nuclei up to Z=36 with an rms mass resolution of ≤0.3 amu[7]. This investigation could be conducted on an Explorer class spacecraft. It could also be lunar based, or on a polar orbiter.

A second approach[8] would utilize measurements of dE/dx vs. E in a high pressure gas ionization chamber for charge and mass determination. A balloon flight version of this instrument (the Tracking Heavy Isotope Spectrometer Telescope for Low Energies, THISTLE) which will have a geometrical factor of 0.8 m^2sr is presently being constructed. A test model of this instrument has been exposed to iron nuclei at the Bevalac and a resolution of 0.35 amu was achieved[12]. The pressure vessel weight in this approach will probably be large enough so that the mission would have to be on a Polar Orbiter or Lunar Based.

Elemental Abundances for 30≤Z≤100--The elemental abundances in the cosmic radiation for Z≥30 have been measured by experiments on the HEAO-3[10] and Ariel-6[13] spacecraft, with earlier exploration of the Z>50 region being done using plastic track detectors on high altitude balloons (See Fig. 1). As shown in Fig. 3, the abundances of even Z nuclei extending from Z=30 to 60 have been individually resolved and are mostly consistent with a source having a solar system composition with a first ionization potential enhancement. For Z>60, the statistics are low and the charge resolution is broadened in the HEAO and Ariel data (Fig. 4), making it necessary to group charges for meaningful abundance measurements. These measurements for Z>60 are consistent with an enhanced r-process at the source. For the charge region ($88 \leq Z \leq 100$), the two experiments combined give a "best estimate" for the ratio of actinides to iron of $(0.13^{+0.12}_{-0.4}) \times 10^{-6}$ (Ref. 10), an enhancement of a factor of about 3 over the solar system value[6].

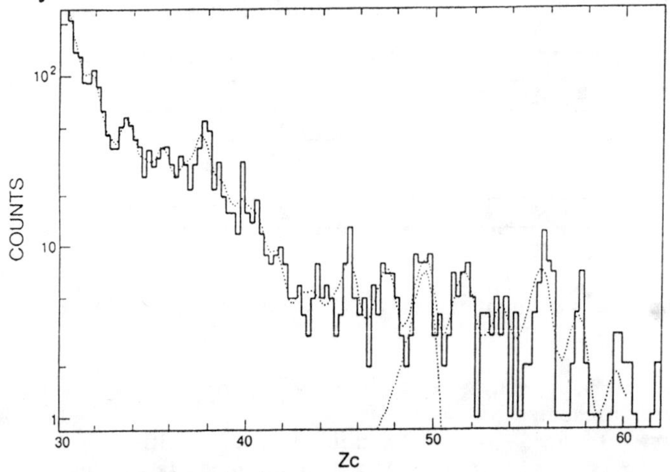

Fig. 3--Histogram of nuclei observed over the charge range $30 \leq Z \leq 62$ by the Heavy Nuclei Experiment on HEA0-3. The plotted nuclei correspond to 7.7×10^6 iron nuclei.

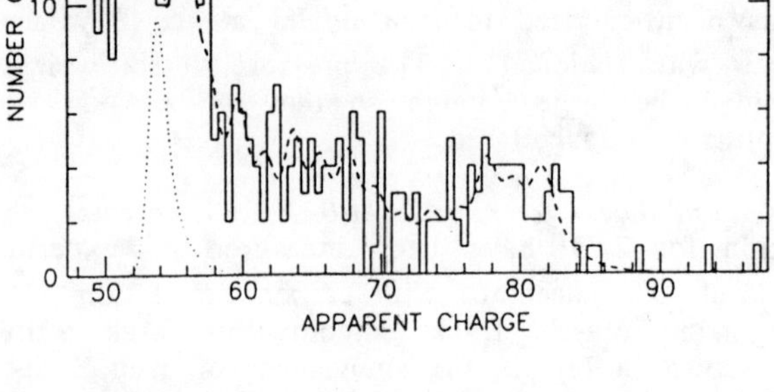

Fig.4--Histogram of nuclei observed over the charge range 50<Z<100 by Ariel-6[13]. The plotted nuclei correspond to 8.7×10^6 iron Nuclei.

There are strong scientific reasons for extending precision abundance measurements to these high charges. A high resolution study with good statistics in the actinide region would yield important new information on the relative contributions of r- and s-process nucleosynthesis and on the evolution of matter. An accurate measurement of the U/Th ratio will give a mean age of the heavy cosmic rays from the time of nucleosynthesis (Fig. 5)[14,15]. The presence of transuranics (e.g. Cm, $\tau_{1/2}$= 1.6 x 10^7 years) would be the signature of a fresh r-process component in the cosmic rays,

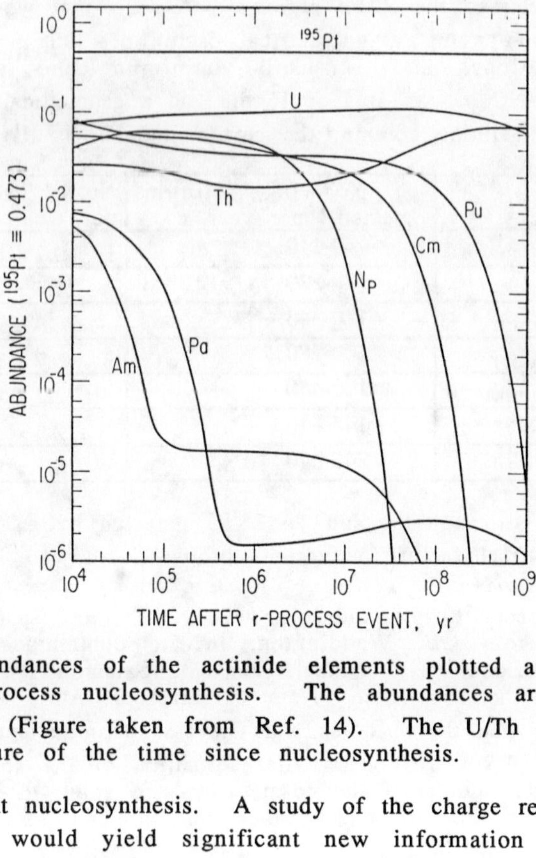

Fig.5--The abundances of the actinide elements plotted as a function of time after r-process nucleosynthesis. The abundances are normalized to ^{195}Pt=0.473. (Figure taken from Ref. 14). The U/Th ratio provides a sensitive measure of the time since nucleosynthesis.

indicating recent nucleosynthesis. A study of the charge region from $_{76}$Os through $_{82}$Pb would yield significant new information for comparison with r- and s-process production. Individual abundance measurements in the secondary region extending from Z= 62 through 74 will give information on propagation for the heaviest nuclei. High statistics measurements with excellent charge resolution in the region from Z=30 through 60 would provide significant new information. Odd Z nuclei, in particular, which are generally less abundant than even Z nuclei, and predominantly secondary, would provide new information on propagation of U.H. cosmic rays. In addition, higher precision measurements of the even Z nuclei would allow more detailed comparison with solar system, special source production of cosmic rays, and r- and s-process abundances.

Measurements in the charge range above Z=60 with a quality which matches the results at lower charge will clearly require a new generation of experiments involving very large arrays of detectors exposed in space for long periods of time. Such an experiment should have a collecting factor of at least one order magnitude larger than the Heavy Nucleus Collector (HNC) experiment, an attached payload planned for flight in the 1990's on Space Station. It is projected that HNC may detect about 40 actinides (Ref. 15 and M. Salamon, Private Communications), though this estimate has a large uncertainty. Table 2 gives the estimated number of actinides and nuclei in the Pt-Pb region (Z=76-83) that would be expected for an instrument having a geometrical factor of 300 m^2sr for various experiment locations, and for solar minimum and maximum (For comparison, the HNC has an area of 16 m^2 and a geometrical factor of about 45m^2sr). In the actinide column, the smaller number is the number

Orbit	No. of Actinides detected per year.	No. of $76 \leq Z \leq 83$ detected per year.
28°	40-110	0.7×10^4
57°, Solar Maximum	80-220	1.4×10^4
57°, Solar Minimum	120-340	2.0×10^4
90°, Solar Maximum	100-280	1.7×10^4
90°, Solar Minimum	160-450	2.7×10^4
Lunar Base Sol. Max	145-410	2.5×10^4
Lunar Base Sol. Min	240-670	4.1×10^4

Table 2--Number of actinides and $76 \leq Z \leq 83$ detected by an instrument with 300 m^2sr geometrical factor for various orbits.

of actinides expected using Anders and Grevesse[6] solar system abundances (Estimates are from C.J. Waddington, private communications, and are based on calculations by J. Adams and M. Salamon), while the larger number assumes the abundances from combined HEAO-3 and Ariel-6 results[10]. (Note that the Anders and Grevesse actinide abundances are for the present time, 4.5 x 10^9 years after formation of the solar system, and would be an underestimate if the cosmic rays are young.)

Two approaches to making these measurements were discussed at this workshop.

The first approach was to consider the assembly of a large electronic detector in space with sufficient collecting power and resolution to accomplish the above objectives. An approach to such an instrument is the Cherenkov Space Heavy Element Leviathan (C-Shell)[16] experiment. The C-Shell would consist of a large array of detector modules consisting of Cherenkov and scintillation counters. The modules would be grouped into two different sets which are separated by ~20m so that time-of-flight information can be obtained. The combination of two coincident Cherenkov signals from opposite modules with TOF measurements from the scintillators would provide the necessary information for charge identification and velocity determination. A prototype module has been constructed and tested at the Bevalac. Based on this test, the estimated charge resolution is ≤0.22 c.u[15], and it is expected that this can be improved. This array would have to be assembled in space into a configuration capable of expansion as more modules become available. The weight of each module is expected to be about 50kg. As many as 1000 modules were suggested as a possibility. The experiment would be operated for a time sufficient to achieve the experiment goals. The operation time would obviously depend upon the number of modules assembled in space and upon the actual actinide event rate. This approach has the advantage that it can cover the full range of charges from Z=30 through the actinides. However, the fact that the array would require assembly in space, and the ability to take the large mass and volume of the detectors up to the assembly area, will likely limit the number of modules that could actually be installed.

A second approach to this problem is the development of a Lunar Based Heavy Nucleus Collector (LBHNC), similar to the HNC for Space Station[15]. An area of 100 m^2 of glass track detectors would provide a geometry factor of about 280 m^2sr and would result in the detection of between about 300 to 700 actinides in a 5 year exposure (See Ref. 15 for more details). The weight of the LBHNC as it was presented is about 10,000 kg. Excellent charge resolution has been demonstrated for these detectors in accelerator tests. However this excellent resolution may be degraded to some extent in the cosmic ray environment over the long period of time required for event collection. The actual resolution of such detectors is expected to be demonstrated in the 1990's on the attached payload HNC. This approach has the disadvantage that the detectors must be recovered for analysis. In addition, they are not sensitive to elements with charge less than about Z=60. At the present time this would appear to be difficult for a lunar based experiment with its projected weight. It was also suggested that the track detectors could be processed on the moon. While this may be possible in the future, it appears unlikely that this will be possible for the period between 2000-2010.

The important point is that the science objectives are very significant, and there are at least two possible approaches for achieving these objectives. Each of these approaches requires further study and would appear to have some difficulty in implementation at the present time, but future work may resolve these problems.

Very High Energy Cosmic Ray Isotope Studies-- Second generation Astromag experiments will be important to extend cosmic ray investigations into regions of the spectrum not previously investigated, and to further utilize the capabilities of the facility. A Ring Imaging Cherenkov (RICH) counter, combined with a magnetic spectrometer to measure isotopic abundances of light isotopes at energies 10 GeV/n≤E≤40 GeV, was described[17]. This experiment would complement, for example, the expected ACE and Astromag data on the $^{10}Be/^{9}Be$ ratio, by making measurements at these high energies, and would help determine the density of matter in the propagation region. It would also measure the ^{3}He and deuterium abundances at these high energies. Since they are mainly secondaries of

^4He, their abundances relative to ^4He would allow us to study propagation, and to compare these results with the propagation of heavier nuclei. The reader is referred to Ref. 17 for further details.

OTHER POTENTIAL MISSIONS

Two other potential missions were described in this splinter group. The first was a magnet spectrometer placed on a polar orbiter to measure light element and isotope abundances, and their energy spectra in the energy range between those covered by the ACE-CRIS and Astromag Isotope Spectrometers, i.e over the energy range from about 0.6 to 2.5 GeV/n[18]. In addition antimatter could be studied in this energy range. A study of individual elements as a function of energy would provide information related to propagation and would complement both the ACE and Astromag programs. It was felt, however, that this possible mission could best be addressed after Astromag has flown.

The second mission described was an investigation of the cosmic ray distribution in our galaxy. This study would be performed by measuring the diffuse gamma ray emission resulting from protons interacting in clouds, thus creating π_0's which then decay into gamma rays. This is an interesting study which overlaps both cosmic ray and gamma ray physics in its objectives and was, in fact, presented to both groups. The instrument used in this study is a gamma ray telescope. Because of this, it was felt that it had the strongest overlap with gamma ray astronomy and that it should be considered by that discipline. For further information the reader should see Ref. 19.

LABORATORY STUDIES AND ACCELERATOR FACILITIES

The scientific significance of the observed abundances of cosmic ray nuclei depends critically on detailed knowledge of the nuclear cross sections. A significant beginning has been made in the measurement of cross sections[20,21,22] of importance in cosmic ray propagation. However, much more work is needed if full advantage is to be taken of present and future observations of cosmic ray abundances. At present, these cross section measurements can only be performed systematically at the LBL Bevalac. For more details on the present and future uses of accelerators in cosmic ray astrophysics see Ref. 22.

In addition, the development and calibration of instruments for balloon and satellite flight demands exposures to beams of ions which can only be obtained at the Bevalac. At present the running time for NASA programs for both instrument calibration and cross section measurements is inadequate and additional running time is needed for this discipline. The anticipated shut down of the Bevalac in 1994-5 will represent a serious loss to the field of cosmic ray astrophysics and it is not clear how it will be replaced. NASA should give serious consideration to obtaining considerably larger blocks of time than are now available for cross section measurements, and should begin a study of how to calibrate future generations of cosmic ray instruments.

THE BALLOON PROGRAM

A vigorous balloon program is essential to progress in this field since it is the only way that instruments can effectively be tested in the same cosmic ray environment that exists on satellite flights. In addition it is a relatively inexpensive way of attaining some important scientific objectives. Thus NASA should continue to give strong support both to the ballooning program, and to balloon payload development. For the current status of the balloon program, the reader should refer to Ref. 23.

REFERENCES

1. P. Ferrando, N. Lal, and F. B. McDonald, AIP Conference Proceedings, "Cosmic Abundances of Matter", Ed. C. Jake Waddington,**183**, 388 (1988).

2. R.A. Mewaldt and E.C. Stone, "Cosmic Ray Studies With an Interstellar Probe", Proceedings of this Workshop.

3. T.E. Holzer, To appear in Ann. Rev. Astron. Ap. (1989).

4. E.E. Stone, L.F.Burlaga, A.C. Cummings, W.C. Feldman, W.E. Frain, J. Geiss, G. Gloeckler, R.E. Gold, D. Hovestadt, S.M.Krimigis, G.M. Mason, D. McComas, R.A. Mewaldt, J.A. Simpson, T.T. von Rosenvinge, and M.E. Wiedenbeck, "The Advanced Composition Explorer", Proc. of this Workshop.

5. W.R.Binns, J. Klarmann, M.H. Israel, T.L. Garrard, R.A.Mewaldt, E.C. Stone, J.F. Ormes, R.E.Streitmatter, I.L. Rasmussen, and M.E.Wiedenbeck, "Large Isotope Spectrometer for Astromag", Proc. of this Workshop.

6. Edward Anders and Nicolas Grevesse, Geochim. et Cosmochim. Acta. **53**, 197 (1989).

7. R.A. Mewaldt, E.C. Stone, W. R. Binns, and J. Klarmann,"The Isotopic Composition of Cosmic Ray Nuclei Beyond the Iron Peak", Proceedings of this Workshop.

8. Barrett Milliken, "Experiments to Study Cosmic Ray Isotopic Composition Up to $_{40}$Zr", Proceedings of this Workshop.

9. N. Prantzos, M. Arnould, J. P. Arcoragi, and M. Casse, Proc. of 19th Intl. Cosmic Ray Conf., (La Jolla), **3**, 167, (1985).

10. W. R. Binns, T. L. Garrard, P. S. Gibner, M. H. Israel, M. P. Kertzman, J. Klarmann, B.J. Newport, E.C. Stone, and C. J. Waddington, To be published, Ap. J, Nov.15, 1989.

11. J. Klarmann, W.R.Binns, M.H. Israel, R. K. Fickle, C.J. Waddington, T.L. Garrard, and E.C. Stone, 17th Intl. Cosmic Ray Conf., **2**, 13 (1983).

12. R. A. Leske, P. Meyer, D. Rubbolo, C. Smith, and M. E. Wiedenbeck, Nucl. Instr. and Meth., **A277**, 627 (1989).

13. P.H. Fowler, R.N.F. Walker, M.R.W. Masheder, R.T. Moses, A. Worley, and A.M. Gay, Ap.J., **314**, 739 (1987).

14. J.B. Blake and D.N. Schramm, Astrophys. and Space Sci., **30**, 275 (1974).

15. M.H. Salamon, P.B. Price, and G. Tarle, "LBHNC: Lunar Based Heavy Nucleus Collector", Proceedings of this Workshop.

16. C. J. Waddington, "The C-Shell, An Active Detector of U.H. Nuclei", Proceedings of this Workshop.

17. Simon P. Swordy, "A Ring Imaging Cherenkov Counter for Astromag To Study the Isotopic Composition of Cosmic Rays", Proceedings of this workshop.

18. J. Wefel, Workshop discussion.

19. J. Adams, "Astrogam Experiment", Proceedings of this Workshop.

20. W.R.Webber, J.C. Kish, and D.A.Schrier, A series of 4 papers on cross section measurements, accepted for publication in Phys. Rev. C.

21. W.R. Binns, J.R. Cummings, T.L. Garrard, M.H. Israel, J. Klarmann, E.C. Stone, and C.J. Waddington, Phys. Rev. C, **39**, 1785 (1989).

22. T. Gregory Guzik, "Current and Future Uses of Accelerators in Particle Astrophysics",Proceedings of this Workshop.

23. W. Vernon Jones, "Status of the Balloon Program", Proceedings of this Workshop.

COSMIC RAYS ABOVE 1 TeV/n AND NEUTRINO ASTRONOMY
(Splinter Group Summary)

Michael L. Cherry
Louisiana State University, Baton Rouge, LA 70803

ABSTRACT

In order to extend space-based measurements of the cosmic ray spectrum and composition into the very interesting $10^{14}-10^{16}$ eV/nucleus region, a step-by-step program can be envisioned (in order of increasing energy) consisting of 1) the SCIN/MAGIC emulsion exposure on the Space Station, progressing through 2) a reflight of the CRN detector and 3) flight of a new, large calorimeter, and eventually leading up to the construction of 4) a very large calorimeter on the moon. The arguments for such a program are described below. The size required for a lunar calorimeter to provide reasonable statistics near 10^{16} eV is 50–150 m^2sr. Since it is not feasible to consider flying such a large detector in orbit, but since most of the calorimeter mass could be obtained locally on the moon by using lunar regolith, such a detector is a natural candidate for a lunar base.

Another candidate detector suggested for a lunar base is a neutrino telescope. Observations of neutrinos from extraterrestrial sources offer a probe with which to study astrophysical regions which are otherwise obscured, and provide information about hadronic acceleration processes. A lunar site would provide significantly less background from atmospheric neutrinos than would a terrestrial location. However, to compete with proposed earth-based neutrino detectors, the mass and area required for a lunar detector would be extremely large.

In summarizing the discussions of the High Energy and Neutrino Astronomy splinter group, I first address the question of the spectrum and composition at energies approaching the "knee" near $10^{15}-10^{16}$ eV. In order to compare detector capabilities in a systematic way, trial high energy spectra based on the CRN and JACEE measurements are used to discuss the expected event rates and required detector sizes. This leads to a suggested step-by-step program to measure spectra and composition with space-borne detectors at energies up to in excess of 10^{16} eV. Finally, the discussion of the potential of a lunar neutrino telescope is briefly summarized.

HIGH ENERGY SPECTRA AND COMPOSITION

The primary cosmic ray spectrum above 10^{12} eV/nucleus is shown schematically in Fig. 1. It appears to be a power law spectrum from 10^{12} eV up to the

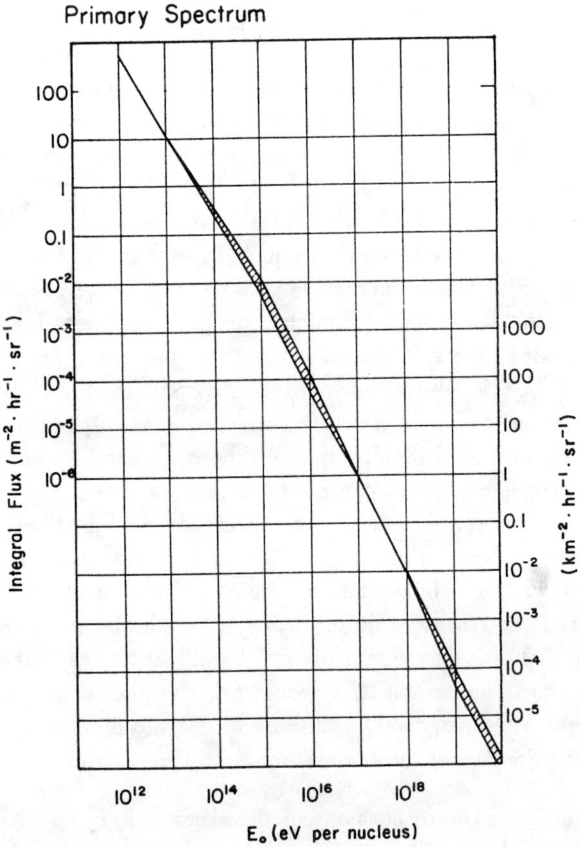

Figure 1: Integral spectrum of all primary cosmic rays at the top of the atmosphere.[1]

"knee" region around $10^{15} - 10^{16}$ eV, and then a somewhat steeper power law up to the "ankle" at energies above 10^{18} eV, where there are suggestions of a flattening. From the point of view of the experimenter, an important (and daunting) aspect of this figure is the units appropriate above the knee: the right–hand scale in Fig. 1 is in units of $km^{-2}hr^{-1}sr^{-1}$. In practical terms this means that, although direct particle–by–particle space experiments may be able to push up from lower energies to the knee, the region above the knee will remain the exclusive bailiwick of the air shower experiments for the foreseeable future.

Fig. 2 shows the results in more detail, and shows the energy range covered so far by balloon and spacecraft experiments (solid bars) together with the energies of the Astromag experiments LISA and SCIN/MAGIC (open bars). Based on the air shower data, the existence of the knee is well established experimentally, although the details of the composition in the knee region are very poorly understood.[3] Theoretically, too, the situation is not at all well defined,[4] largely because it has so far been difficult to extend acceleration mechanisms efficiently up to these energies. In addition, escape of cosmic ray particles from the Galaxy may become important somewhere near $10^{15} - 10^{16}$ eV, and a new high-energy source component may be appearing here, as well.

These effects can be expected to modify the composition, but there are also indications of changing particle physics. The JACEE group[5] has shown evidence for a population of anomalous high-p_t events in their collection of high-energy, high-Z, high-multiplicity cosmic ray nucleus-emulsion interactions (Fig. 3). Also, the underground muon observers[6] and PeV air shower experimenters[7] have suggested the presence of transient sources (Cyg X-3, Her X-1) of high energy gamma rays which produce muons as copiously as hadrons. The positive observations are still typically at the $2 - 4\sigma$ level, but if they are taken at face value, then (at least when the sources are "on") a small number of PeV sources must be major contributors to the high energy cosmic ray energy budget of the Galaxy.

In addition to the problems of measuring the composition in a region where the underlying particle physics may be changing and where the fluxes are so low, further difficulties come from the indirect nature of the air shower technique itself. In order to interpret the results, measurements made at the back end of a thick calorimeter (500 - 1000 g cm^{-2} of atmosphere) must be coupled to a complicated Monte Carlo program describing the interactions of the heavy high-energy primaries and the propagation of the shower through the atmosphere. At least some of the difficulties in interpreting the experimental results come from the disagreements between different groups' codes.

Given the inherent difficulties with the air shower technique, and both the astrophysics and particle physics interest in the cosmic ray spectrum near the knee, it is essential to push direct space–based cosmic ray observations up as close to 10^{16} eV as possible. The Grigorov et al.[8] calorimeter measurements and the JACEE emulsion exposures[9] now stop at just about the lower end of the

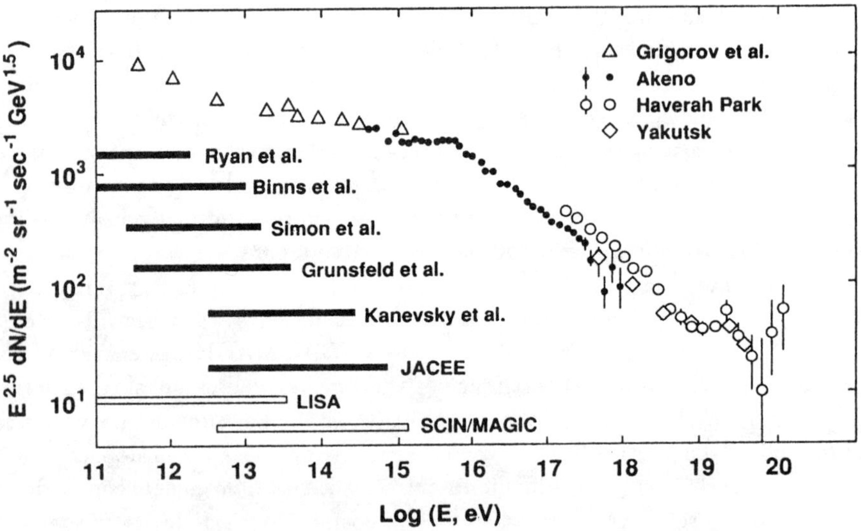

Figure 2: Differential spectrum showing results of individual measurements and the energy range of ground-based, balloon, and space-borne experiments[2].

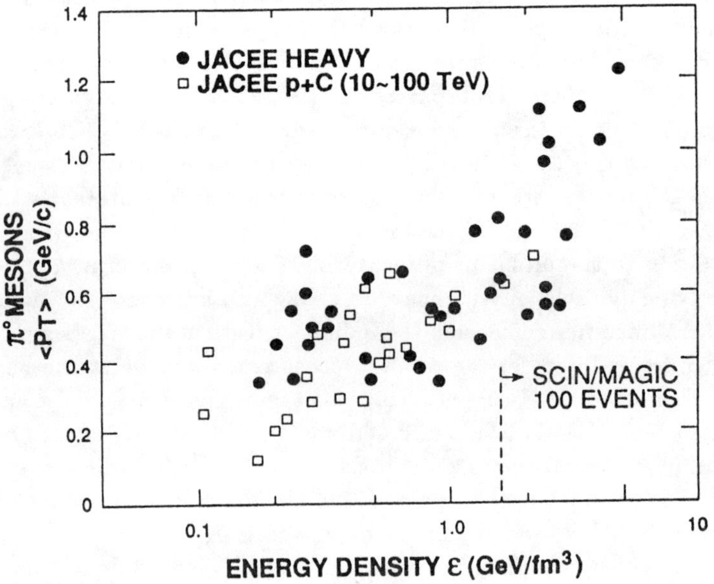

Figure 3: Transverse momentum distribution measured in JACEE emulsions, plotted against central energy density.[5]

air shower range. Larger space-based experiments capable of both answering the particle physics questions and determining the composition on a direct particle-by-particle basis up to 10^{16} eV would provide an overlap with the air shower measurements over the $10^{14} - 10^{16}$ eV energy range. Anchoring the air shower results and supplying a much-needed calibration at the lower end of their regime would then enable a much more confident application of the air shower technique to still higher energies.

Ian Axford, in his summary at the 1987 Cosmic Ray Conference in Moscow[10], noted that "the problem of the higher energy particles above the 'knee' remains. Since the spectrum is continuous it is likely that these particles are also galactic in origin ... It is evident that the main pressure from the observational point of view will come with detailed measurements of the composition spectra ... particularly in the region $10^{14} - 10^{17}$ eV. There is a need for a new large space experiment to provide this information."

ESTIMATES OF EVENT RATES

In order to estimate the detector sizes required for a realistic program to measure spectra and composition up to the knee, the JACEE[9] and CRN[11] data provide a starting point. The recent JACEE proton and helium results of Burnett et al.[9] above 10^{13} eV are shown in Fig. 4 and are used up to a rigidity of 5×10^6 GV. Following Fichtel and Linsley[12], the assumption is made that all species bend over and approach a differential spectral index of -3.05 above a common rigidity of 5×10^6 GV. The resulting proton and helium spectra, as shown in Fig. 4, are then

$$\left.\frac{dN}{dE}\right|_{\mathbf{P}} = \begin{cases} 1.75 \times 10^4 E^{-2.76} \quad m^{-2}s^{-1}sr^{-1}GeV^{-1} & for\ E < 5 \times 10^6\, GeV \\ 1.53 \times 10^6 E^{-3.05} & for\ E > 5 \times 10^6\, GeV \end{cases}$$

$$\left.\frac{dN}{dE}\right|_{\mathbf{He}} = \begin{cases} 4.20 \times 10^3 E^{-2.87} \quad m^{-2}s^{-1}sr^{-1}(GeV/n)^{-1} & for\ E < 2.5 \times 10^6\, GeV/n \\ 5.95 \times 10^4 E^{-3.05} & for\ E > 2.5 \times 10^6\, GeV/n \end{cases}$$

(It should be noted that at 1 TeV/n the measured JACEE helium points, on which the helium spectrum here is based, are about a factor of 2 higher than the extrapolation of lower energy data.)

For iron, the JACEE points are based on a very small number of events. The fit to the iron spectrum shown in Fig. 4 is therefore based on the slope derived from the Chicago CRN results and normalized to the JACEE points. Again, the spectrum is allowed to bend over at R = 5×10^6 GV. The resulting spectrum is

$$\left.\frac{dN}{dE}\right|_{\mathbf{Fe}} = \begin{cases} .137 E^{-2.55} \quad m^{-2}s^{-1}sr^{-1}(GeV/n)^{-1} & for\ E < 2.5 \times 10^6\, GeV/n \\ 217 E^{-3.05} & for\ E > 2.5 \times 10^6\, GeV/n \end{cases}$$

The all-particle spectrum of Grigorov et al.[8] and the Akeno air shower results[13] are also shown in Fig. 4 for comparison.

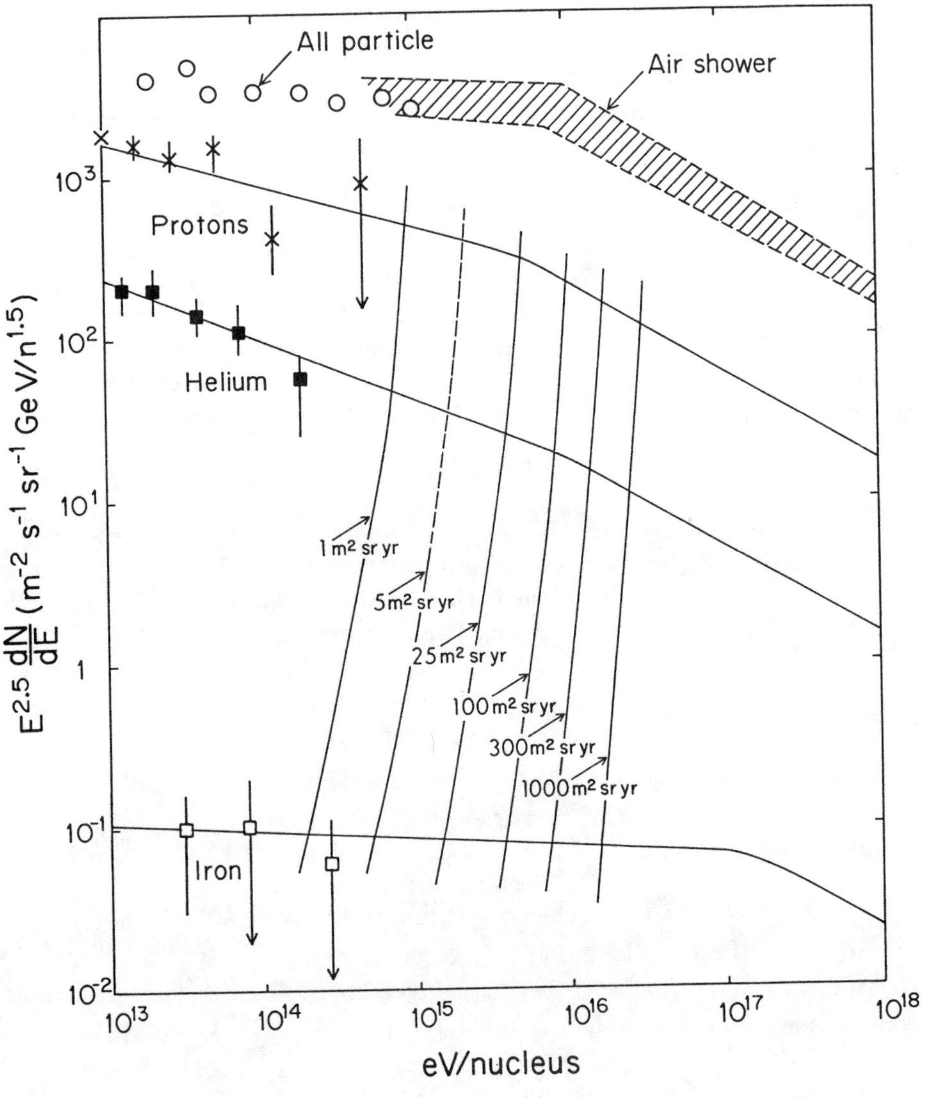

Figure 4: Assumed spectra for protons, helium, and iron based on JACEE and CRN results. The highest energy JACEE points are shown, as are the all-particle Grigorov et al. satellite calorimeter data and the region of the Akeno air shower results. E_{max} values corresponding to $N(> E_{max}) = 10$ are calculated for representative choices of $A\Omega T$.

Given these assumed spectra, we can evaluate the required collecting power $A\Omega T$ for proposed experiments. We take as a proposed figure of merit the energy E_{max} above which an experiment with a given $A\Omega T$ can still collect 10 events (i.e., we require $N(E > E_{max}) = 10$). For values of $A\Omega T$ ranging from 1 to 1000 m^2sr yrs, the resulting energies E_{max} are plotted in Fig. 4 and tabulated below.

m^2sr-yrs:	1	5	25	100	300	1000
Protons	9.2×10^5	2.3×10^6	5.2×10^6	1.0×10^7	1.8×10^7	3.2×10^7
Helium	7.4×10^5	1.8×10^6	4.1×10^6	8.4×10^6	1.4×10^7	2.6×10^7
Iron	1.8×10^5	5.2×10^5	1.5×10^6	3.6×10^6	7.2×10^6	1.6×10^7

Table 1. E_{max}(GeV/nucleus) such that $N(> E_{max}) = 10$.

A PROGRAM TO MEASURE HIGH ENERGY SPECTRA AND COMPOSITION

Table 1 suggests a possible step-by-step program to extend the existing measurements of energy spectra and composition:

- **1 m^2-sr-yr:** The SCIN/MAGIC emulsion experiment on Astromag,[2] with two 2.4 m^2sr exposures for three months each, will conduct composition and interaction studies for $Z \geq 1$ with E_{max} ranging from nearly 10^{15} eV for protons to 2×10^{14} eV for iron. The results should provide an interesting test of confinement models by measuring the Li–O abundances near 10^{14} eV, where the predictions of the closed galaxy model are significantly different from those of the leaky box model. The emulsions will also provide data on primary electrons above several TeV.

  The high spatial resolution ($\sim 1\mu$) of emulsions makes it possible to study secondary particle multiplicity, rapidity, and transverse momentum distributions, and to look for effects of spatial clustering on very fine scales. With a 3-month exposure of a single 2.4 m^2sr emulsion stack in the Astromag field, approximately 100 interactions are expected with energy density $\epsilon > 2$ GeV/fm^3. No matter what the result of a search through these data for a nuclear matter phase change at high energy density, the study of these interactions will certainly yield valuable information for the analysis and interpretation of air shower data.
- **5 m^2-sr-yr:** The CRN experiment on Spacelab 2 successfully measured spectra from boron through iron up to 2×10^3 GeV/nucleon, and could provide 5 m^2sr exposure for a 1-year flight, giving improved statistics for charges beginning at about 4. The instrument is flight-tested and well-understood. Also, the transition radiation detectors depend on an electromagnetic effect and so their calibration is not subject to uncertainties

in the nuclear or particle physics of the interaction. The combination of SCIN/MAGIC and CRN could be very useful, especially for relatively high charges.

The transition radiation detectors may need modification in order to reach higher energies, but retuning them for Lorentz factors up to at least $\gamma \sim 3 \times 10^4$ should be straightforward.[14] It should be noted, however, that for medium Z nuclei, the saturation of the transition radiators may determine the highest energy at which the spectrum can be measured. In this case, an integral point can be measured above $\sim 6Z \times 10^{13}$ eV; only for relatively high Z does the transition radiation saturate high enough so that the detector can begin to reach energies where it becomes flux-limited in a 1-year exposure. It is for this reason that the upper portion of the 5 m^{2}sr yr curve in Fig. 4 is drawn as a dashed rather than a solid line.

With a weight of approximately 2.5 tons, CRN could be flown either as a Large Attached Payload on the Space Station, on an equatorial mission, or on an available polar platform.

- **75–100 m^2-sr-yrs:** A 3–4 year exposure of a large (25 m^{2}sr) calorimeter sensitive from protons up through iron[15] would provide a major improvement over SCIN/MAGIC for Z = 1 − 5 and, for iron, a factor ~ 7 in E_{max} beyond a 1-year CRN reflight. Until a lunar base becomes feasible, a large space-based calorimeter will probably provide the only opportunity to approach the knee in the spectrum with reasonable statistics (other than with air shower arrays).

  Although calorimeters are reasonably well-understood objects, detailed study must be done to address the questions of charge resolution, albedo (which has caused major problems in previous instruments), and overall detector design. With a weight estimated to be 25 tons for 25 m^{2}sr, the instrument might be configured as two modules brought to the Space Station separately and assembled as a single Large Attached Payload. Alternately, given the demonstrated Soviet interest and experience in flying large calorimeters in space, it might be possible to arrange a joint US-USSR mission using the large lift capacity of a Soviet launch vehicle.
- **200–1000 m^2-sr-yr:** A 4–6 year exposure of an extremely large 50–150 m^{2}sr calorimeter will require a lunar base. The bulk of the calorimeter material would be compressed lunar regolith; only relatively light plastic scintillators, electronics, and support structure would be carried up from the earth. Based on the rough initial estimates by Swordy[16], a 10-layer, 3.6 m deep (10 interaction lengths) calorimeter with an area 8 m $\times$ 8 m would use 640,000 kg of lunar soil as the calorimeter material, together with $\sim$ 17,000 kg of scintillator, electronics, and mechanical support structure to be transported from the earth. The detector could be modular, so that

the pieces could be lifted from the earth and assembled on the moon in stages. (These estimates are based on scaling Swordy's initial numbers up by a factor of 4 to go to a larger area, and making the somewhat arbitrary assumption that his initial scintillator thickness can be decreased by a factor of two to 1 cm.)

PROSPECTS FOR NEUTRINO ASTRONOMY FROM A LUNAR BASE

Extraterrestrial neutrinos have been detected at low energies (below 50 MeV) from the sun and the 1987 supernova. At high energies, no sources have been clearly detected. In principle, however, the observation of such neutrinos from astronomical objects offers the potential of probing deep into otherwise obscured regions, and studying the details of high energy hadronic acceleration mechanisms. Because of the small neutrino cross section, proposed neutrino telescopes on the earth must be free of background and exceptionally large: DUMAND is to be a 20,000 m^2 (effective area) array of photomultiplier tubes 4.8 km down in the Pacific Ocean, and GRANDE is to be a 100,000 $m^2 \times$ 50 m deep detector in a fresh-water pit in Arkansas.

Shapiro and Silberberg[17] have emphasized the potential value of neutrino astronomy, and have discussed the advantages of going to the moon to eliminate the background from the earth's atmosphere. In the GeV–TeV energy range, the lunar background can be $\sim$ 3 orders of magnitude lower than on the earth[18,19]. At low energies, too, nuclear reactors provide a troublesome terrestrial background which can also be eliminated on the moon (assuming a lunar base does not have its own power reactor)[20].

Unfortunately, despite the background advantages, a lunar-based neutrino telescope will still need an enormous mass and/or collecting area in order to compete with the sensitivity of earth-based detectors[19]. Active detector masses in excess of 100 tons would need to be transported to the moon, and (at high energies) a detector with area $\sim 10^6$ m^2 would need to be constructed. At this scale, neutrino astronomers will probably have to wait for a more advanced lunar base than is presently contemplated.

REFERENCES

1. T.K. Gaisser and G.B. Yodh, Ann. Rev. Nucl. Part. Sci. **30**, 475 (1980).
2. T. Parnell, this meeting.
3. T.K. Gaisser, this meeting.
4. R. Silberberg, this meeting; R. Jokipii, this meeting.
5. T.H. Burnett et al., Phys. Rev. Lett. **57**, 3249 (1986).
6. M.L. Cherry, *Proc. 1986 Summer Study on the Physics of the SSC*, R. Donaldson and J. Marx, eds., Snowmass, p. 640 (1986).
7. J. Goodman, Nucl. Phys. B, to be published (1989); and this meeting.
8. N.L. Grigorov et al., 12th Intl. Cosmic Ray Conf., Hobart **5**, 1760 (1971); V.V. Akimov et al., Acta Phys. Suppl. **29**, 517 (1970).
9. T.H. Burnett et al., Ap. J. Lett., to be published (1989).
10. W.I. Axford, 20th Intl. Cosmic Ray Conf., Moscow **8**, 120 (1987).
11. J.M. Grunsfeld et al., Ap. J. Lett. **327**, 31 (1988); S. Swordy, this meeting.
12. C.E. Fichtel and J. Linsley, Ap. J. **300**, 474 (1986).
13. M. Nagano et al., J. Phys. **G10**, 1295 (1984).
14. T.A. Prince et al., Nucl. Instr. Meth. **123**, 231 (1975); M.L. Cherry, Phys. Rev. **D17**, 2245 (1978).
15. D. Berley and R.W. Ellsworth, this meeting.
16. S.P. Swordy, *Physics and Astrophysics from a Lunar Base*, Stanford Univ., May 1989; and this meeting.
17. M.M. Shapiro and R. Silberberg, *Lunar Bases and Space Activities of the 21st Century*, ed. by W.W. Mendell, Lunar and Planetary Inst., p. 329 (1985); and this meeting.
18. M.L. Cherry and K. Lande, *Lunar Bases and Space Activities of the 21st Century*, ed. by W.W. Mendell, Lunar and Planetary Inst., p. 335 (1985).
19. M.L. Cherry, this meeting.
20. A.K. Mann and W. Zhang, *Physics and Astrophysics from a Lunar Base*, Stanford Univ., May 1989.

Papers Contributed to Splinter Group

THE ABUNDANCES OF ULTRAHEAVY NUCLEI IN SOLAR ENERGETIC PARTICLES

Mark E. Wiedenbeck
Enrico Fermi Institute and Department of Physics
University of Chicago, Chicago, Illinois 60637 USA

ABSTRACT

Measurement of the elemental and isotopic composition of nuclei accelerated in solar flares is important both as a probe of physical conditions at the acceleration site and as a source of data on the composition of solar surface material and of the presolar nebula from which it was derived. Present data extend up to $Z \simeq 30$. In order to obtain information on the rare, "ultraheavy" (UH, $Z > 28$) elements, it will be necessary to develop new large area detectors. An attractive approach for extending isotope measurements to the heaviest nuclei involves the use of time-of-flight vs. total energy, with the flight paths required being comparable to the typical detector dimensions needed to obtain adequate event statistics. Sites suitable for carrying out these kinds of measurements, include a lunar base, a dedicated satellite outside the Earth's magnetosphere, and a polar platform in low Earth orbit.

INTRODUCTION

The elemental and isotopic composition of solar system material forms an essential part of the observational basis for our understanding of stellar nucleosynthesis. Until recently, the available composition measurements relied on analysis of samples of solid bodies—the Earth, the Moon, and meteorites—and on spectroscopic measurements of certain elements in the solar photosphere or corona.[1] Energetic nuclei accelerated in solar flares can be detected in space near the Earth, and comprise all of the stable nuclides. Breneman and Stone[2] showed that deviations of these solar energetic particle (SEP) abundances from spectroscopically-measured coronal abundances are well correlated with the charge states of the individual elements, presumably because the injection and/or acceleration processes at the Sun depend on the particles' magnetic rigidity, which is proportional to M/Q (the ratio of mass to atomic charge) at a fixed energy per nucleon. They also derived photospheric abundances by applying a correction for the observed correlation of ratios of coronal and photospheric abundances with first ionization potential (FIP). While the physical origin of this correlation is not yet known, the magnitude of the effect has been reasonably well determined empirically and can be used to obtain photospheric abundances even for elements which have not been measured spectroscopically. Thus SEP measurements provide a means for deriving coronal and photospheric abundances of a wide range of elements using a technique which is complementary to traditional approaches, and which is subject to a different set of systematic uncertainties.

To date accurate SEP elemental abundance measurements are available up to $Z \simeq 30$, with isotopic measurements only for the most abundant elements.

It is of great interest to extend this work to include nuclides with $Z \gtrsim 30$ and fully exploit this new channel of information on solar abundances.

INSTRUMENTATION FOR MEASURING UH NUCLEI IN SEP EVENTS

The natural abundances of the elements drop precipitously as one goes above $Z = 28$, due to the fact that all of these species have lower binding energy per nucleon than ^{56}Fe and therefore are only synthesized in endothermic reactions. As a result of the low fluxes of these nuclides, very large instruments are needed to collect an adequate data sample for composition studies. The recent interest in establishing a permanent manned base on the Moon with capabilities for carrying out large scientific experiments has led us to examine whether these developments could make possible the experiments required for measuring UH nuclei in SEP events. Thus scientific requirements rather than instrument size constraints were treated as the major consideration in defining the sensor approach discussed below.

In addition to requiring a large geometrical acceptance, the instrumentation must have resolution sufficient to resolve heavy isotopes. As an ultimate goal one would like to achieve mass resolution $\sigma_M \lesssim 0.2$ amu for particle masses up to $M \simeq 200$, that is $\sigma_M/M = 10^{-3}$. Detector systems used to study elements $Z \lesssim 30$ above a few MeV/nucleon have identified particle charge and mass using dE/dx vs. total energy. This technique is not readily extended to particles with significantly higher Z, since for a heavy nucleus at low energies electron attachment reduces the particle's effective charge and, therefore, the sensitivity of the dE/dx–E technique to the particle's charge and mass.

At energies below ~ 10 MeV/nucleon particle identification has been accomplished using time-of-flight (TOF) vs. total energy. This approach has the advantage that the TOF measurement yields a direct determination of the particle's velocity unaffected by details of the ionization process. However, with the time resolution presently being realized in space experiments the flight paths needed tend to be long. The possibly of deploying large structures on the lunar surface promises to significantly extend the range of applicability of the TOF vs. E technique for SEP studies by allowing the use of long flight paths.

In a TOF vs. E instrument the particle velocity, v, is obtained as $v = d/t$ where d is the flight path and t is the measured time-of-flight. Combining this result with the measured total energy, $E = \frac{1}{2}Mv^2$, one obtains the particle's mass and the corresponding fractional uncertainty as

$$M = \frac{2Et^2}{d^2} \quad \text{and} \quad \frac{\sigma_M}{M} = \left[\left(\frac{\sigma_E}{E}\right)^2 + \left(\frac{2\sigma_t}{t}\right)^2 + \left(\frac{2\sigma_d}{d}\right)^2\right]^{1/2} . \qquad (1)$$

Thus to achieve the desired mass resolution one needs to measure E to $\sim 0.1\%$, while t and d must be determined with a relative accuracy $2\times$ better than this.

If we assume a time resolution $\sigma_t \simeq 100$ psec and limit consideration to energies below 10 MeV/nucleon ($v/c = 0.15$) we find that the required flight path is approximately 10 m. In addition, if one wants to collect several thousand UH nuclei from an SEP event of moderate size, a geometrical acceptance of at least several m^2sr is needed. This requirement, together with the above flight path, determines the minimum size of the sensors which provide the start and stop signals for the TOF measurement.

Figure 1 illustrates general features of an instrument for measuring UH composition in SEP events. It consists of two sensor arrays separated by $d = 10\,m$. Detectors in the upper array provide a start time signal, while those on the bottom provide the stop signal and the total energy. In order to achieve the desired accuracy in the length of the flight path it is necessary to determine the position at which each particle penetrates the start and stop detectors to an accuracy

$$\sigma_x/d \lesssim \left(\sqrt{2}/\sin 2\theta\right)\left(\sigma_{\sec\theta}/\sec\theta\right) ,$$

where θ is the maximum angle of incidence measured from the detector normal and $\sigma_{\sec\theta}/\sec\theta$ is the maximum allowable fractional uncertainty in the flight path. For the geometry shown in Figure 1 we find $\sigma_x \lesssim 7\,mm$.

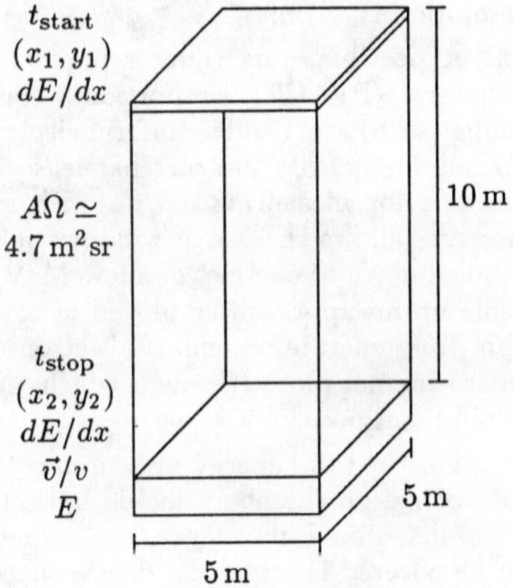

Figure 1. Sensor configuration for measuring UH nuclei in solar energetic particles.

An important requirement for a large SEP sensor system is the capability for measuring the particles of interest in the presence of a background of lighter nuclei which are much more abundant. Iron group nuclei present the most serious background for a UH experiment, since the signals which they produce are comparable to those from nuclei with $Z \simeq 30$, but their flux is more than a thousand times greater. Adequate rate capability can be achieved by using sensors with a relatively fast response and by constructing the sensor layers as mosaics of modules with independent electronics. For example, using a basic module size of $10 \times 10\,cm^2$ a sensor array covering a $25\,m^2$ area would require 2500 modules. An additional advantage of modular construction is that failure of a small number of modules will not significantly degrade the instrument performance and, if facilities for servicing are available, individual modules can be replaced.

The modules from which the start and stop signals are derived should provide both the time and the position measurements. Further reduction of the background due to accidental coincidences can be achieved by making coarse dE/dx measurements in the start and stop detectors, thereby helping identify which of several hits in the start counter actually corresponds to a UH nucleus detected in the stop counter. In addition, one (or both) of the counters could be designed to provide two sets of position measurements of higher accuracy than required for pathlength determination, with a modest separation (e.g., 10 cm) between them, in order to provide a vector indicating the particle's direction of travel with sufficient accuracy to predict which module it should have passed through in the other counter. For the dimensions indicated above the required resolution is $\sim 1\,\mathrm{mm}$.

A possible start counter implementation utilizes parallel plate avalanche counters[3] (PPAC's), proportional counters operated with a low pressure gas filling ($< 0.1\,\mathrm{atm}$) and a uniform electric field. Such counters provide fast timing signals for heavily ionizing particles, can be made position sensitive, and are suitable for measuring dE/dx with moderate resolution. The low operating pressure allows the use of windows sufficiently thin so that UH nuclei could be studied down to energies of a few MeV/nucleon. Alternatively one could greatly scale up an approach employed at lower energies: particles passing through a thin foil generate secondary electrons which are electrostatically imaged on a microchannel plate (through which the particles do not pass), providing both timing and position signals.

For the total energy measurements, arrays of silicon solid state detectors or gas ionization chambers should be suitable. The only stringent requirement for these detectors is an energy resolution of 0.1% or better. Thus although the area to be covered is large, the devices needed are relatively simple and fabrication should not present any fundamental difficulties.

PLATFORMS SUITABLE FOR UH COMPOSITION STUDIES

An instrument similar to that described above could be ideally accommodated on the lunar surface. However, other platforms would also be suitable. Possibilities include an Explorer-class satellite in an orbit with apogee $\gtrsim 20$ Earth radii, which would spend most of its time outside the Earth's magnetosphere, or a polar platform in low Earth orbit, to which low energy particles can penetrate during $\sim$ 30% of the orbit when the spacecraft is at high magnetic latitudes. For these latter two options it may be necessary to scale down some of the detector dimensions discussed above. For example, one might reduce both the flight path and the sensor areas by a factor ~ 3, thereby retaining approximately the same geometrical factor and degrading mass resolution by up to a factor of 3. Such an instrument could still make significant progress in the exploration of UH nuclei in solar flare events and, with some improvement of timing resolution, might rival the instrument envisioned for a lunar base.

REFERENCES

1. N. Grevesse and E. Anders, "Solar-system Abundances of the Elements: A New Table", in *Cosmic Abundances of Matter*, ed. C. J. Waddington (Am. Instit. of Phys., New York), 1989, pp. 1–8.
2. H. H. Breneman and E. C. Stone, *Astrophys. J. Lett.*, **299**, L57 (1985).
3. G. F. Knoll, *Radiation Detection and Measurement*, 2nd Edition (Wiley, New York), 1989.

A Ring Imaging Čerenkov Counter for Astromag To Study the Isotopic Composition of Cosmic Rays

Simon P. Swordy

Enrico Fermi Institute, University of Chicago.

LASR, 933 E56th St, Chicago, Il 60637

We describe a ring imaging Čerenkov counter, (RICH), for the Astromag facility. This counter can be used in combination with a rigidity measurement from Astromag to determine the isotopic composition of cosmic ray particles over an energy range extending from ≈ 10 GeV/amu to ≈ 100 GeV/amu. Such measurements are important to study the lifetime of cosmic rays in our galaxy, and their history of propagation. They also determine if the sources of cosmic rays are similar in composition to the Solar System, or if the contributions of more evolved or exotic matter must be included. In order to resolve isotopes at these energies, the Lorentz factor and rigidity of the particle must be measured at the few percent level The magnetic field of Astromag permits a rigidity determination to this level of accuracy. We propose to perform an accurate measurement of the Lorentz factor by imaging the cone of Čerenkov emission from the particle as it passes through a gas volume above the Astromag rigidity detector.

Introduction

The objective of the instrument presented here is to provide a means for measurement of the isotopic composition of cosmic ray nuclei at energies above 10 GeV/amu. We propose to accomplish this with an accurate velocity measuring device, RICH, which is to be combined with a rigidity analyzer using the Astromag facility. The RICH technique is being actively developed by our group at the University of Chicago for balloon measurements of cosmic ray energy spectra. This type of instrumentation is intended as a natural follow on to the presently selected Astromag experiments.

The isotopic composition and its variation with energy will provide major new insight into the origins and history of these particles in our galaxy. At these energies, the measurements are not affected by solar modulation and consequently accurately represent the particle populations of interstellar space. Further, the cross sections for fragmentation of these nuclei have become essentially energy independent at such energies. This greatly aids the interpretation of observed abundances, since propagation calculations become far more reliable and less affected by uncertainties in the cross sections, unlike the case at lower energies.

The history of the cosmic rays in the galaxy can be studied by measuring the abundances of radioactive isotopes. For example, the ^{10}Be isotope, with a half life of 1.5x10^6 years is produced entirely by the fragmentation of heavier nuclei during their passage through the galaxy. The measurements of these radioactive isotopes at high Lorentz factors, as we propose, will provide lifetimes that are comparable with the loss timescales from escape and spallation , because of relativistic time dilatation. High energy measurements should therefore definitely show an appreciable abundance of both ^{10}Be and ^{26}Al if our present understanding of cosmic ray propagation is correct.

RICH Principle.

Traditional integrating Čerenkov counters measure a signal proportional to the total amount of light L produced by an energetic particle along its path through the radiator, while imaging counters determine, in addition, the emission angle θ. An *ideal* imaging counter would require the measurement of the emission angle θ of not more than one photon with high precision in order to determine the value of γ. Thus a major advantage is its inherent immunity to statistical fluctuations in the number of photoelectrons detected. An important additional feature of imaging counters is their discrimination power against sources of background. This is due to the fact that most of these background photons will be emitted in directions very different from those of the true Čerenkov photons, and thus will not be properly focussed onto the detector image.

In these devices either a spherical mirror (for long gas radiators), or simple proximity focusing (for thin liquid or solid radiators), is used to focus the Čerenkov light produced by a particle traveling parallel to the optical axis, into a circular ring on a position sensitive photon detector. This is shown schematically in Fig. 1. The radius r of the ring on the detector is directly related to the Čerenkov angle θ. The detection of the position of photons with high accuracy has become possible in the UV region with drift chambers filled with a gas containing a small admixture of tetrakis (dimethylamine) ethylene (TMAE) vapor. The photoelectric absorption length is about 1-2 cm at the TMAE vapor pressure of about 1 torr at room temperature. These chambers have very high quantum efficiency in the wavelength range 170-220 nm. Photoelectrons produced in the chamber by UV photons are drifted to proportional counter cells on one wall of the drift chamber where they are detected.

Unlike accelerator RICH devices, the isotropic nature of the cosmic rays prevents us from aligning the optical axis of the mirror with

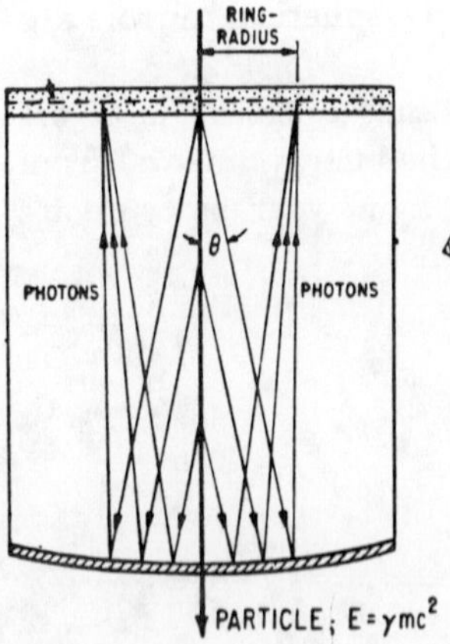

Figure 1.

RICH

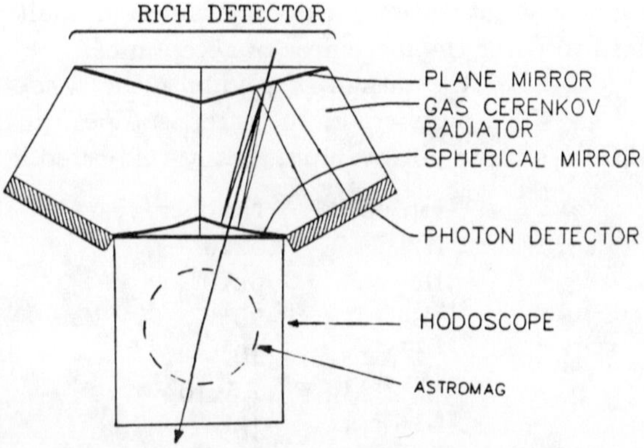

Figure 2. Possible implementation of RICH on Astromag.

the direction of incidence. For an off axis particle, the ring image becomes elliptical but is well defined and still provides a precise measure of the Čerenkov angle.

RICH on Astromag

A RICH can be combined with a rigidity measurement provided by an already existing Astromag experiment. The tracking detectors in the existing expermient will determine the particle trajectories, and the acceptance aperture will be unlikely to extend further than 30 degrees from the instrument axis.

A possible implementation of RICH with an Astromag experiment is shown in Figure 2. The hodoscope of the existing experiment will measure the particle rigidity. The RICH is low in mass, and could be located above the existing experiment as shown. This design would use a gas with a threshold Lorentz factor, $\gamma_0 = 12$ and UV-sensitive drift chambers as photon detectors. The drift chambers have been positioned out of the acceptance aperture by the use of additional, flat mirrors. By scaling from measured yields of RICH counters at accelerators, a radiator length of 1m would provide $\approx$40 photoelectrons from a singly charged particle at saturation yield. In this design, multiple spherical mirrors are used to cover the large area of acceptance.

For the design of Figure 2 the acceptance aperture is ≈ 0.1 m^2 str. The numbers of the rarer isotopes, in the energy interval from 12GeV/amu to 40Gev/amu, that are expected in a one year exposure are

Isotope	Number/year
^{2}H	$5x10^{6}$
^{3}He	$5x10^{6}$
^{10}Be	10^{4}
$^{22,20}Ne$	10^{5}
$^{24,25,26}Mg$	$1.5x10^{5}$
$^{28,29,30}Si$	10^{5}
^{26}Al	$1.2x10^{3}$

Expected results from a combination of this RICH with Astromag are shown in Figure 3. Here the rigidity of an event, as measured by Astromag is plotted against the Lorentz factor measured by RICH. The shaded regions indicate the FWHM separation of isotopes for $^{26}Al/^{27}Al$, $^{10}Be/^{9}Be$, $^{3}He/^{4}He$, and $^{2}H/^{1}H$. The maximum detectable rigidity for Astromag has been assumed to be 2TV, The position resolution in the photon detector is 1mm. Clearly, isotopes can be resolved over a wide range of energies using this technique. As has already been mentioned, the RICH device has an inherent immunity to background sources of light emission. In fact, any photon that is not travelling within a cone of angle $1/\gamma_0$ of the particle path will not fall in the ring image. This severely depresses isotropic background. The only potential source remaining is high energy knock-on electrons, closely aligned to the particle path. These knock-ons may be produced directly in the radiator or, more likely, spill into the counter from the solid material surrounding the counter. A knock-on may contribute to the RICH signal as long as it is not scattered away from the particle trajectory. For the counter described here we estimate the bakcground of photon hits generated by knock-ons to be $\approx 10^{-4}$ of the Čerenkov yield, negligible for practical purposes.

Conclusions

We conclude that the addition of a RICH is a scientifically important and technically feasible endeavour that will be an important contribution to later stages of the Astromag facility. An instrument of this type could be used to explore the isotopic composition at high energy which is unknown at present. We believe the RICH technique is the only known practical method to make reliable measurements of this type.

This instrument would seem to provide a natural follow-on experiment to the initial investigations selected for the Astromag facility.

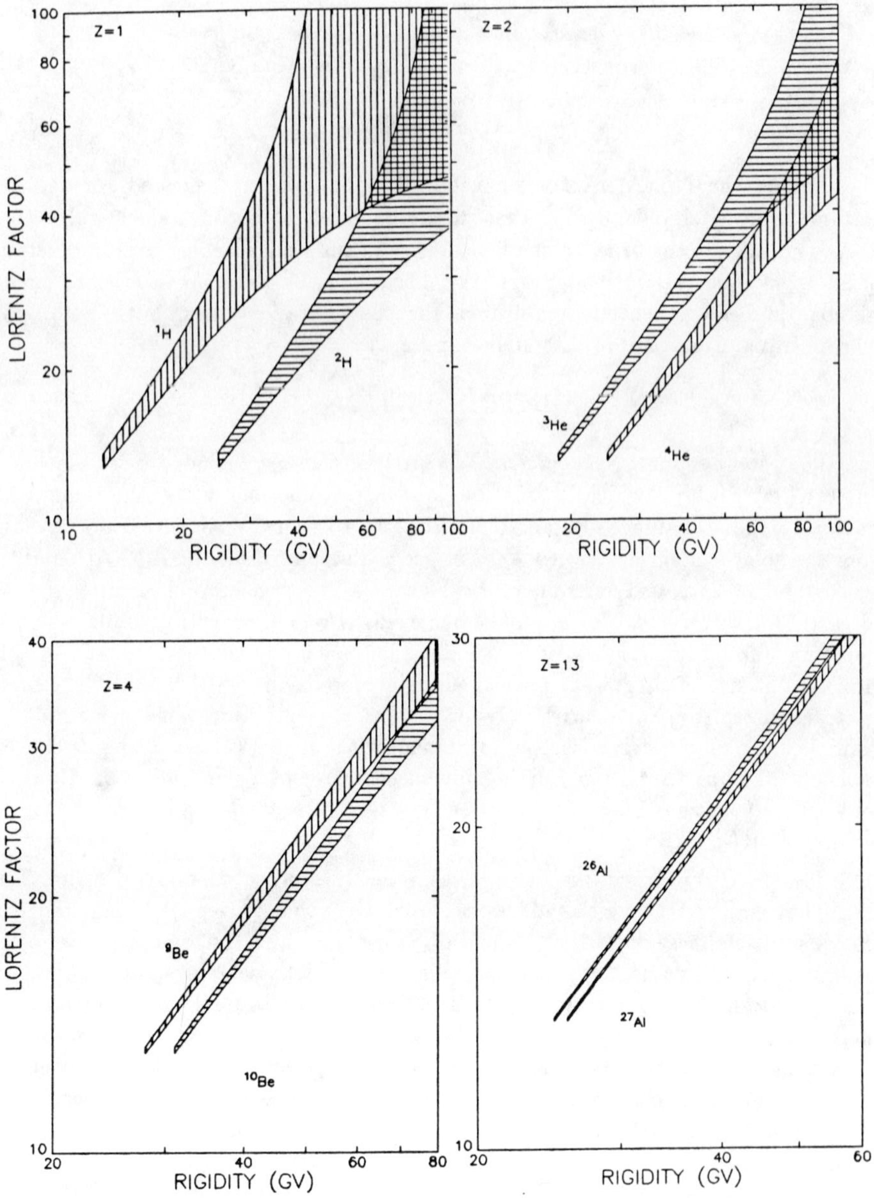

Figure 3. The calculated isotopic separations for the RICH/Astromag combination shown in Figure 2. The shaded regions indicate the FWHM of the distribution of events expected in rigidity versus Lorentz factor scatter plots.

COSMIC RAY STUDIES WITH AN INTERSTELLAR PROBE

R. A. Mewaldt and E. C. Stone
California Institute of Technology, Pasadena, CA 91125

ABSTRACT

Among the NASA mission concepts that have been suggested for the 21st century is an Interstellar Probe that might be accelerated to a velocity of ~10 to 20 AU/yr, allowing it to leave the heliosphere, ultimately reaching a radial distance of ~500 to 1000 AU in ~50 years. We discuss briefly previous studies of such a mission, and its potential significance for cosmic ray studies, both within the heliosphere, and beyond, in interstellar space.

INTRODUCTION

The recent report[1] *Space Science in the 21st Century* included discussion of an "Interstellar Probe", which might be launched sometime after the year 2000, reaching a radial distance of 100 AU within 5 to 10 years. With a velocity of escape from the solar system of ~50 to 100 km/sec a distance of 500 to 1000 AU could be reached in an extended mission of ~50 years. The acceleration required for the necessary spacecraft velocity might be achieved by one of several possible schemes[2], including Jupiter gravity assist and subsequent swingby of the Sun at ~4 solar radii, or the use of megawatt nuclear electric propulsion. Data could, if necessary, be transmitted to Earth with advanced optical communications systems. For example, a 1-m optical telescope combined with a 10-W laser transponder can transmit 20 kbps to a 10-m Earth-orbit-based telescope from 1000 AU. However, conventional systems, such as those on Voyager, are adequate for lower bit rate requirements (<1 kbps).

The Jet Propulsion Laboratory has recently studied a mission called TAU (for Thousand AU), which addresses objectives that cover a broad range of scientific disciplines, including:[3] optical astronomy and astrometry; cosmology; stellar evolution; gamma-ray astrophysics; gravitation; solar system studies; and space physics including cosmic ray studies. The strawman payload included several energetic particle and cosmic ray detectors, as well as an optical/IR telescope, UV spectrometer, VLF radio astronomy antenna, magnetometer, and instruments to detect plasma and plasma waves, dust, and γ-ray bursts. Total spacecraft plus payload masses in the range from several hundred to 1000 kg have been discussed[1,2].

HELIOSPHERIC AND INTERSTELLAR COSMIC RAY STUDIES

In our present view of the large-scale structure of the heliosphere (see, e.g., ref. 4), the solar wind (SW) flows radially outward to a termination shock, surrounded at a somewhat greater distance by a contact surface called the "heliopause", the boundary between the SW plasma and the interstellar gas.

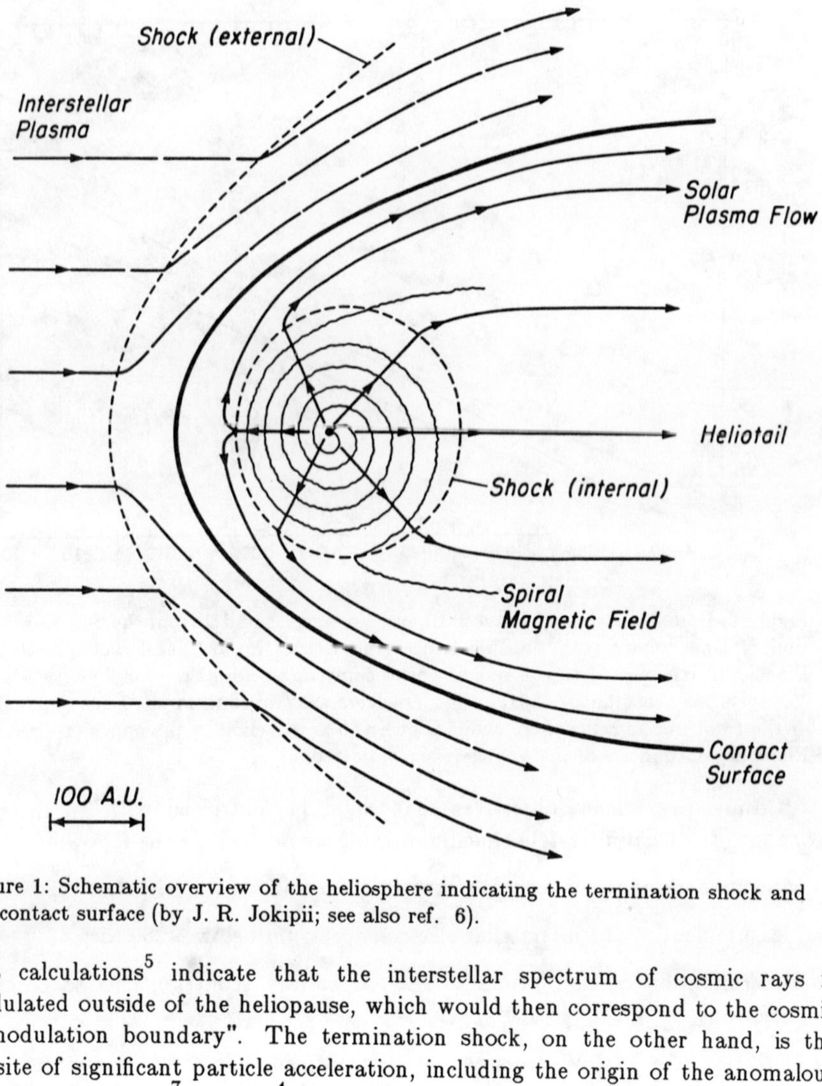

Figure 1: Schematic overview of the heliosphere indicating the termination shock and the contact surface (by J. R. Jokipii; see also ref. 6).

Recent calculations[5] indicate that the interstellar spectrum of cosmic rays is unmodulated outside of the heliopause, which would then correspond to the cosmic ray "modulation boundary". The termination shock, on the other hand, is the likely site of significant particle acceleration, including the origin of the anomalous cosmic ray component[7]. Holzer[4] has recently estimated a distance of 50 to 60 AU to the termination shock, with an estimated uncertainty of a factor of ~2. The heliopause is estimated to be approximately twice as distant as the termination shock. Some indirect estimates of the distance to the cosmic ray modulation boundary are as small as ~50 AU (e.g., refs. 8, 9).

Figure 2 shows the radial distance of Pioneer 10 and Voyagers 1 and 2 from the Sun as a function of time, along with two examples of possible trajectories for an Interstellar Probe. Note that while it is possible that one (or more) of the Voyager or Pioneer spacecraft may cross the termination shock within its operating lifetime, it is much less likely that one will cross the heliopause.

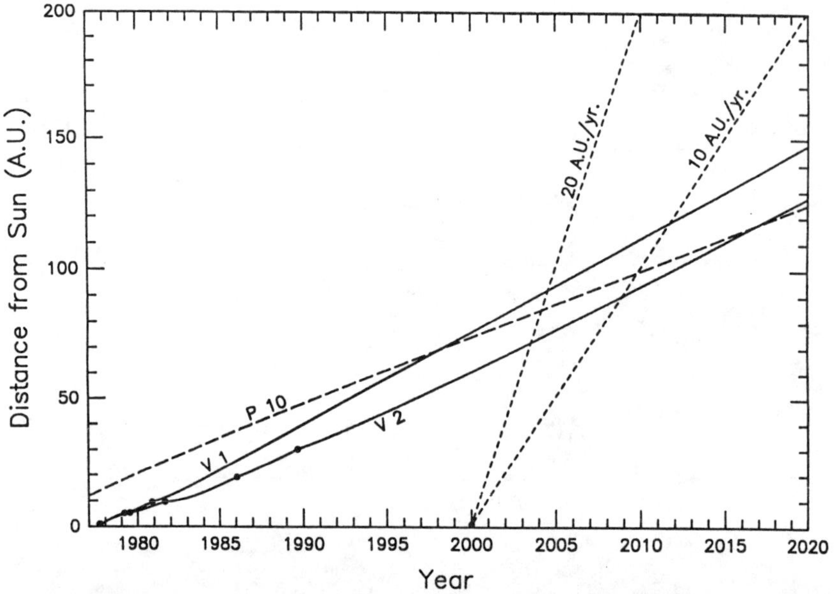

Figure 2: Radial distance of Pioneer 10 and Voyagers 1 and 2 from the Sun as a function of time, along with possible trajectories for an Interstellar Probe. Note that Pioneer 10 is expected to run out of power sometime before the year 2000, while the Voyagers may last until ~2015. The Voyagers are traveling in the general direction of the front of the heliosphere, while Pioneer-10 is traveling in the opposite direction. Planetary encounters of the Voyagers are indicated.

Cosmic ray science objectives that could be addressed with an Interstellar Probe into the local interstellar medium (ISM) would include the following:

- Measurement of the interstellar spectra of cosmic ray nuclei below ~1 GeV/nuc.
- Measurement of the interstellar electron spectrum below ~200 MeV.
- Determination of the "primary" and "secondary" contributions to cosmic ray electrons.
- Detection of ~1 MeV interstellar positrons from ^{56}Co decay in supernovae, the likely source of the galactic 0.511 MeV γ-ray line.
- *In situ* study of the acceleration of the anomalous cosmic ray component, including both seed and accelerated nuclei.
- *In situ* measurement of acceleration effects of the SW termination shock on GCR nuclei and electrons.
- Measurement of spatial gradients of galactic, solar, and interplanetary particles within the heliosphere.
- Correlation of interplanetary cosmic ray, SW, and magnetic field parameters in the outer heliosphere.

- Identification of the cosmic ray modulation "boundary".
- Measurement of local ISM abundances for comparison with SEP, SW, GCR, ACR, and spectroscopic abundances.
- Determination of the cosmic ray contributions to the "cosmic abundances" of nuclei such as ^{6}Li and ^{7}Li, of interest to early universe studies.
- Measurement of interstellar parameters to normalize cosmic ray propagation models.
- Determination of the cosmic ray contribution to the heating, pressure, and energy balance of the local ISM.
- Search for interstellar cosmic ray anisotropies due to nearby/recent supernovae.

An Interstellar Probe admittedly poses significant technical challenges, including development of the propulsion and data systems, and the need for a spacecraft and instruments (not to mention investigator teams) with ~50 year lifetimes. Because of the broad range of objectives for such a mission, it would be an obvious candidate for cooperation with NASA scientific disciplines outside of Space Physics. While a broad-based mission such as TAU^3 would clearly be a very major scientific undertaking, it is also worth considering less ambitious (and less expensive) options that address a subset of these objectives. In the end, mass and power limitations always force some compromises on instrumentation, but it would appear that an Interstellar Probe could address a large fraction of the outstanding problems in cosmic ray physics.

Acknowledgements: This work was supported in part by NASA under grant NGR 05-002-160. We thank M. D. Looper for producing Figure 2 and R. Ramaty for suggestions.

REFERENCES

1) *Space Science in the Twenty-First Century: Imperatives for the Decades 1995 to 2015,* Task Group on Solar and Space Physics, National Academy Press, Washington D.C. (1988).
2) Jaffe, L. D. et al., *An Interstellar Precursor Mission,* JPL Publication 77-70 (1977).
3) Etchegaray, Maria Ines, *Preliminary Scientific Rationale for a Voyage to a Thousand Astronomical Units,* JPL Publication 87-17 (1987).
4) Holzer, T. E., to appear in *Ann. Rev. Astron. Ap.* (1989).
5) Jokipii, J. R., as quoted in ref. 6.
6) Mewaldt, R. A., A. C. Cummings, J. H. Adams Jr., P. Evenson, W. Fillius, J. R. Jokipii, R. B. McKibben, and P. A. Robinson Jr., in *Interplanetary Particle Environment,* Joan Feynman and Stephan Gabriel editors, JPL Publication 88-28, p. 14 (1988).
7) Pesses, M. E., J. R. Jokipii and D. Eichler, *Ap. J. (Letters)* **246,** L85 (1981).
8) Randall, B. A., and J. A. Van Allen, *Geophys. Research Lett.* **13,** 628 (1986).
9) Webber, W. R., *Astr. Ap.,* **179,** 277 (1987).

THE ISOTOPIC COMPOSITION OF COSMIC RAY NUCLEI BEYOND THE IRON PEAK

R. A. Mewaldt and E. C. Stone
California Institute of Technology, Pasadena, CA 91125

W. R. Binns and J. Klarmann
Washington University, St. Louis, MO 63130

ABSTRACT

Isotope measurements of cosmic ray nuclei beyond the Fe peak are considered, using the charge region from Z=29 to Z~40 as an example. Such studies can address a number of important questions that bear on cosmic ray origin, acceleration, and propagation. One possible approach for measuring isotopes with Z≥30 is based on large-area arrays of silicon solid state detectors combined with scintillating optical fiber trajectory detectors.

INTRODUCTION

Studies of the isotopic composition of cosmic rays heavier than Fe (Z=26), including Co, Ni, and Cu (Z = 27 to 29) and the ultraheavy (UH) nuclei (Z≥30) although complicated by the relative rarity of these nuclei, can address a number of important objectives. Over the next decade we can expect missions such as the Advanced Composition Explorer[1] (ACE) to provide precision measurements of cosmic ray isotopes from H to nickel (Z = 1 to 28), and the HNC experiment[2] on the Space Station to provide greatly improved information on the composition of UH elements with Z≥60. We consider here, as an example, the significance of isotope measurements of nuclei with Z = 29 to Z~40, the next most accessible region for further isotope studies. Similar considerations apply to heavier nuclei.

COSMIC RAY NUCIEI HEAVIER THAN IRON

One basic question that can be addressed by UH isotope studies is whether cosmic ray nuclei in the upper 2/3 of the periodic table have a common origin and history with that of lighter nuclei. The nucleosynthesis of nuclei with Z>30 is dominated by the rapid and slow (r and s) neutron capture processes, but also includes contributions from equilibrium burning (e-process) and the proton capture (p) process. HEAO-3 and Ariel-VI *elemental* composition studies suggest that cosmic ray nuclei with 32≤Z≤60 contain a somewhat higher proportion of r-process material than does the solar system, while the r-process contribution to nuclei with Z>60 apparently substantially exceeds that in the solar system[3].

Such questions are much more easily addressed with *isotope* studies, since these nucleosynthesis processes do not in general produce the same isotopes of a given element. Figure 1 shows "solar system" abundances[4] of isotopes with 30≤Z≤40, including an indication of the dominant nucleosynthesis process contributing to each isotope. Note that while the production of some elements such as Ga, Se, and Sr is dominated by a single process, others, such as Zn, Ge, Kr, and Rb contain significant contributions from two processes. While there are uncertainties in these assignments, it is clear that abundances in this charge region represent a superposition of contributions from several processes (and therefore sites) of nucleosynthesis. In addition, the "Wolf-Rayet" model[5] for cosmic ray origin, designed to explain anomalies in the elemental and isotopic composition of lighter nuclei[6], also predicts a number of sizeable isotopic anomalies in these elements, including enhanced abundances of both isotopes of Cu and Ga, and of ^{67}Zn, ^{70}Ge, ^{82}Kr, and ^{86}Sr. The charge region from Z=29 to Z=40 is relatively free of "secondary" production because there are no abundant heavier nuclei, and HEAO-3 studies[7] indicate that "primaries" should account for ~70% of the observed flux of nuclei with $31 \leq Z \leq 40$. This charge region therefore presents an opportunity to isolate the contributions of several nucleosynthesis processes to cosmic ray source material.

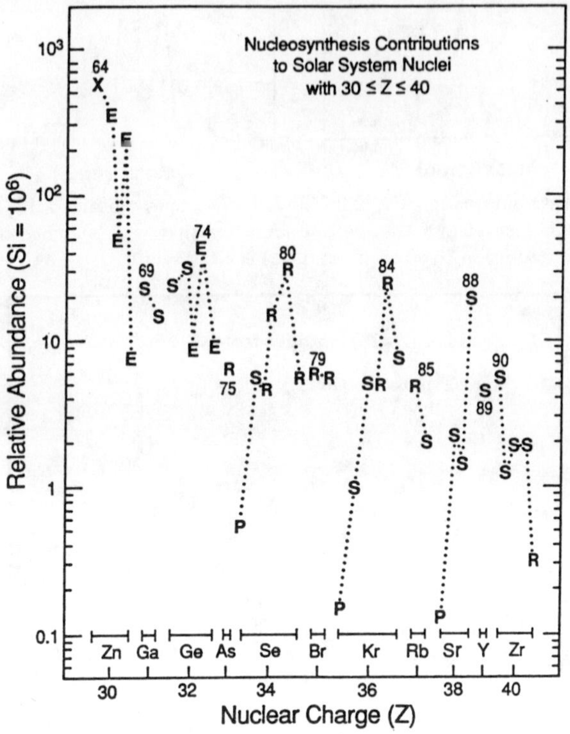

Figure 1: The relative abundances of the nuclides with $30 \leq Z \leq 40$ as compiled for "solar system" material[4] are shown on a scale where Si $= 10^6$. Each isotope is labeled with a letter representing the process that produces the largest fraction of its solar system abundance, as tabulated by Anders and Grevesse[4] [r, s, e, and p processes and explosive nucleosynthesis (x)]. The isotopes of a given element are connected by a dotted line with the mass number of the most abundant isotope labeled.

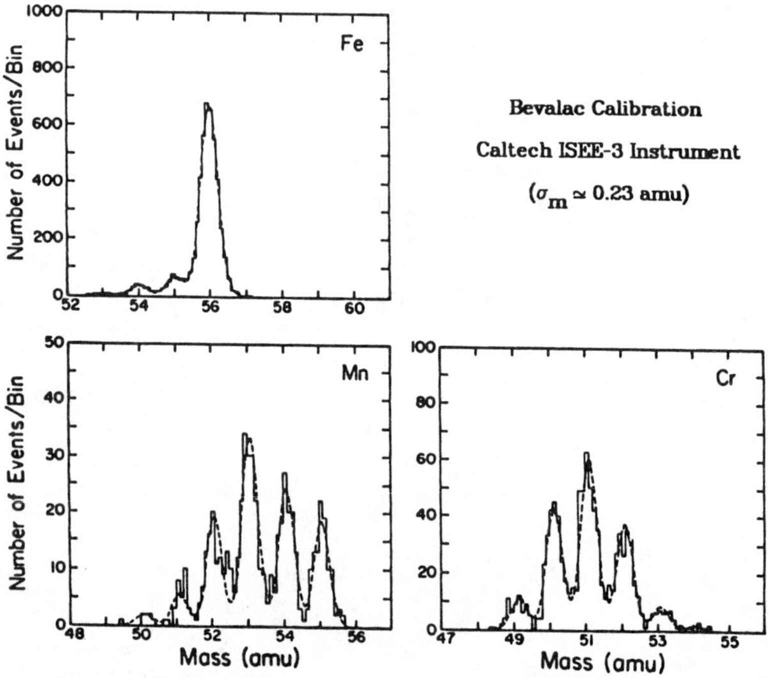

Figure 2: Mass distributions of Cr, Mn, and Fe isotopes obtained in a Bevalac calibration of the Caltech Heavy Isotope Spectrometer that was launched on ISEE-3. The observed mass resolution (dashed fit to data) is ~0.23 amu.

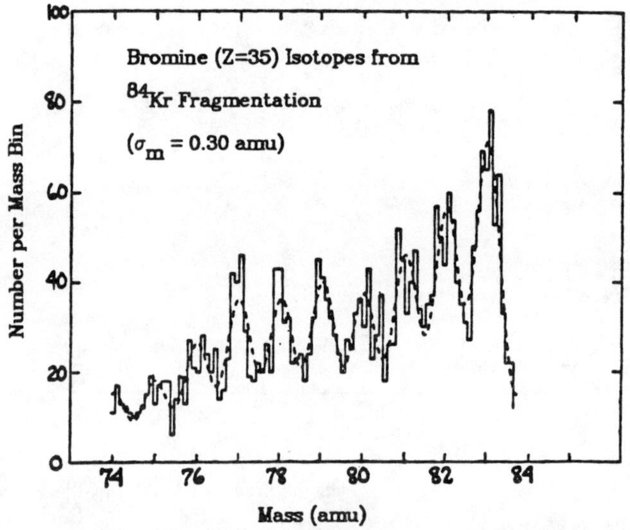

Figure 3: Mass histogram of Br isotopes obtained in a calibration of a large area solid state detector telescope at the Bevalac. The observed mass resolution of 0.30 amu (dashed fit to data) is expected to be improved when corrections for detector non-uniformity are applied.

Trans-Fe nuclei also present several new opportunities for studying cosmic ray propagation. There are pure electron capture nuclei with $31 \leq Z \leq 40$ that might be used as probes of the density of the cosmic ray propagation region(s)[8], and of possible energy-changing processes in the interplanetary and interstellar medium. In addition the element technicium (Z=43), which has no stable isotopes, has been suggested as a cosmic ray clock[9].

Isotope studies of UH nuclei require an instrument with both sufficient collecting power and good mass resolution. One possible approach is based on a combination of multi-detector arrays of large-area silicon solid state detectors combined in a module with a scintillating optical-fiber trajectory (SOFT) detector hodoscope, a concept that is under study for ACE. Silicon detectors have already demonstrated the capability to separate isotopes of nuclei up to Z = 36 with an rms mass resolution of $\leq$0.3 amu (see, e.g., Figures 2 and 3), while 200-μm thick SOFT detectors have demonstrated a position resolution of ~70 μm for Fe nuclei[10]. With a combination of these devices it should be possible to resolve isotopes at least through Z=40 over the approximate energy range from ~100 to ~700 MeV/nuc, and also abundant isotopes of heavier nuclei. The collecting power of such an instrument would depend on the number of modules, but a reasonable module size would yield $>10^6$ Fe/yr, more than 10 times that of ACE, and sufficient so that in a 3-year mission each module would give several thousand events each of Z=29 and Z=30, and >1000 nuclei total with Z=31 to 40. A test of this concept is planned for the Bevalac in 1990.

Acknowledgements: This work was supported NASA grant NGR 05-002-160 at Caltech, and by NASA grants NGR 26-008-001 and NAG 5-846 and the McDonnell Center for Space Sciences at Washington University. We appreciate the assistance of B. W. Gauld and T. L. Garrard.

REFERENCES

1) Stone, E. C., *et al.*, this volume (1989).
2) Price, P. B., this volume (1989).
3) Binns, W. R., T. L. Garrard, M. H. Israel, J. Klarmann, E. C. Stone, and C. J. Waddington, in *Cosmic Abundances of Matter,* Ed. C. J. Waddington, AIP Conference Proceedings #183, New York, p. 147 (1989).
4) Anders, E., and N. Grevesse, *Geochimica et Cosmochimica Acta* **53,** 197 (1989).
5) Prantzos, N., M Arnould, J. P. Arcoragi, and M. Casse, *Proc. 19th Internat. Cosmic Ray Conf.* (La Jolla), **3,** 167 (1985).
6) Mewaldt, R. A., in *Cosmic Abundances of Matter,* Ed. C. J. Waddington, AIP Conference Proceedings #183, New York, p. 124 (1989).
7) Garrard, T. L., private communication (1989).
8) Letaw, J. R., J. H. Adams, Jr., R. Silberberg, and C. H. Tsao, *Astrophys. Sp. Sci.* **114,** 365 (1985).
9) Drach, J., and M. H. Salamon, *Ap. J.* **319,** 237 (1987).
10) Davis, A. J., P. L. Hink, W. R. Binns, J. W. Epstein, J. J. Connell, M. H. Israel, J. Klarmann, and V. Vylet, *Nuc. Instrum. Meth.* **A276,** 347 (1989).

EXPERIMENTS TO STUDY COSMIC RAY ISOTOPIC COMPOSITION UP TO $_{40}$Zr

Barrett Milliken

Enrico Fermi Institute, The University of Chicago
933 East 56th Street, Chicago, Illinois 60637 USA

Abstract

Instruments being currently developed for balloon–borne cosmic ray isotopic composition measurements shoul lead to instruments with the precise mass resolution and large geometry factor needed to study isotopes in the range $30 \leq Z \leq 40$ with long duration exposure on a polar orbiting platform or on the moon.

Introduction. The importance of the isotopic composition of the galactic cosmic radiation for understanding a wide range of astrophysical topics, including stellar nucleosynthesis, galactic chemical evolution, and particle acceleration and transport has been discussed in detail in several recent review articles (Mewaldt 1989, Wiedenbeck 1984). Measurements that will be carried out in the coming decade promise to provide detailed information for isotopes in the range of $1 \leq Z \leq 28$. In particular, two experiments proposed for the late 1990's, the Advanced Composition Explorer (ACE) and the Large Isotope Spectrometer for Astromag (LISA) promise to provide 2 orders of magnitude more data, reducing statistical uncertainties to a few percent for the iron isotopes.

The much rarer ultraheavy elements are extremely interesting as well. Neutron capture reactions replace fusion of charged nuclei as the process responsible for the production of these elements, thus their isotopic abundances provide information on qualitatively different aspects of stellar nucleosynthesis than that addressed using abundances of the lighter elements. Because of the very small fluxes of nuclides heavier than iron (see figure 1, taken from Binns *et al.* 1989), it is clear that the generation of isotopic composition experiments following ACE and LISA must have significantly larger collection power. For example, to collect 100 $_{40}$Zr events with a full year exposure, one would have to increase the geometry factor of the cosmic ray instrument planned for ACE by about two orders of magnitude.

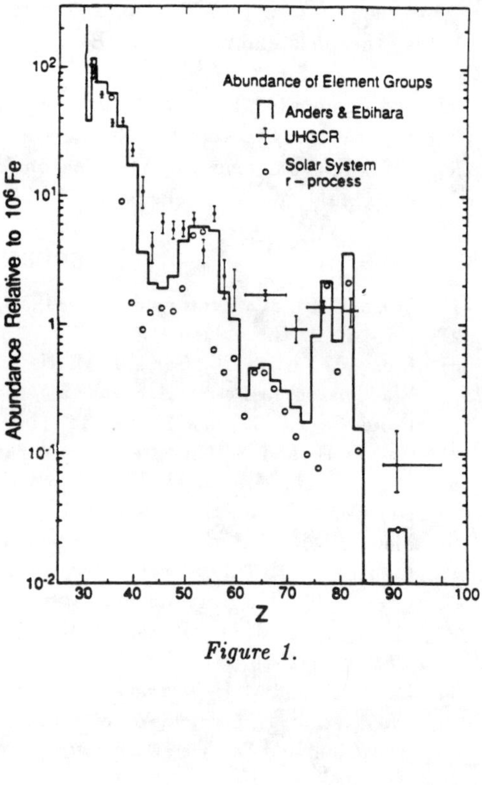

Figure 1.

There have been several instruments developed with very large geometrical factors, for use in high altitude balloon experiments (Webber *et al.* 1985, Esposito *et al.* 1985, Christian *et al.* 1987, Connell *et al.* 1987, Leske *et al.* 1989). Instruments designed to have mass resolution sufficient to resolve the iron isotopes and geometry factors 10 to 50 times that of ACE are being constructed, and should be flown in the next few years. With further developement such instruments should be able to achieve the resolution needed to resolve the ultraheavy isotopes.

For such instruments to acquire the statistics needed to study elements in the range $30 \leq Z \leq 40$ they require several year exposure times. This exposure time can be provided on a polar orbiting platform (where collection efficiency for low energy charged particles will be ~ 30–40%), or at a lunar base.

One example of a large instrument with excellent mass resolution is the Tracking Heavy Isotope Spectrometer Telescope for Low Energies (THISTLE), now being developed at the University of Chicago. The sensor system of this instrument utilizes measurements of dE/dx *vs.* E in a high pressure gas ionization chamber for charge and mass identification. Figure 2 shows the mass resolution achieved by a prototype chamber at the BEVELAC acelerator (Leske *et al.* 1989). The flight instrument now being constructed has a geometry factor of roughly 0.8 m^2 sr. The spatial uniformity and the long term stability of the gas ionization chamber technology should make it an ideal candidate for a long exposure time, large collection area instrument capable of studying the rare ultraheavy isotopes.

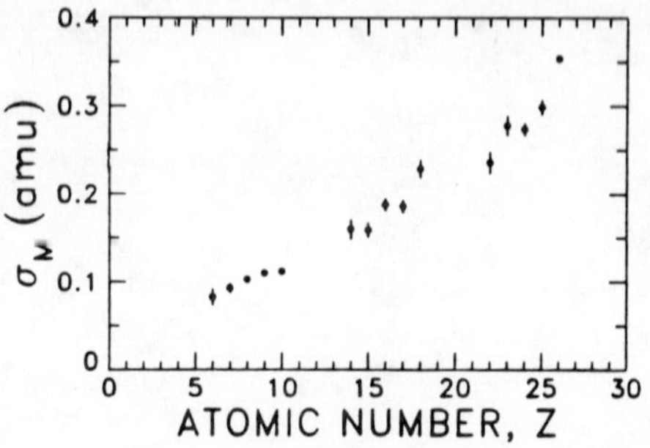

Figure 2.

Conclusion. The next step in cosmic ray isotopic composition experiments after ACE and LISA must include a significant increase in collecting power. Instruments using sensor technology currently being developed for high altitude balloon flights should be able to provide geometry factors of 1 m^2 sr or greater. Intensive developement work is now under way to achieve the necessary mass resolution. Long duration exposure of such instruments, on a polar orbiting platform or at a lunar base will be necessary to make the first measurements of the isotopic composition of ultraheavy cosmic rays.

ACKNOWLEDGEMENTS. I would like to thank M. E. Wiedenbeck for many helpful discussions. This work was supported by NASA grants NGL 14–001–005.

REFERENCES.

Wiedenbeck, M. E., *Adv. Space Res.* **4**, 15 (1984).

Mewaldt, R. A., "The Abundances of Isotopes in the Cosmic Radiation", in *Cosmic Abundances of Matter*, (C. J. Waddington, ed.), American Institute of Physics, New York (1989(, p.124.

W. R. Binns, T. L. Garrard, P. S. Gibner, M. H. Israel, M. P. Kertzman, J. Klarmann, B. J. Newport, E. C. Stone, and C. J. Waddington, "The Abundances of Ultraheavy Elements in the Cosmic Radiation: Results from HEAO–3", to be published in *Ap. J.* (1989)

W. R. Webber, J. C. Kish, and D. A. Schier, *Proc. 19th Internat. Cosmic Ray Conf.* (La Jolla) **2**, 88 (1985).

J. A. Esposito, B. S. Acharya, V. K. Balasubrahmanyan, B. G. Mauger, J. F. Ormes, R. E. Streitmatter, W. Heinrich, M. Henkel, M. Simon, and H. O. Tittel, *Proc. 19th Internat. Cosmic Ray Conf.* (La Jolla), **3**, 278 (1985).

E. R. Christian, J. E. Grove, R. A. Mewaldt, S. M. Schindler, T. Zukowski, J. C. Kish, and W. R. Webber, *Proc 19th Internat. Cosmic Ray Conf.* (Moscow) **2**,382 (1987).

J. J. Connell, J. W. Epstein, M. H. Israel, J. Klarmann, W. R. Binns, and J. C. Kish, *Proc 19th Internat. Cosmic Ray Conf.* (Moscow) **2**,398 (1987).

R. A. Leske, P. Meyer, D. Ruffolo, C. Smith, and M. E. Wiedenbeck, *Nucl. Instr. and Meth.*, **A277**, 627 (1989). 627.

CURRENT AND FUTURE USES OF ACCELERATORS IN PARTICLE ASTROPHYSICS

T. Gregory Guzik
Louisiana State University, Baton Rouge, LA 70803

ABSTRACT

Beams of artificially accelerated heavy ions, p, $\bar{p}$, e^- and e^+ currently available at (and planned for) numerous facilities around the world are a valuable resource to the Cosmic Ray community. Such beams have been used to test detector concepts, calibrate balloon-borne and space flight experiments and to measure fundamental nuclear physics parameters necessary for the interpretation of Cosmic Ray data. As new experiments are flown the quality and extent of Cosmic Ray measurements will continue to improve. It will be necessary to increase activity at ground based accelerators in order to test/calibrate these new instruments and to maintain (or possibly improve) our ability to interpret these data. In this area, the newly formed *Transport Collaboration*, supported by NASA, will be providing new nuclear interaction cross section measurements for beams with $Z \leq 58$ and supporting new instrument calibrations at the Lawrence Berkeley Laboratory Bevalac accelerator.

INTRODUCTION

Even before the 1950's, when the elemental composition of Galactic Cosmic Rays (GCR) was just beginning to emerge[1,2], it was clear that fundamental nuclear and atomic processes were critically important not only for understanding these energetic particles but also as input to simulations of new cosmic ray instrument designs and to determine corrections for instrument and atmospheric induced backgrounds. Indeed, long before ground based accelerators the "Cosmic Ray Beam" was of critical importance in the study of elementary particles[3,4]. Today, it is still crucial to understand the nuclear and atomic processes, since this understanding has lagged behind our technical ability to measure the GCR composition. The variety of nuclear/atomic physics information required is quite broad. We need cross sections as a function of energy for the secondary production of pions (which decay into gamma rays, electrons and positrons), muons, neutrinos, anti-protons (and heavier anti-nuclei), alphas, heavy-ions and fission products. For radioactive species we need the mode of decay (beta, alpha, e^- capture, fission), the branching ratio of each mode and its lifetime. We need to know the electron capture and loss cross sections as a function of energy and to understand the ionization energy loss process in solids, liquids and gases.

Most of these processes can be measured in the laboratory and much information can be obtained from sifting through the existing literature. On the one hand, it is perhaps not surprising that researchers have used simplifying approximations to, for example, secondary production cross sections or radioactive decays rather than devoting scarce resources to the considerable effort necessary for such a literature search. On the other hand, data analysis simplifications may have contributed to several "controversies" in the literature (e.g. N isotope source abundances[5,6,7,8]; Constant value for λ_e and/or PLD truncation[9,10]). As instrumentation improves, resulting in data over an extended energy/charge range with increased precision, such simplifications will limit the reliability of astrophysical conclusions.

Over the last ten years there has been some movement to improve this situation where various members of the community have summarized some of these important data (e.g. Ahlen on ionization energy loss[11]) or have provided new measurements (e.g. Crawford et al. with electron capture and loss cross sections[12], Binns et al., Guzik et al., Webber et al. with heavy ion fragmentation cross sections[13,14,15]). In addition, beam time at the Lawrence Berkeley Laboratory (LBL) Bevalac accelerator has been made available for the community to perform such measurements and/or to test and calibrate new instruments.

Recently, a NASA supported collaboration has been formed involving researchers at Goddard Space Flight Center (F. Jones), Louisiana State University (T.G. Guzik), University of California (Berkeley) - Space Science Laboratory (H.J. Crawford), University of Minnesota (C.J. Waddington) and the University of New Mexico (W.R. Webber). The principal goals of this *Transport Collaboration* are to: 1) Perform measurements of important fragmentation cross sections and study cross section systematics, 2) Improve the theoretical development of the Astrophysical context of cosmic rays, 3) Apply the knowledge gained to an extensive interpretation of the available measurements, and 4) Provide supervision of the cosmic ray instrument test/calibration beam time at the LBL Bevalac.

Since the scope of this effort is quite broad, I will focus the discussion on three topics; available accelerator resources, a summary of the current fragmentation cross section database, and a brief description of the new measurements planned by the *Transport Collaboration*.

ACCELERATOR RESOURCES

The new space missions discussed during this workshop, which hopefully will be carried out over the next decade, will push our ability to measure the charged particle radiation to higher energies and charges as well as improve the precision in determining the charge state and isotopic composition. The following generation of instrumentation will presumably move these boundaries even further. Such experiments will need to fly proven technology and have available accurate calibrations. At accelerators the charged particle beam energy, composition and trajectory can be precisely controlled allowing new detector concepts to be thoroughly tested and new flight instruments to be calibrated. For the last ~14 years about 150 to 200 hours of beam time per year has been available at the LBL Bevalac for this purpose, and the cosmic ray community has grown to depend upon this service. The Bevalac, however, does have a limited lifetime and capability, such that the community should examine the possibility of establishing similar programs at other accelerators.

A number of accelerators, such as FERMILAB, BNL, SLAC, CERN and KEK, around the world can provide beams of protons, anti-protons, neutrons, pions, electrons, positrons, and muons from a few MeV or 10's of MeV up to 1 TeV.[16] Some of these facilities have "Test Beam" areas where a detector can be exposed to a mixture of hadrons and leptons. Such exposures may be ideal for experiments, such as POEMS and/or WIZARD, to test particle discrimination capability. There are also areas where relatively pure beams could be obtained for precise instrument calibrations. The principal objective of these accelerators, however, is for the study of elementary particle physics and as a result, the newest and highest energy beams are part of collider systems. Thus, fixed target facilities, necessary for the "Cosmic Ray" work, may not be available for all beam species and energies.

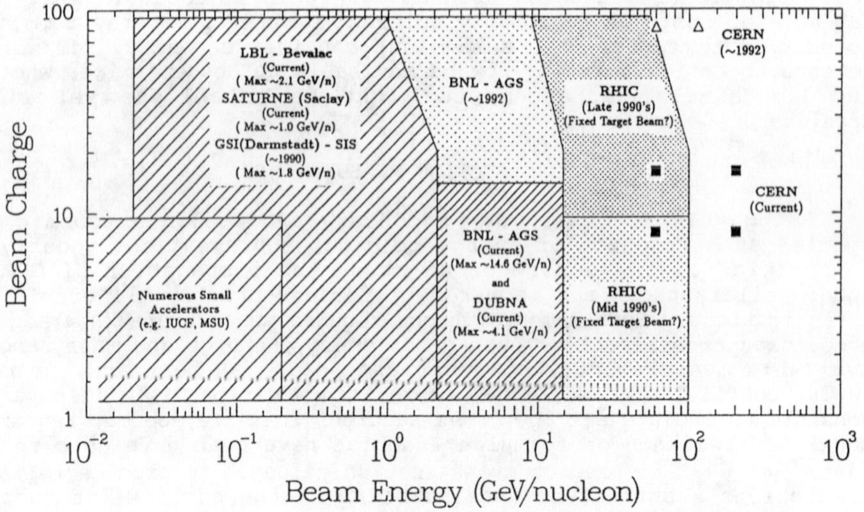

Figure 1: Current, planned and proposed heavy ion accelerator capabilities.

The beams possible at various heavy ion accelerators around the world are indicated in Figure 1. Machines which are currently available are shown by the diagonal hashing and the filled squares. Planned extensions or proposed machines are indicated by the speckled regions and the open triangles. For light ions ($Z_b < 8$) and low energy ($E_b < 200$ MeV/nucleon) various small cyclotrons are available. Other accelerators such as SATURNE, Dubna and BNL can provide beams up to ^{28}Si - ^{40}Ar. CERN has the highest energies currently available and can deliver precisely ^{16}O and ^{32}S at 60 and 200 GeV/nucleon. For higher charges the mainstay is still the LBL-Bevalac, though the highest energies attainable are somewhat less than that for the lower charges as the ultra-heavies are difficult to fully strip of electrons prior to acceleration. Shortly, the Bevalac will be joined by a machine with roughly the same capability located at GSI (Darmstadt). The SIS is expected to be online by next year and capable of accelerating up to Uranium with a maximum energy somewhat less than the Bevalac. By about 1992 the Booster ring at the BNL-AGS will be completed and this should allow that machine to accelerate nuclei at least up to Gold and possibly heavier. Also in the same time frame CERN should have completed its Lead beam. Beyond this is the proposed Relativistic Heavy Ion Collider (RHIC) and details about this machine are still being developed. It is expected, however, that RHIC may be operational with light ions up to 100 GeV/nucleon by the middle 1990's and eventually move up to heavier beams. It is uncertain, however, whether RHIC will have a fixed target beam capability, without which its usefulness for measurements of fragmentation cross sections or instrument tests at high energy is questionable.

It appears that the capability of currently available and planned ground based accelerators is sufficient to enable tests/calibrations and measurements of support data for most of the cosmic ray instrumentation planned for the next decade. While the LBL-Bevalac has been reasonably tolerant of the cosmic ray community using beam time, for what some might consider secondary (or worse) science objectives, it may now be advisable to be determined whether and how similar arrangements might be established at other accelerators.

CROSS SECTION MEASUREMENTS

Being able to unfold the secondary production of cosmic ray species as a function of energy in order to reveal the underlying composition of the source or to study the propagation of GCR in interstellar space is a requirement for essentially all cosmic ray investigations. Such deconvolutions require an accurate knowledge of production cross sections over broad energy, charge and mass ranges. Available measurements, however, are sparse at best and, in many cases, conflict or have large errors. In addition, it will never be possible to measure all the cross sections that are needed, requiring continued reliance on formulae, such as have been developed in the past,[15,17,18,19] to predict cross section values. To produce reliable predictions a sufficient understanding of the nuclear interaction process needs to be developed and new cross section measurements as a function of beam, target species and energy are necessary to improve this understanding. In this section, I will discuss the *Transport Collaboration's* effort in this area for species with $Z > 3$ and refer the reader to a number of other references for discussions of the secondary production of anti-protons,[20] positrons,[21,22] and light isotopes.[23,24]

a. The Cross Section Database

Recently a database of measured cross sections compiled from over 100 references has been established. This database, described by Crawford et al.,[25] contains cross sections from proton-induced as well as heavy ion nuclear interactions and can be used for investigations of cross section systematics, for checking the accuracy of any cross section prediction routines or for developing the nuclear interaction parameters needed as input to propagation calculations. Part of the *Transport Collaboration* effort is to maintain this database (e.g. correcting errors, providing public access) and to update it as new measurements become available.

Figure 2 illustrates the current contents of the database and, thus, indicates the extent of available measurements. The format of the figure is in terms of fragment charge (Z_f) versus beam energy (E_b) and the darkness of a square is proportional to the number of measurements in the database. One should note that the cross sections are actually functions of the beam (Z_b, A_b, E_b), target (Z_t, A_t) and fragment (Z_f, A_f), so the number of values attributed to a single cell in Figure 2 may actually be for measurements involving different beams and/or targets. Figure 3 shows a restricted subset of the database where the target is ^1H (i.e. p-induced, CH_2 - C subtraction, or actual H target) and the experiment reports values for fragment isotopes with an uncertainty less than 20%.

From the figures we can see that there appear to be fragments which are well measured and, conversely, vast regions with little or no data. Before the age of heavy ion accelerators, cross section measurements were accomplished by irradiating a target with high energy protons and then analyzing that target using mass spectrographic or radioactive decay counting methods (see review by

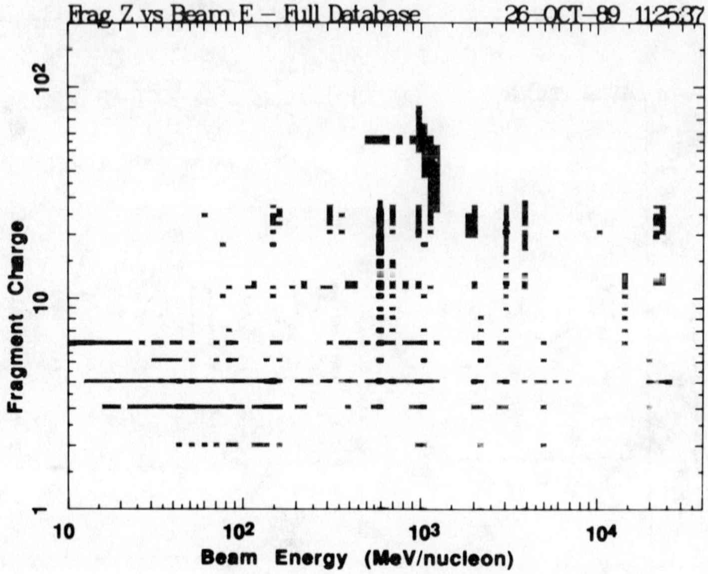

Figure 2: Fragmentation cross sections contained within the database.

Raisbeck and Yiou[26] and references in Garcia-Munoz et al.[10]). As a result, cross sections for the production of radioactive fragments such as ^7Be, ^{10}Be, ^{22}Na, ^{26}Al, and unstable isotopes of Sc - Ni account for much of the data (including those at energies around 20 to 30 GeV/nucleon). In addition, the intense interest by the cosmic ray community in using ^{10}Be as a cosmic ray chronometer resulted in the Be isotopes being well studied during the first half of the 1970's.

By 1975 the LBL Bevalac was accelerating light ions and direct measurements of the fragmentation of ^{12}C and ^{16}O started to become available.[27] This technique has several advantages. First, the incident beam can be selected to be a particular isotope. Second, the resulting fragments are observed directly and include stable as well as unstable species and, finally, the measured interaction is similar to the process occuring in interstellar space. In the intervening years, the Bevalac capability has improved such that now isotopes of Uranium can be accelerated. However, the Bevalac is devoted mainly to Nuclear Physics investigations which often do not result in cross section measurements of direct use to cosmic ray studies. As a result, there has been little improvement in the database until recently. Over the last several years Webber published values for a variety of fragment isotopes produced by beams from ^{12}C to ^{58}Ni at 600 MeV/nucleon[15] and Waddington,with colleagues, have been providing charge changing measurements of the ultra-heavy (Z>30) elements.[18,19] It is mainly because of work by these and other members of the cosmic ray community that our knowledge of fragmentation cross section results have been improved.

There remains, however, significant areas where cross section measurements are needed. For example, little data is available for

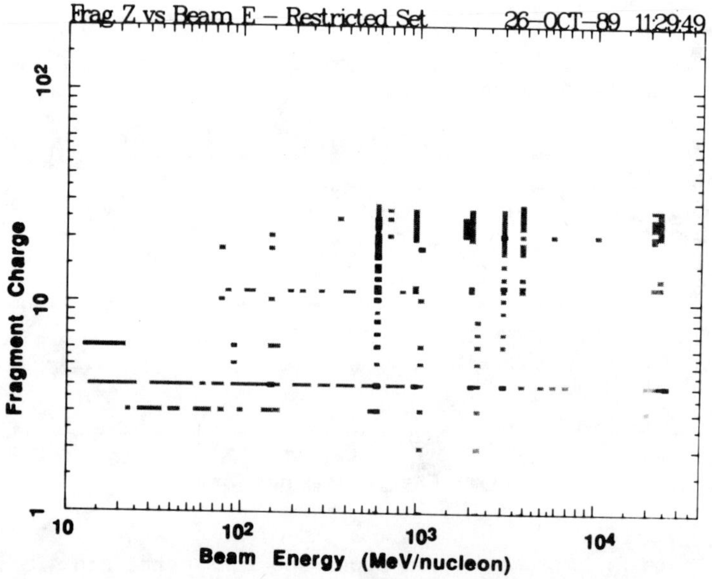

Figure 3. Restricted subset of database cross section measurements.

He and heavier targets. The He cross sections are needed for GCR propagation in a realistic interstellar media and the cross sections on heavier targets are necessary for producing a reliable scaling method that could be used in correcting atmospheric or instrument induced backgrounds. Further, the fragmentation cross sections can be energy dependent below a few GeV/nucleon. The very highest energies (< 10 GeV/nucleon) also need further investigation in order to determine whether the cross sections are indeed independent of energy.[28] In addition, beams commonly run at heavy ion accelerators may not be the isotope of interest for cosmic ray work. For example,

Argon fragmentation cross sections are typically for ^{40}Ar, rather than for the isotopes ^{36}Ar and ^{38}Ar which may dominate the cosmic ray Argon composition. Finally, no measurements in the ultra-heavy region are yet available where the isotopic composition of the secondary fragments are identified. Such measurements are important even if the cosmic ray instrument only has elemental resolution in order to correctly predict the GCR secondary component due to radioactive species. It is expected that the new measurements to be performed by the *Transport Collaboration* over the next several years will address some of these deficiencies and lead to a significant improvement in our knowledge of heavy ion fragmentation cross sections

b. Planned Cross Section Measurements

One of the immediate goals of the *Transport Collaboration* is to provide new fragmentation cross section measurements which will aid in the interpretation of Ne to Fe GCR data. These measurements will take place at the Lawrence Berkeley Laboratory (LBL) Bevalac using the Heavy Ion Spectrometer System (HISS) facility.[29] In particular, we have applied for and received 300 hours of accelerator beam time (independent of the "NASA Calibration" time) which we will use to measure ^{32}S, ^{36}Ar, ^{40}Ca and ^{56}Fe fragmentation in liquid H_2 and He targets within the energy range ~400 MeV/nucleon to 1.6 GeV/nucleon. Our first run, currently scheduled for Nov/Dec 1989, will concentrate on the low energy (~400 - 900 MeV/n) measurements and the high energy data (~900 - 1600 MeV/n) will be taken during our second run planned for May/June 1990. After the successful completion of these runs it is our intention to move on to higher charges and measure cross sections for fragments of ^{90}Zr and/or ^{140}Ce beams.

The HISS facility configuration to be used during the Fall, 1989 runs is shown in Figure 4. The beam enters the facility from the left hand side and passes through a series of detectors (TOF0, V1, UDEW, DT1, FISH, V2, E) which define the incoming beam geometry and characteristics. The beam strikes the target (TGT) and the trajectory of any fragment produced is measured (DT2) prior to entering the field of the large aperture superconducting HISS magnet.

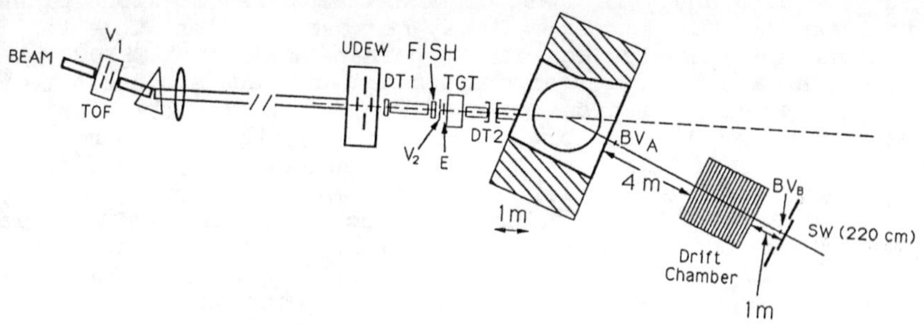

Figure 4: HISS facility configuration for fall 1989 runs.

The fragments are separated by the magnetic field and measured in detectors downstream of the magnet to determine the particles velocity, charge and final trajectory. From these measurements the particles' rigidity (R) can be determined and the particular isotope identified using the rigidity versus velocity technique. Depending upon the charge and energy range to be studied, the detectors used downstream will be somewhat different. For the Fall, 89 runs the downstream trajectory will be measured by the HISS Multiplane Drift Chamber (DC), velocity will be determined by time-of-flight and the charge will be determined from the scintillator signals in the small TOF wall (SW). During the higher energy runs, we will replace the DC with MUSIC, a multiple sampling ionization chamber, which will provide both charge identification and particle tracking and we will use a set of total-internal-reflection Cherenkov counters as the velocity measuring device (VMD). For these configurations we expect rigidity to be determined with an accuracy of $\Delta R/R < 0.01$, the time-of-flight to have an uncertainty of $\Delta t = 50 - 90$ ps and the VMD to have a velocity resolution comparable to a 20 ps resolution in a TOF system.

With this kind of performance, we expect that the uncertainty in a majority of the cross sections measured will be dominated by statistical errors. The liquid H_2 and He targets used will be thin; resulting in only 3% to 5% of the beam interacting. While this is a low yield, the thin targets greatly simplify the data analysis as multiple-scattering, multiple-interactions and energy losses in the target are reduced and the thin-target approximation can be used to determine cross sections. Background in this experiment is dominated by interactions of the beam in the detectors and air gaps, and will be assessed with target out runs and eliminated during data analysis by accepting only those fragments with a trajectory that can be traced to a single interaction point in the target and matched to a single beam particle within the accepted geometry. We will also use a beam veto (BV) detector to identify and eliminate from the data stream most of the uninteracted beam. Using these methods the beam time allocated to us will allow measurements of 1.0 mbarn cross sections to ~10% statistical accuracy.

CONCLUSIONS

Detailed knowledge of a number of nuclear and atomic physics processes and parameters will be needed from the initial development through the data analysis phase of most cosmic ray missions planned for the next several decades. It may be most efficient to establish a "Particle Astrophysics Database", perhaps under the NSSDC, where such support data is collected, maintained and made available to the community. The new missions planned would also greatly benefit by exposing prototype detectors or completed flight instruments to controlled beams of heavy ions, protons, anti-protons, electrons and positrons available at accelerators world wide. Currently, however, only the LBL-Bevalac has an established mechanism where NASA programs can obtain such beam time and developing similar procedures at other accelerators may prove necessary. Our understanding of heavy ion fragmentation continues to improve, but further measurements as a function of beam species, target and energy are still required. The *Transport Collaboration* has undertaken a program of new measurements using the HISS facility and of investigations into cross section systematics which will be of great importance to the analysis of data from future cosmic ray missions.

ACKNOWLEDGEMENTS

I would like to thank my colleagues in the *Transport Collaboration,* (H.J. Crawford, C.J. Waddington, W.R. Webber) and at LSU (J.W. Mitchell, J.P. Wefel, G.C. Sutton) for their contributions to this paper. This work has been supported in part by NASA (NAGW-1526) and the U.S. Department of Energy (DE-FG05-84ER40147).

REFERENCES

1. Freier, P. S., Lofgren, E. J., Oppenheimer, R., Dradt, H. L., and Peters, B., Phys. Rev., 74, 213 (1948).
2. Bradt, H. L. and Peters, B., Phys. Rev., 80, 943 (1950).
3. Meyer, P., "Cosmic Rays" in *Encyclopedia Britannica,* (Helen Benton, 15th Edition), 200 (1974).
4. Simpson, J. A., *Early History of Cosmic Studies,* Eds. Y. Sekido and H. Elliot (Dordrecht:Reidel), 385 (1985).
5. Guzik, T. G. , Ap. J., 244, 695 (1981).
6. Meyer, J. P. , Proc. 19th Int. Cosmic Ray Conf. (LaJolla), 9, 141 (1985).
7. Krombel, K. E. and Wiedenbeck, M. E., Ap. J., 328, 940 (1988).
8. Gupta, M. and Webber, W. R., Ap. J., 340, 1124 (1989).
9. Ormes, J. F. and Protheroe, R. J., Ap. J., 272, 756 (1983).
10. Garcia-Munoz, M., Simpson, J. A., Guzik, T. G., Wefel, J. P., and Margolis, S. H., Ap. J. Suppl., 64 269 (1987).
11. Ahlen, S. P., Rev. Mod. Phys., 52, No. 1, 121 (1980).
12. Crawford, H. J., Grenier, D. E., Heckman, H. H., Lindstrom, P. J., Wilson, L. W. and Raisbeck, G. M., LBL Report - 8807 (1979).
13. Binns, W. R., Cummings, J. R., Garrard, T. L., Israel, M. H., Klarmann, J., Stone, E. C., and Waddington, C. J., Phys. Rev. C, 39, 1785 (1989).
14. Guzik, T. G., Wefel, J. P., Crawford, H. J., Greiner, D. E., Lindstrom, P. J., Schimmerling. W., and Symons, T. J. M., Proc. 19th Int. Cosmic Ray Conf. (LaJolla), 2, 80 (1985).
15. Webber, W. R., Kish, J. C., and Schrier, D. A., Phys. Rev. C (in press) (1989).
16. Particle Data Group, "Current Experiments in Elementary Particle Physic," LBL-91, UC-34D, p. 153 (1987).
17. Silberberg, R., Tsao, C. H., and Letaw, J. R., Proc. 20th Int. Cosmic Ray Conf. (Moscow), 2, 133 (1987).
18. Binns, W. R., Garrard, T. L., Israel, M. H., Kertzmann, M. P., Klarmann, J., Stone, E. C., and Waddington, C. J., Phys. Rev. C, 36, 1870 (1987).
19. Cummings, J. R., Ph.D. Thesis, University of Minnesota, 1989.
20. Stephens, S. A., Adv. Space Res., 9, 55 (1989).
21. Kozlovsky, B., Lingenfelter, R. E. and Ramaty, R., Astrophys. J., 316, 801 (1987).
22. Protheroe, R. J., Astrophys. J., 254, 391 (1982).
23. Meyer, J. P., Astron. Astrophys. Suppl., 7, 417 (1972).
24. Beatty, J. J., Astrophys. J., 311, 425 (1986).
25. Crawford, H. J., Engelage, J., Ferrando, P., Guzik, T. G., Jones, F. C., Lindstrom, P. J., Mitchell, J. W., Soutoul, A., Waddington, C. J., Webber, W. R., Wefel, J. P., Proc. 21st Intl. Cosmic Ray Conf. (Adeliade, Australia), OG 8.3-14 (1989).

26. Raisbeck, G. M. and Yiou, F., *Spallation Nuclear Reactions and Their Applications*, Ed. B.S.P. Shen and M. Merker (Dordrecht: Reidel), 83 (1975).
27. Lindstrom, P. J., Greiner, D. E., Heckman, H. H., Cork, B. and Bieser, F. S., LBL Report, LBL-3650, (1975).
28. Waddington, C. J., Freier, P. S., Holynski, R., Jones, W. V., Jurak, A., Moon, K. H., Olszweski, A., Pruet, O. E., Szarska, M., Wefel, J. P., Wilczynska, B., Wilczynski, H., Wolter, W., Wosiek, B. and Wozniak, K., Proc. 21st Int. Cosmic Ray Conf. (Adelaide) HE1.3-15 (1989).
29. Engelage, J., Nucl. Instr. Meth., A277, 431 (1989).

DEVELOPMENT OF LONG-DURATION BALLOONING IN ANTARCTICA

W. Vernon Jones
National Aeronautics and Space Administration, Code ES
Space Physics Division, Washington, DC 20546

ABSTRACT

Following four successful long-duration test flights from Australia to South America, a new flight support system with global tracking, command, and telemetry capability is being developed to support long-duration balloon flights of relatively sophisticated instruments at both mid-latitudes and in Antarctica. The first test flight for the joint NASA-NSF program to support flights in Antarctica is scheduled from McMurdo in December 1989, with operational capability two years later.

INTRODUCTION

The traditional role of the National Aeronautics and Space Administration (NASA) Balloon Program has been to provide rapid turn-around, low cost science investigations, as well as the development of new technology and innovative instrumentation for follow-on space experiments. An important adjunct of the program has been the training of graduate students and post-doctoral researchers, so they can subsequently play leading roles in space investigations. With the apparent shortage of near-term space flight opportunities, there has been a significant trend toward ballooning becoming a recognized surrogate for space missions. This has, in turn, resulted in requests for a greater number of flights, longer duration flights, and support for more sophisticated investigations.

Part of the increased interest in balloon-borne science payloads can be attributed to the success rate of the flight program, which has experienced only two balloon failures in about 110 launches over the past two-three years. In spite of the new-found confidence accompanying this unprecedented success rate, there has been an increase in concerns about flight safety in the wake of the Challenger accident. The 1988 re-evaluation of ballooning safety resulted in even greater restrictions on launches from the National Scientific Balloon Facility (NSBF) in Palestine, Texas.[1] In particular, turnaround flights are no longer permitted from Palestine, because their calculated risk exceeds the accepted casualty expectation, 1×10^{-6}. To mitigate the impact on the flight program, Fort Sumner, New Mexico has been designated the semi-permanent NSBF site for conducting turnaround flights in the western U.S.

The development of a long-duration ballooning capability in Antarctica has been undertaken in response to these changing roles of the balloon program. Antarctica offers a unique opportunity for studies at high latitudes, such as continuous, week-long observations of solar flares during solar maximum or cosmic ray investigations requiring low geomagnetic cutoff. Obviously, there are no safety concerns associated with balloons overflying populated regions in Antarctica.

CURRENT BALLOON CAPABILITIES

The NASA balloon program currently supports approximately 50 flights per year for the astrophysics, space physics and upper atmosphere research disciplines. In recent years the program has relied on five, standard balloons that offer the nominal range of weights and altitudes shown in Fig. 1. The minimum and maximum design limits for these balloons are based on calculations of a stress-index for the ascending balloon.[2] Test flights ballasted to suspended weights outside the nominal ranges are currently being undertaken to check whether the range of standard weights can be extended to both smaller and larger weight limits. Simultaneously, a development program has been initiated to provide larger balloons that would be capable of carrying heavy payloads to higher altitudes. Figure 2 shows the nominal stress-index design limits for 40 MCF two-cap (light) and three-cap (heavy) balloons. The first successful test of a 40 MCF two-cap balloon was carried out in early November, 1989. Future studies will check the effect of reducing the standard 0.8 mil balloon film thicknesses currently in use, in order to reduce the weight of the large balloons.

PRECURSER LONG-DURATION FLIGHTS

Two long-duration campaigns, one in January-February, 1987 and the other in January-February, 1988, have tested the feasibility of long-duration balloon flights between Australia and South America.[3] Each campaign consisted of two flights, one of which had daily ballasting whereas the other utilized the unballasted RACOON (Radiation Controlled Balloon) mode. Figure 3 shows the altitude profiles for the two ballasted flights, which remained above the expected minimum nighttime altitude of 110 kft. Both of the unballasted flights experienced severe decay in the altitude profile after the first few days. The lessons learned from these test flights have helped focus the efforts for developing a reliable long-duration flight support system for global flights.

Concurrently with the 1988 NASA flights from Australia to South America, the Air Force Geophysical Laboratory (AFGL) launched an 11.8 MCF balloon from McMurdo, Antarctica.[4] The balloon traveled in a circumpolar direction nearly along the the launch latitude (~78° S) for about three days before its termination; during this time it covered almost 25% of the distance around the globe. The balloon's altitude profile, shown in Fig. 4, tended to track the Sun's declination angle. The initial float altitude was about 115 kft, with daily altitude deviations of about 5 kft; ballasting occurred only during the first day of the flight.

In January 1989, a series of three "pathfinder" flights were carried out by AFGL, in order to collect meteorological data for future ballooning activities in Antarctica.[5] These relatively small (0.25 MCF) balloons reached float altitudes of

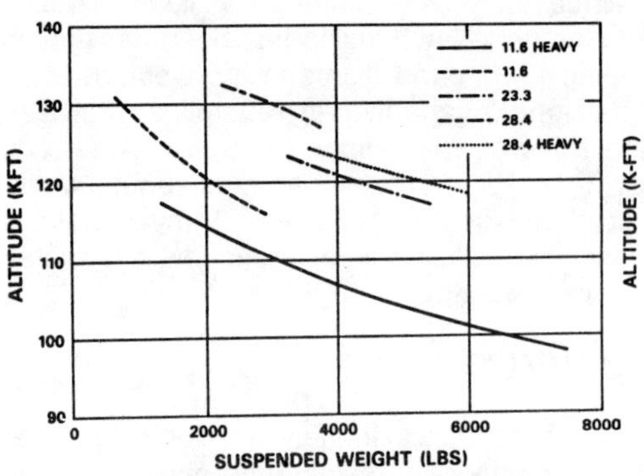

Fig. 1. Capabilities of existing balloons (Courtesy of Wallops Flight Facility).

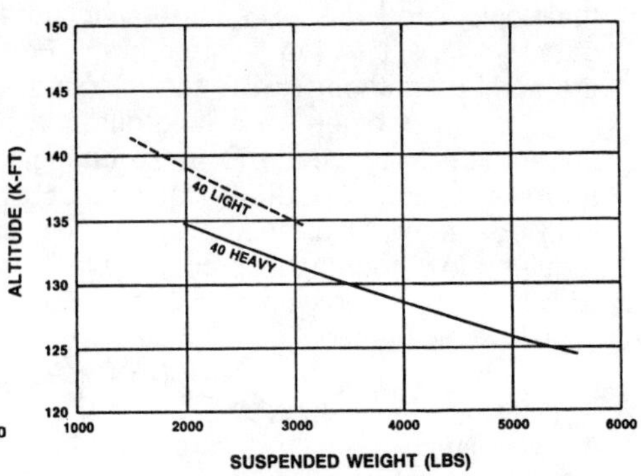

Fig. 2. Capabilities of 40 MCF balloon (Courtesy of Wallops Flight Facility).

about 125 kft, but even with ballasting the altitudes decayed unexpectedly to about 50 kft within two-three days. None of the balloon trajectories followed the 78th parallel of the launch site. It is not known whether the balloons had unusual characteristics or whether they may have encountered abnormal environmental conditions that caused them to experience random, rather than circumpolar, trajectories.

LONG-DURATION FLIGHT SUPPORT SYSTEM

The NASA objective is to develop the capability for supporting two global, long-duration campaigns each year, with at least two flights per campaign. The need for stable winds requires that the campaigns be carried out during the local summer season, so the goal is to have one southern hemisphere campaign and one northern hemisphere campaign approximately six months apart. The two prime candidates for the southern hemisphere, Australia-Brazil and Antarctica, have more-or-less common time windows. Currently, there is no capability in place for northern hemisphere long-duration flights, although candidate routes frequently mentioned are U.S.-China and Sicily-U.S.

The systems needed for Antarctica and mid-latitudes are similar, although there are some crucial differences, e.g., the satellites used for telecommunications would generally be different. For that reason, a modular system is being developed.[3] The generic flight support system, shown schematically in Fig. 5, consists of the support instrumentation package (SIP) with electronics for controlling the flight, antennas for bi-directional telecommunications, solar panels with rechargeable batteries for power, and a crude (~ 5°) pointing capability to ensure that the solar panels will function adequately. If the scientific payload employs a pointing gondola, the pointing capability can generally be met with the user-provided pointing system.

NASA-NSF ANTARCTIC TEST FLIGHTS

The Antarctic balloon program is a joint effort with the National Science Foundation (NSF), and potential users are expected to submit proposals to both NASA and the NSF. Selections for flight will be made jointly by NASA and NSF.

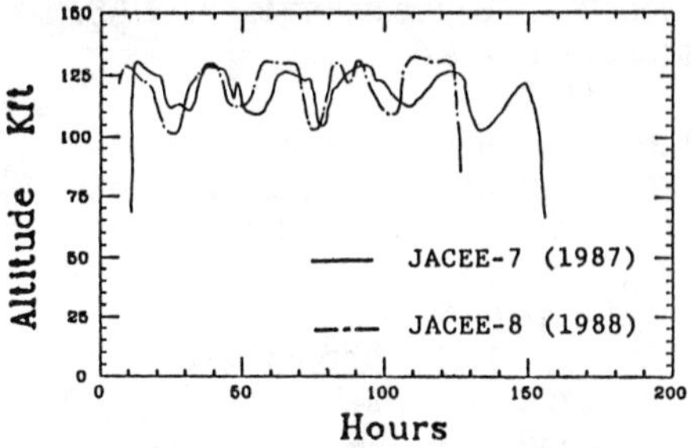

Fig. 3. Altitude profile of ballasted flights from Australia to South America (Courtesy of J. P. Wefel).

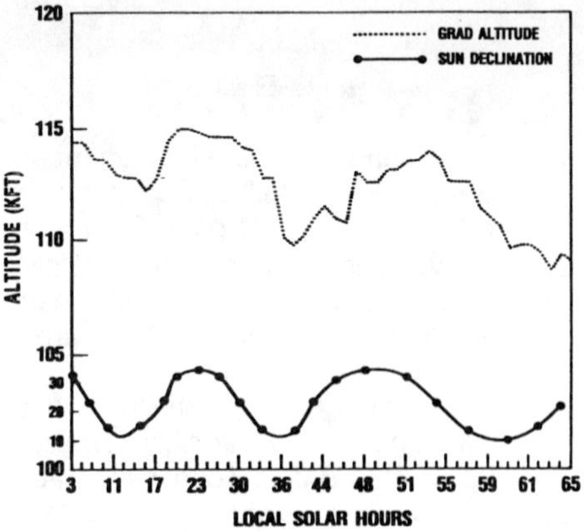

Fig. 4. Altitude of Antarctic flight with Sun declination angle at bottom (Courtesy of J. Grounds).

The first test campaign to Antarctica in December 1989-January 1990 will focus on tests of the long-duration subsystems and studies of the wind circulation patterns. The latter will involve launching four small (~0.25 MCF) pathfinder balloons, in addition to the 28 MCF balloon used for the primary flight. The campaign will check the Antarctic launch technique and, hopefully, demonstrate the ability to circumnavigate the globe at about 78° S latitude. A second test campaign with the first full-up flight test of the long-duration SIP is planned for the December 1990-January 1991 launch window. An operational capability is expected to be in place by the December 1991-January 1992 season.

CONCLUDING REMARKS

The NASA Balloon Program is a valuable asset for the scientific community, not only for the scientific data it provides, but also for the rapid turnaround flight opportunities it provides young researchers. The latter will be especially important in developing the careers of young scientists involved in the major cosmic physics missions selected for the 1990's and beyond. The Antarctic capability being developed jointly with the NSF is a significant enhancement to the program, as is the capability for long-duration flights at mid-latitudes. These new capabilities help mitigate the effective loss of the Shuttle/Spacelab program to the observing sciences for ton-class payloads for week-long observations in a space-like environment.

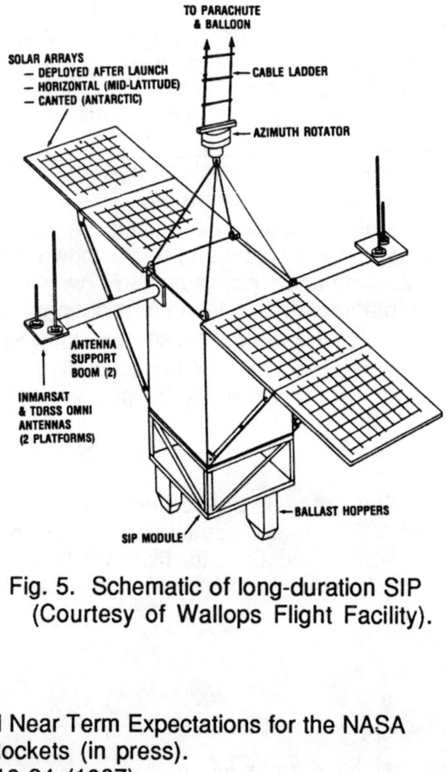

Fig. 5. Schematic of long-duration SIP (Courtesy of Wallops Flight Facility).

REFERENCES

1. W. V. Jones, "Recent Developments and Near Term Expectations for the NASA Balloon Program," J. Spacecraft and Rockets (in press).
2. I. S. Smith, Adv. Space Research 7, 19-24 (1987).
3. W. V. Jones, "Long-Duration Ballooning at Mid-Latitudes and in Antarctica," (Proceedings of the Astrophysics in Antarctica Conference, Bartol Research Institute, University of Delaware, June 8-10, 1989), AIP Conference Proceedings (in press).
4. J. Ground, "A Large Balloon, Antarctica, and Supernova 1987A," in *Environmental Research Papers, No. 1015*, pp. 1-38; Preprint AFGL-TR-0265, Air Force Geophysical Laboratory, Hanscom AFB, MA 01731.
5. J. Ground, Air Force Geophysical Laboratory, Hanscom AFB, MA 01731 (private communication).

LBHNC: A LUNAR-BASED HEAVY NUCLEUS DETECTOR

M.H. Salamon
Physics Department, University of Utah, Salt Lake City, UT 84112

P.B. Price
Physics Department, University of California, Berkeley, CA 94720

G. Tarlé
Physics Department, University of Michigan, Ann Arbor, MI 48109

ABSTRACT

A passive, large-area experiment for the detection of cosmic ray actinides on the lunar surface is discussed. Due to the absence of a geomagnetic cutoff, a 100 m^2 array of nuclear-track-detecting glass plates in 5 years will detect ~300-1000 U and Th cosmic ray nuclei of energies > 0.85 GeV/u (compared to the present world's total of 4 actinides). With a charge resolution at uranium of ~0.25e, the U/Th ratio can be accurately determined, thereby dating the r-process component of the cosmic rays; the presence of a fresh r-process component would be corroborated by the likely detection of transuranics as well. In addition, abundances in the Pt/Pb and sub-Pt/Pb regions and abundances of secondary actinides would provide detailed data on the 0-1 g/cm^2 region of the cosmic ray path length distribution, hence on the astrophysical site of origin of these cosmic rays. Finally, should a fresh r-process component exist, the detection of postulated superheavy nuclei is conceivable. With an analysis station at the Lunar Base, glass plates could periodically be harvested, analyzed, annealed/remelted, and replaced onto the lunar surface.

INTRODUCTION

Data from the Danish-French instrument[1] and the Heavy Nuclei Experiment[2] on board HEAO-3, and the Ariel 6 satellite[3] show that for $Z < 60$ the galactic cosmic ray (CR) elemental distribution is very similar to that of the solar system (SS), after the effects of cosmic ray propagation have been taken into account[4]. However, for $Z > 60$, data from the Heavy Nuclei Experiment and Ariel-6 instruments suggest that there may be a significant enhancement of r-process elements in the ultra-heavy component of the CR's[2], raising the possibility that a fraction of this component may be freshly synthesized nuclei from recent supernovae, accelerated to CR energies by the supernovae shocks.

A high-statistics measurement of the CR actinide abundances would definitively answer this question. Figure 1 (taken from Ref. 5) shows the abundances of the actinide elements from Z=90 to 96 as a function of time after explosive nucleosynthesis. Since the propagation time of CR actinides

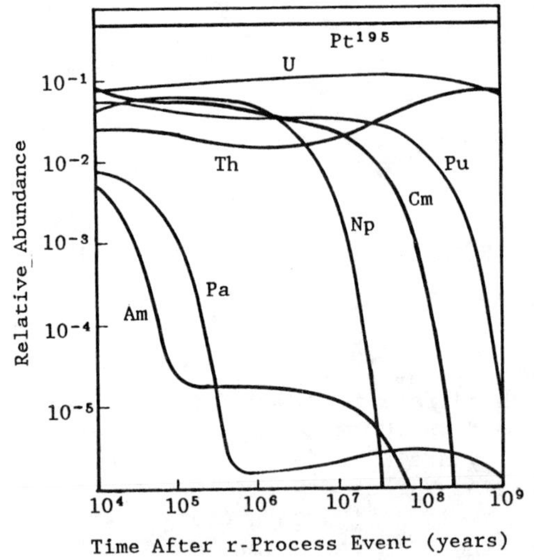

Fig. 1

should be less than 10^7 yr (nuclear interactions reduce the lifetime from the CR lifetime of ~10^7 yr derived from ^{10}Be data), the CR U/Th ratio is an excellent indicator of CR age since nucleosynthesis; the measurement of several hundred U + Th CR's would give a U/Th ratio to ~10% accuracy, sufficient to determine the epoch of CR nucleosynthesis. In addition, the observation of short-lived transuranics such as Cm and Pu would be an unambiguous signature of a fresh r-process enhancement of the CR's, and would provide strong evidence for the role of shocks in CR acceleration.

A detector that is also capable of cleanly separating even and odd charges in the (Pt to Pb)/sub(Pt to Pb) region and beyond would dramatically improve our knowledge of the CR path length distribution (PLD) in the critical region of 0-1 g/cm^2, thereby yielding information on the astrophysical sites of origin of the CR's. To simultaneously fit CR secondary/primary ratios B/C and sub-Fe/Fe, an energy-dependent deficit in short (0-1 g/cm^2) pathlengths in the CR PLD is required[6], supporting the picture of nested leaky box sources, such as OB associations in dense molecular clouds. The secondaries of Pt-Pb and U-Th are ideally suited to study this region of the PLD because their mean nuclear interaction length is of the order of 1 g/cm^2 in hydrogen. An extraordinary charge resolution is required for this purpose: < 0.3e at a Z of ~90.

Finally, a high-statistics measurement of the ultraheavy CR's could conceivably result in the first detection of superheavy nuclei (Z > 100). Should an island of nuclear stability exist beyond the transuranics with lifetimes of > 10^7 yr, then a fresh r-process CR component might carry with itself the only extant superheavy nuclei accessible to us.

At the present time, the total number of observed CR actinides is no more than four, detected with satellite instruments[2,3] of charge resolution > one charge unit at uranium. A lunar base, by comparison, would allow the deployment of an array capable of detecting several hundred actinides with a charge resolution of ~0.25e.

LBHNC

The recent discovery[7] of the extraordinary properties of certain phosphate glasses as nuclear track detectors has made possible the construction of an extremely simple and powerful detector of ultraheavy CR's of Z>60. Schott's BP-1[8] phosphate glass has remarkable immunity to many of the problems inherent in a long-duration space experiment involving nuclear track detectors. These include fading of latent tracks over exposure times of months or years at even fairly low temperatures[9], the variation of sensitivity with registration temperature[10] and oxygen pressure[11], and the growth of latent track reactivity with time[12]. BP-1 glass suffers no fading of latent tracks up to a test limit of 5 months at 50 °C, ensuring negligible fading over several years at a lunar base site. At the design operating temperature of -20 °C, the dependence of track etch rate on registration temperature is remarkably weak, with a fractional shift in etch rate of only $\sim 1 \times 10^{-3}/°C$. Finally, the response of BP-1 is independent of oxygen pressure, so pressure vessels with their unavoidable mass overburden are unnecessary; also, there is no measurable variation of latent track reactivity with time.

The charge resolution of BP-1 glass is unequalled in the high-Z regime. Figure 2 shows data on the fragmentation of 1 GeV/u Au ions at LBL's Bevalac. A stack of 5 sheets of glass (10 surfaces) sufficed to yield a charge resolution of 0.06e (note that the ordinate in Fig. 2 is logarithmic)! With CR energies initially unknown, the charge resolution versus energy also depends upon the ability to uniquely fit response curves of individual elements to the measured track etch rate versus stack depth. Figure 3 shows the results of a Monte Carlo calculation for a stack of 14 sheets of 0.2 cm-thick BP-1 glass. It is seen that a majority of events have their charge resolved to < 0.3e.

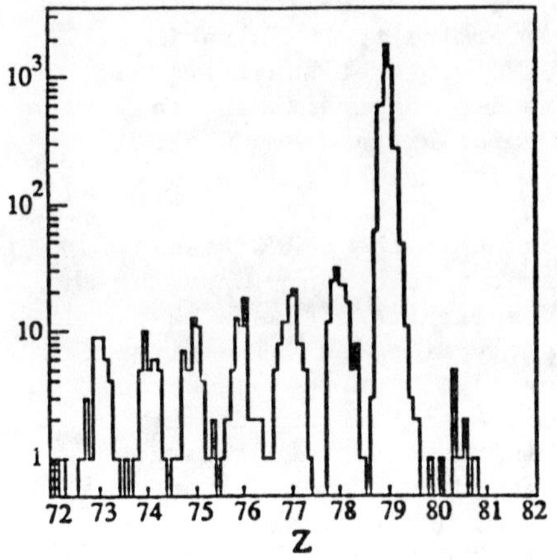

Fig. 2

A Lunar-Based Heavy Nucleus Detector (LBHNC) would consist of an array of BP-1 glass stacks totalling 100 m² in area (and acceptance solid angle of 2.8 sr), deployed for an exposure period of 5 years, with each stack consisting of 16 0.2 cm-thick sheets. A lunar site is arguably the ideal location for such an experiment. The absence of a lunar magnetic field implies the absence of any rigidity cutoff, resulting in substantially higher CR fluxes on the Moon than in a low-inclination orbit about the

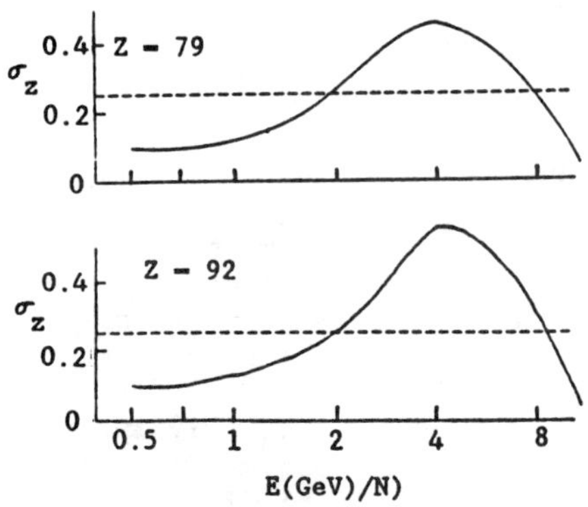

Fig. 3

Earth. The flux of CR U and Th can be estimated from the measured CR Fe flux by scaling with the SS U+Th/Fe ratio, and accounting for propagation effects. The CR Fe flux measured by the Danish-French HEAO-3 experiment[1] has been corrected for the effects of its (variable) geomagnetic cutoff, resulting in a spectrum corresponding to that at 1 AU outside the Earth's magnetosphere[13]. This is then scaled by the SS abundance ratio[14] U+Th/Fe = 5.1×10^{-8}. Propagation effects are taken into account by calculating the survival fraction of both Fe primaries and U+Th primaries after propagation through a truncated exponential PLD using the energy-dependent parameters of Garcia-Munoz et al.[6], and correcting for first-ionization-potential selection effects[4]. Imposing an energy cut of > 0.85 GeV/u to ensure adequate range within a glass stack, we estimate that a total of ~350 U+Th nuclei (this does not include actinide secondaries) will be detected by LBHNC. Should the r-process enhancement suggested by the HEAO-C and Ariel-6 data be real, this number might be as high as ~1000. In either case, these statistics would be sufficient to determine the U/Th ratio to within ~10%, and they imply that LBHNC would be sensitive to transuranics even if the fresh r-process component were to comprise only a few percent of the total ultraheavy CR flux.

The absence of a lunar shielding magnetosphere, while increasing CR flux, raises the question of whether solar flares would generate unacceptable backgrounds in the LBHNC stacks, which are protected only by multilayer insulation. Choosing the June 3, 1982 (1130-1800 UT) Fe-rich solar flare[15] as a test case, we find that LBHNC is not significantly affected. The first few sheets bear the brunt of the exposure: for our test case, the third glass sheet in the stack acquires an Fe-track density of ~25/cm^2. LBHNC will be completely insensitive to the larger fluxes of protons and alphas, as these do not even register as tracks in BP-1 glass (as extensive exposure to radioactive α sources has shown). The above Fe-track density will not obscure actinide tracks; it only places a hardship on the track imaging system by increasing the number of tracks requiring analysis. The background density drops off roughly as a power law in stack depth, so several sheets into the stack (corresponding to a small fraction of an actinide's mean interaction length of 11 g/cm^2 in BP-1) the background becomes insignificant. Thus, at worst, the first few sheets of a stack will be sacrificed, with minimal impact on statistics or charge resolution.

To minimize the effects of accruing background tracks, an etching facility and automated track imaging system could be established at the lunar base. Glass stacks could then be periodically harvested, analyzed, remelted to eliminate all track structures, and placed back onto the lunar surface; in effect, LBHNC could indefinitely be recycled at little additional payload cost.

ACKNOWLEDGEMENTS

We wish to thank Dick Mewaldt for helpful conversations.

REFERENCES

[1] P. Goret et al., Proc. 18th ICRC (Bangalore) 9, 139 (1983), and references therein.
[2] W.R. Binns et al., Ap. J. 346, 997 (1989).
[3] P.H. Fowler et al., Ap. J. 314, 739 (1987).
[4] J.P. Meyer, Ap. J. Suppl. 57, 173 (1985).
[5] J.B. Blake and D.N. Schramm, Astrophys. and Sp. Sci. 30, 275 (1974).
[6] M. Garcia-Munoz et al., Ap. J. 280, L13 (1984).
[7] P.B. Price et al., Nucl. Instr. Meth. B21, 60 (1987).
[8] Shicheng Wang et al., Nucl. Instr. Meth. (in press, 1989).
[9] M.H. Salamon et al., Nucl. Instr. Meth. B17, 173 (1986).
[10] A. Thompson et al., Proc. 11th Inter. Conf. on Solid State Nuclear Track Detectors (Bristol, 1981).
[11] J. Drach et al., Nucl. Instr. Meth. B23, 367 (1987).
[12] J. Drach and P.B. Price, Nucl. Instr. Meth. B28, 275 (1987).
[13] E. Juliusson et al., Proc. 18th ICRC (Bangalore) 2, 21 (1985).
[14] A.G.W. Cameron, in Essays in Nuclear Astrophysics, ed. C.A. Barnes et al., Cambridge University Press (1982).
[15] M.A.I. Van Hollebeke et al., Proc. 19th ICRC (La Jolla) 4, 209 (1985).

NEW MEASUREMENTS OF THE CROSS SECTIONS OF ^{4}HE INTO ^{2}H AND ^{3}HE AND THEIR IMPLICATION FOR ^{2}H AND ^{3}HE PRODUCTION IN COSMIC RAYS

W.R. Webber
Space Science Center
University of New Hampshire
Durham, NH 03824

ABSTRACT

We have recently measured the cross sections for the production of ^{2}H and ^{3}He fragments from ^{4}He interactions in hydrogen targets at six energies between ~140 MeV/nuc and 2GeV/nuc. In this paper we present these cross sections and their implication for the production of ^{2}H and ^{3}He by cosmic rays in the interstellar medium, and compare these predictions with currently available data. The changes introduced by these cross sections as compared with earlier predictions are small below ~200 MeV/nuc but much larger at energies near 1 GeV/nuc. Sharp decreases in both the ^{3}He/^{4}He and ^{2}H/^{4}He ratios are predicted below ~1 GeV/nuc which may be utilized to understand features of both galactic propagation and energy gain and solar modulation.

1. INTRODUCTION

Here we will present the new cross sections and examine the implications of these cross sections on the production of secondary ^{2}H and ^{3}He by cosmic rays traversing the interstellar medium. Most of the secondary ^{2}H and ^{3}He is produced by interactions of cosmic ray ^{4}He and previously these production cross sections were characterized by very few measurements and uncertainties ~25% or larger. As a result, it has not been possible to fully utilize the measurements of cosmic ray ^{2}H and ^{3}He abundance to determine an accurate cosmic ray path length as a function of energy and to examine possible differences in this path length for ^{4}He and for heavier nuclei such as C and Fe.

2. THE CROSS SECTIONS

During the period from June to November 1988 we measured the cross sections for the production of ^{2}H and ^{3}He fragments from ^{4}He interactions in CH_2, C and He targets at six energies from ~140 MeV/nuc to ~2 GeV/nuc at the BEVALAC. The 2 GeV/nuc measurements were made with the HISS array and analysis is still in progress. The lower energy measurements were made with a modified version of the charge-isotope telescope we have used previously for measuring the cross sections of heavier nuclei fragmentation (Webber and Brautigam[1]). The charge and isotope resolution was achieved by a combination of techniques which included dE/dx xE, CxE, CxC and Cx dE/dx, depending upon the energy. The preliminary results of these runs are summarized in Table I. This data for hydrogen targets is also shown in Figure 1a along with other available data and Figure 1b shows the data summarized by total ^{3}He and ^{2}H production along with the total inelastic cross section for ^{4}He interactions in hydrogen.

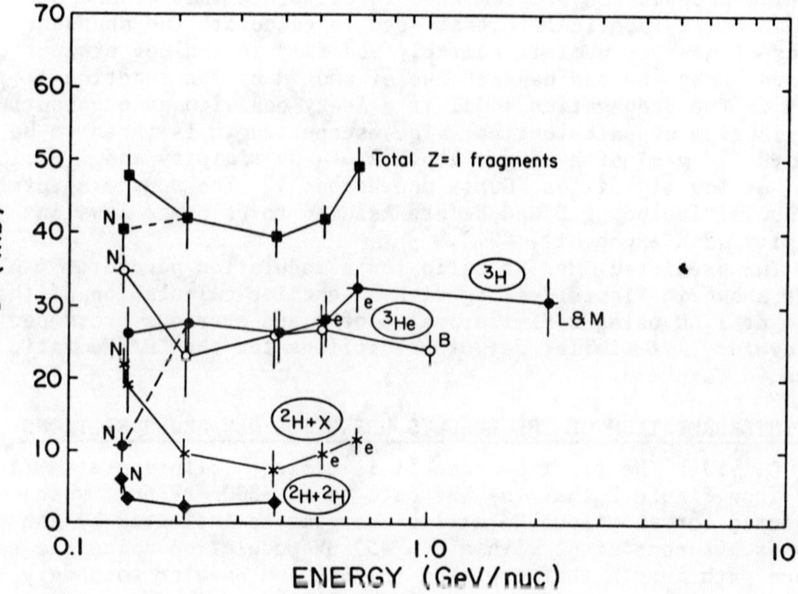

Figure 1a Partial cross sections of ^4He into ^3He, ^3H and ^2H in hydrogen targets. Total cross section into Z=1 fragments is also shown as are other measurements labelled N, B and L & M.

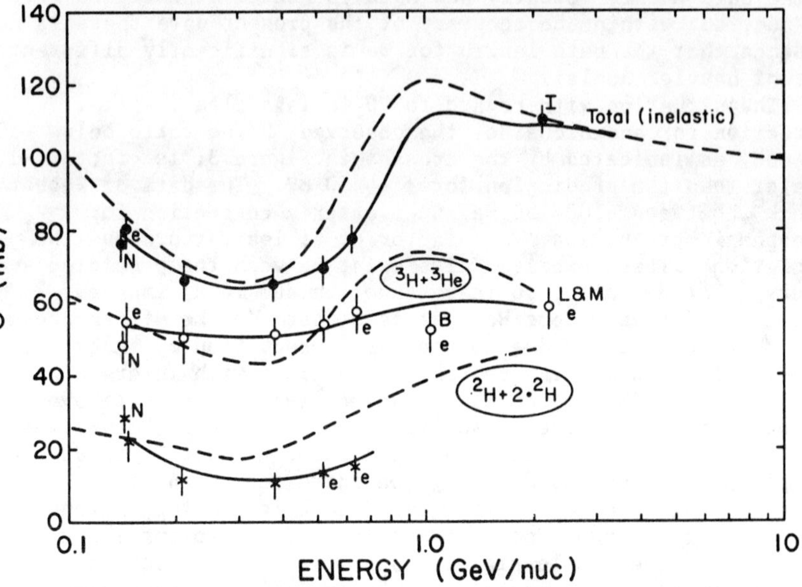

Figure 1b Total inelastic cross sections for ^4He in H targets. Also shown are the summed cross sections into ^3He and ^2H after

3. CALCULATED ^2H AND ^3HE ABUNDANCES AT EARTH

The propagation program used to calculate the ^3He and H^2 abundances is identical to that used to calculate the abundance ratios of heavier nuclei, suitably modified to include production of ^3He and ^2H by ^4He and heavier nuclei and ^2H by the reaction $p + p \rightarrow d + \pi^+$. The propagation model is a leaky box with an exponential distribution of path lengths. The escape length is taken to be $\lambda_e = 24.9\,\beta R^{-0.6}$ g/m^2 of hydrogen above R = 4 GV rigidity and $\lambda_e = 10.8\beta$ g/cm^2 at low rigidities (Gupta and Webber[2]). The source spectra of all nuclei including H and He are assumed to be power laws in ridigity with exponent = -2.3.

The predicted $^3He/^4He$ ratio for a modulation parameter $\phi = 450$ MV is shown in Figure 2 along with an earlier calculation of this ratio at 1 AU using a similar value of ϕ and previous cross sections by Mewaldt[3]. A similar set of predictions for the $^2H/^4He$ ratio is shown in Figure 3.

4. INTERPRETATION OF THE RESULTS USING THE NEW CROSS SECTIONS

Consider ^3He first because it is easiest to interpret. It is seen from Figure 2 that the ^3He data up to ~300 MeV/nuc, when corrected for anomalous He at low energies as indicated by the crosses, is consistent with a $\phi = 450$ MV modulation using the same escape path length that explains the B/C and heavier secondary to primary ratios. At higher energies only one measurement exists (Jordan[4]). This measurement, reinterpreted by Webber et al.[5], also agrees with this escape length within the very large uncertainties in the data point. Clearly new data is needed at energies >300 MeV/nuc, but within the accuracy of the present data there is no evidence that the path length for He is significantly different from that of heavier nuclei.

The situation with regard to ^2H is less clear. After correction for anomalous He, the observed $^2H/^4He$ ratio below ~100 MeV/nuc, as indicated by the crosses in Figure 3, is factor ~1.5 greater than the prediction for $\phi = 450$ MV. The data of Webber and Yushak[8], between ~100-150 MeV/nuc, after a correction for atmospheric production, is a factor of at least two above the prediction. These excesses also exist between the predicted and observed $^2H/^3He$ ratio, so they cannot be due to an improper correction for anomalous He. At least part of the difference below ~100 MeV/nuc could be due to increased production of ^2H by a factor of ~2.5 from cosmic ray proton interactions with interstellar He (the so-called IM component). However, the difference above ~100 MeV/nuc can only be due to additional production of ^2H via the p + p reaction. In order to explain these observations, ^2H production via the p + p reaction would have to be increased by ~4 times, corresponding to either an increased path length for protons at low energies, or a larger intensity of low energy protons above that predicted by the simple rigidity spectra used. In any case essentially all of the data on ^2H below ~150 MeV/nuc suggests an excess production of ^2H by cosmic ray protons beyond that predicted by our model.

It is essential that more precise measurements of ^2H be made in the entire energy range up to and beyond ~1 GeV/nuc. There is a

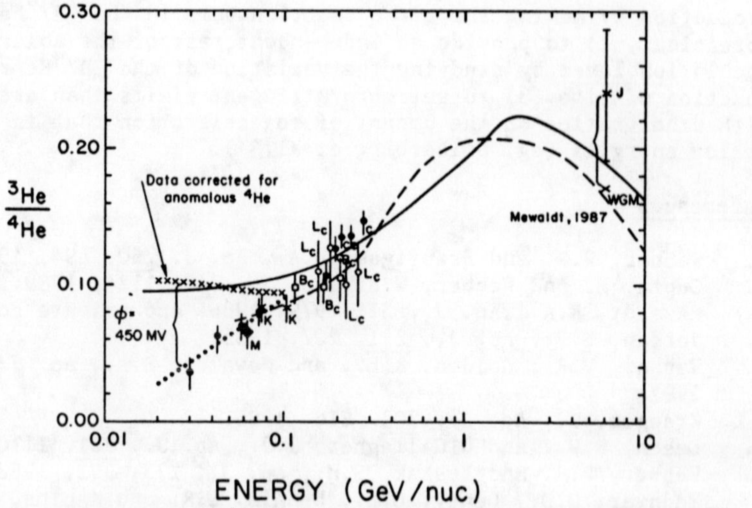

Figure 2 The predicted $^3He/^4He$ ratio as a function of energy using new cross sections for ^4He fragmentation. Data points are: K = Kroeger[6]; WGM = Webber et al.[5], modified Jordan[4] measurement; M = Mewaldt[3]; L = Leech and O'Gallagher[7]; O = Webber and Yushak[8]; B = Badhwar, et al.[9]. The subscript indicates measurement corrected for atmosphere production.

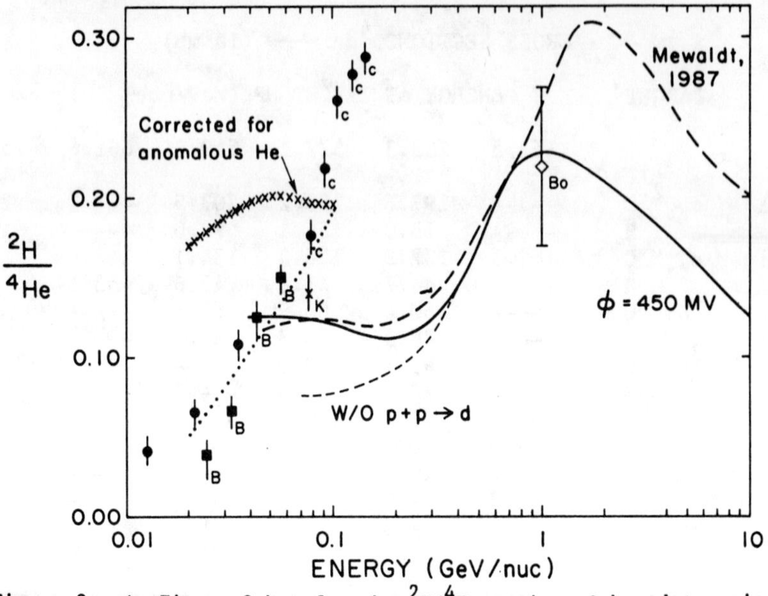

Figure 3 As Figure 2 but for the $^2H/^4He$ ratio. Other data points are: B = Beatty[11]; Bo = Bogomolov et al.[12].

sharp decrease in the $^2H/^4He$ ratio below ~1 GeV/nuc that may be used to: 1) better understand the role of cosmic ray protons in ^2H production by making the p + p reaction bump in the $^2H/^4He$ ratio more clear, 2) to provide an independent test of the solar modulation level by studying the variation of the $^2H/^4He$ ratio as a function of time, 3) to set more stringent limits than are possible with other ratios on the amount of reacceleration that is occurring at low energies (eg. Silberberg et al.[10]).

References

1. Webber, W.R. and Brautigam, D.A., Ap. J. 760, 694, 1982.
2. Gupta, M. and Webber, W.R., Ap. J., 340, 1124, 1989.
3. Mewaldt, R.A., Ap. J., 311, 979, 1986, and private communication.
4. Jordan, S.P., Ap. J., 291, 207, 1985.
5. Webber, W.R., Golden, R.L., and Mewaldt, R.A., Ap. J., 312, 178, 1987.
6. Kroeger, R., Ap. J., 303, 816, 1986.
7. Leech, H.W., and O'Gallagher, J.J., Ap. J., 221, 1110, 1978.
8. Webber, W.R. and Yushak, S.M., Ap. J., 275, 391, 1983.
9. Badhwar, G.D., Deney, C.L., Dennis, B.R. and Kaplon, M.F., Phys. Rev., 196, 1327, 1967.
10. Silberberg, R., et al., Proc. 19th ICRC, 3, 238, 1985.
11. Beatty, J.J., Ap. J., 311, 425, 1986.

TABLE I

CROSS SECTIONS $^4He \longrightarrow$ (in mb)

TARGET			ENERGY AT TARGET MP (MeV/nuc)					
			152.5	203.3	377.1	519.8	661.6	1500
Total	$\sigma_{\Delta A}$ [2]	C	-----	197.6	193.2	203.5	-----	260[3]
		H	-----	73.0	68.1	72.0	-----	110[3]
Total	$\sigma_{\Delta z}$ [2]	C	182.5	137.2	126.3	134.1	148.6	
		H	53.8	44.7	41.8	45.6	53.6	
	^4He	C	-----	60.4	66.9	69.4	-----	
		H	-----	28.3	26.3	26.4	-----	
	^3H	C	122.4	93.1	79.0	-----	-----	
		H	29.0	28.3	26.3	-----	-----	
	^2H	C	47.2	25.4	30.8	-----	-----	
		H	19.6	10.2	7.4	-----	-----	
	^1H	C	4.6	10.7	6.1	-----	-----	
		H	0.8	2.8	2.9	-----	-----	
	$2x^2H$	C	4.2	4.3	5.3	(6)	(7)	
		H	2.2	1.7	2.6	(4)	(5)	

1) Estimated errors $\sim\pm10\%$ for $\sigma > 10$ mb and $\pm20\%$ for $\sigma < 10$ mb.
2) Includes $2x^2H$ cross section.
3) Total cross section from Jaros, 1976.

NEUTRINO ASTRONOMY ON THE MOON — PRACTICAL CONSIDERATIONS

Michael L. Cherry
Louisiana State University, Baton Rouge, LA 70803

ABSTRACT

Observations of neutrinos from extraterrestrial sources offer the opportunity to study astrophysical regions which are otherwise unobservable, and provide (in principle) an extremely clean probe of high energy hadronic acceleration processes. Detectors must have low background and high sensitivity. Due to the absence of an atmosphere, the moon provides a very low background environment. However, in order to compete with proposed earth-based experiments, the mass and area required are large.

INTRODUCTION

Observations of neutrinos from astronomical objects offer the possibility of probing deep into regions which cannot otherwise be observed, and enable one to study the details of hadronic acceleration processes in energetic sources[1-4]. Because of the low neutrino cross section, however, neutrino detectors must be enormous: the Homestake ^{37}Cl detector uses 615 tons of C_2Cl_4 to detect neutrinos from the sun; the Kamiokande detector has used 3000 tons of water to see neutrinos from the sun and SN1987a; and the IMB detector has used 8000 tons of water to observe the burst of SN1987a neutrinos. Several very large detectors are now being proposed as telescopes to search for high-energy neutrinos from discrete sources: the DUMAND Octagon array is to be a 20,000 m^2 effective area water Cerenkov detector at a depth of 4.8 km in the Pacific Ocean off the coast of Hawaii[5]; and GRANDE is to be a 90,000 m$^2\times$ 50 m deep water detector in a quarry pit near Little Rock, Arkansas[6].

Shapiro and Silberberg[3], in particular, have pointed out the potential value of neutrino astronomy, and have emphasized the importance not only of detector size but also of reducing atmospheric background. As an example, the galactic plane is known to be a source of γ-rays from π°-decay. One can calculate the expected flux of neutrinos from the accompanying π^+- and π^--decays, and compare this predicted cosmic flux to the background of neutrinos produced by cosmic ray interactions in the earth's atmosphere. It turns out that at energies below ~ 1 TeV, the earth's atmospheric background is too large to see the galactic plane source.

On the moon, however, the absence of an atmosphere means that this background is reduced significantly[3,4]. On the assumption that a lunar detector is covered by 1000 g cm^{-2} of dirt (equivalent in thickness to the terrestrial atmosphere), the ratio of lunar to terrestrial ν_μ fluxes is shown in Fig. 1: the lunar flux is suppressed by up to a factor of 1000 near 1–10 GeV. At energies above ~ 1 TeV, prompt decays in the lunar soil bring the lunar/terrestrial ratio back up toward unity; and at low energies, the absence of a geomagnetic cutoff on the moon (and the corresponding higher primary cosmic ray flux at the moon) likewise bring the ratio up. In the energy range near 1–1000 GeV, however, the lunar neutrino background is significantly lower than that on earth. It is for this reason that the moon has a potentially large advantage over a terrestrial laboratory as a site for a neutrino telescope.

HIGH ENERGY NEUTRINOS (≥ 1 GeV)

The expected ν rates in terrestrial detectors have been calculated by scaling from observed 10^{15} eV γ-ray rates. Assuming the reported 10^{15} eV signals from Cyg X-3, Her X-1, LMC X-1, and Vela X-4 are real and are due to photons from π° decay, then one can calculate the corresponding ν event rates. In GRANDE, with a proposed full area of 9×10^{4} m^{2}, one expects in excess of tens of events/year from each of several Southern Hemisphere sources[2]. The detailed predictions depend on the specific value of spectral index assumed. The DUMAND detector, which is to be somewhat smaller ($\sim 2 \times 10^{4}$ m^{2}), will see lower rates but will be sensitive to a somewhat different region of the sky.

GRANDE will have angular resolution slightly better than 1°, with an expected background rate ~ 5 counts per year per 1° bin on the sky from atmospheric neutrinos. With an area $\sim 10^{5}$ m^{2} and for Southern Hemisphere objects similar to the observed 10^{15} eV γ-ray sources Cyg X-3, Her X-1, LMC X-1, and Vela X-4, therefore, *GRANDE will not be background-limited.* Only if one wants to look at sources a factor of ten weaker than the standard GRANDE sources (which will require a detector area of 10^{6} m^{2}) will the terrestrial background be large enough to warrant going to the moon.

One can nevertheless ask what a lunar detector might look like. GRANDE and DUMAND use the mass of the earth and ocean as a neutrino target. The secondary muons are then detected by measuring the Cerenkov light produced as the muons cross the active water volume. A similar technique could be used on the moon: lunar material would be used as the target, and active detector components and electronics would be the bulk of the material to be transported from earth to the moon.

Just as a terrestrial detector is shielded from the primary cosmic ray flux by 1000 g cm^{-2} of atmosphere, a lunar detector would need to be shielded by ~ 10 m of lunar soil. It may, however, be possible to accomplish this (even on a scale of 10^{6} m^{2}), with a minimum of digging and rock-moving, by using natural lunar lava tubes[7]. On earth, hollow lava tubes, formed by channels of hot, flowing magma, have produced natural caverns in basaltic volcanic formations. Based on the evidence from lunar surface photographs, much larger subsurface natural caverns apparently exist on the moon. One of the most prominent of the lunar lava tubes, seen in photographs of the Northern Oceanus Procellorum region, measures 40 km in length and in places 500 m across. In most places, the roof of the cavern has collapsed; but uncollapsed sections several hundred meters long can be seen in the photographs. One can estimate the thickness of the uncollapsed roof sections by observing the effects of meteorite impacts and calculating the minimum roof thickness required to withstand a direct meteorite impact. The minimum thicknesses of lava tube roofs derived in this manner are a few tens of meters.

Large natural caverns with 10 m thick roofs have been suggested as suitable environments for a lunar base, since they provide the protection against both meteorites and radiation that astronauts will need. If a large enough (2 km $\times$ 500 m) cavern is available, then one can envision a relatively light-weight detector consisting of two layers of gas-filled counters, one laid out along the floor and one along the roof of the cavern. Just as in earth-based detectors, neutrinos that come upward through the rock interact in the material beneath the detector and produce secondary muons that then are detected as they traverse the two active detector layers. If these active detectors are considered to be segmented plastic balloons, prefabricated on earth with wires strung down the middle, then one must carry the weight of the plastic and the compressed filling gas from the earth. Although it is difficult to envision a lighter detector than this, the total mass of plastic which must be carried up from the earth (assuming 4 layers, each 1/4 mil thick, each with an area of 10^6 m^2), is still $\sim$ 200 tons.

LOW ENERGY NEUTRINOS ($\leq$ 100 MeV)

Based on the observation of electron antineutrinos from SN1987a by IMB and Kamiokande, Mann and Zhang[8] have pointed out the possibility of using a lunar detector to search for the flux of relic $\bar{\nu}_e$ from the totality of all supernova collapses in the past. The SN1987a observations provided a detailed confirmation of many of our ideas about gravitational collapse. Armed with this knowledge about a (presumably) standard event, it now may be possible to study evolutionary and cosmological effects by looking for the redshifted and broadened flux of relic $\bar{\nu}_e$ summed over all supernovas at different distances and different past epochs. The relic flux is expected to peak between 6 and 10 MeV, so that a much more fine-grained detector is required than was discussed for high energy observations. On the earth, the atmosphere again provides a troublesome background. But at energies up to about 10 MeV, an even worse background comes from nuclear reactor–generated $\bar{\nu}_e$'s. Assuming a lunar base does not have a power reactor of its own, this terrestrial reactor background, together with the atmospheric background, would be eliminated on the moon. A 200 ton liquid scintillation detector on the moon would then be expected to be sufficiently free of background and have sufficient sensitivity to study the flux of relic $\bar{\nu}_e$'s at levels close to the lowest theoretically expected fluxes.

Similarly, on the assumption of a factor of 200 background reduction taken from Fig. 1, Cline[9] has suggested that a 10^6 ton detector of compacted lunar regolith (with perhaps 100 tons of active sensor mass brought from the earth) might provide a possibility for dramatic improvements both in measurements of proton decay and in the sensitivity of supernova burst detection.

CONCLUSIONS

Although neutrino backgrounds will unquestionably be lower on the moon than on the earth (at least in some energy ranges), the detector masses and areas required to do significant science and to compete with earth-based detectors will be large: $\sim$ 200 tons of material to be brought from the earth and (in the case of a high energy telescope) $\sim 10^6$ m^2 of detector area. Deployment, operation, and maintenance of the electronics will most likely be large–scale enterprises characteristic of earth-based accelerator and large underground experiments. However, preliminary NASA deployment scenarios for a Lunar Outpost Astrophysics Program call for a total of 12 crew members in the "Utilization Phase" of a lunar base (years 2009 – 2015 and beyond) and a total science mass of 40 tons. It is difficult to see how a lunar neutrino detector can satisfy these kinds of constraints.

REFERENCES

1. V.S. Berezinsky and G.T. Zatsepin, Sov. J. Nucl. Phys. **11**, 111 (1970) and *Proc. XIth Intl. Conf. on Neutrino Physics and Astrophysics*, Nordkirchen (K. Kleinknecht and E.A. Paschos, eds.), p. 589 (1984); F.W. Stecker, Ap. J. **228**, 919 (1979); C.E. Fichtel, *Proc. 1978 DUMAND Summer Workshop* (A. Roberts, ed.), p. 289 (1979); R. Silberberg and M.M. Shapiro, *Proc. 1978 DUMAND Summer Workshop*, (A. Roberts, ed.) p.237 (1979); M.M. Shapiro, *Neutrino 81* (R.J. Cence et al., eds.) **1**, 44 (1981); V.J. Stenger, Ap. J. **284**, 810 (1984) and Il Nuovo Cimento **9C**, 479 (1986); H. Lee and S.A. Bludman, Ap. J. **290**, 28 (1985); V.S. Berezinsky et al., Il Nuovo Cimento **8C**, 185 (1985) and Ap. J. **301**, 235 (1986); G. Auriemma et al., Il Nuovo Cimento **9C**, 451 (1986).
2. M.L. Cherry, Nuclear Physics B, to be published (1989).
3. M.M. Shapiro and R. Silberberg, *Lunar Bases and Space Activities of the 21st Century* (W.W. Mendell, ed.), p. 329 (1985) and these proceedings.
4. M.L. Cherry and K. Lande, *Lunar Bases and Space Activities of the 21st Century* (W.W. Mendell, ed.), p. 335 (1985).
5. P. Bosetti et al. (DUMAND Collaboration), *DUMAND II: Proposal to Construct a Deep–Ocean Laboratory for the Study of High Energy Neutrino Astrophysics and Particle Physics*, Hawaii DUMAND Center Publ. HDC-2-88 (1988).
6. A. Adams et al., *Proc. 21st Intl. Cosmic Ray Conf.*, Adelaide, to be published (1990).

7. F. Hörz, *Lunar Bases and Space Activities of the 21st Century* (W.W. Mendell, ed.), p. 405 (1985).
8. A.K. Mann and W. Zhang, NASA Workshop on Physics and Astrophysics from a Lunar Base, Stanford Univ., May 1989.
9. D.B. Cline, NASA Workshop on Physics and Astrophysics from a Lunar Base, Stanford Univ., May 1989.

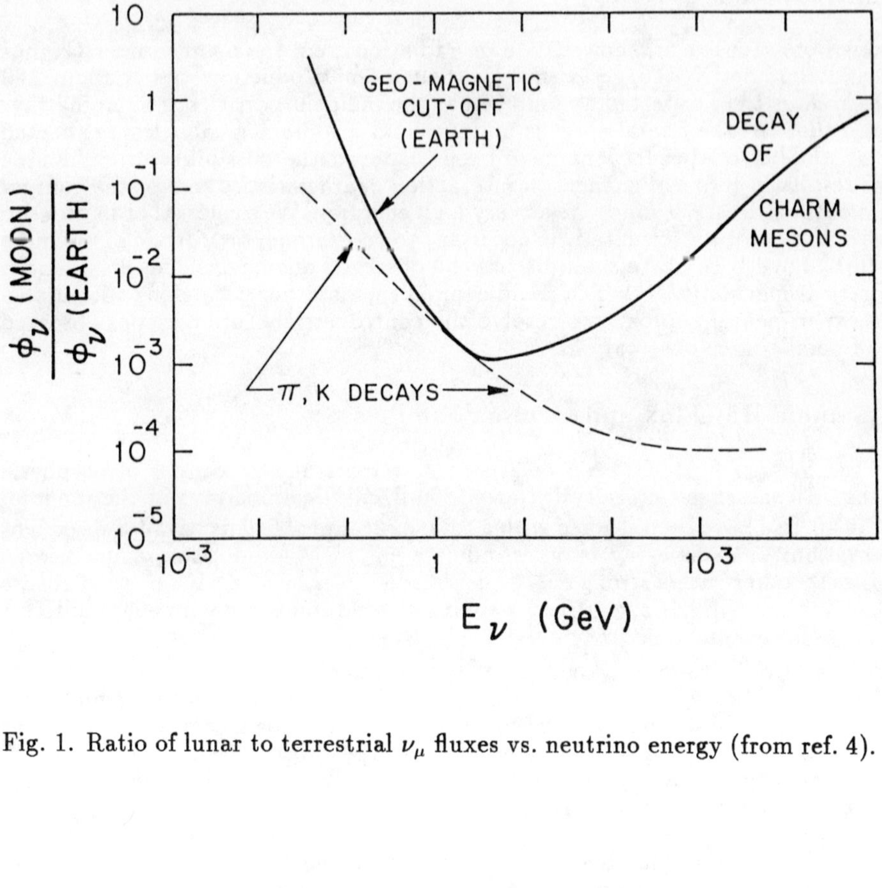

Fig. 1. Ratio of lunar to terrestrial ν_μ fluxes vs. neutrino energy (from ref. 4).

Gamma Rays at Airplane Altitudes

J. Iwai, T.Koss, J. Lord, S. Strausz, J. Wilkes, J. Woosley
University of Washington, Seattle, WA 98195 USA

Abstract

An examination of the gamma ray flux above 1 TeV in the atmosphere is needed to better understand the anomalous showers from point sources. Suggestions are made for future experiments on board airplanes.

1 Introduction

Analysis of extensive air shower data on radiation from the point sources Cygnus X-3 and Hercules X-1 suggests that (i) their muon production rate is about 100 times greater than expected assuming that the incident particles are gamma rays [1] and that (ii) the lateral spreads underground are much greater than expected [2]. A number of speculations have been made on the possibility of explaining these results in terms of changes in interaction characteristics and/or the composition of the primary cosmic rays at very high energies. We suggest that in order to explain the point source data it is necessary to re-examine the cosmic ray gamma ray flux above 1 TeV. We shall discuss the observed anomalous showers[3] and a theoretical speculation[4] which could explain the observed data along with a possible experimental approach to resolve the controversial characteristics observed in the point source observations.

2 Gamma Ray Flux and Gamma Families

The gamma ray flux has been measured systematically at various atmospheric depths. At passenger aircraft altitudes (260 g/cm^2) gamma rays in the range of 0.04 to 40 TeV have been studied with a total exposure of 640 m^{2}hr [5]. Subsequent observations of gamma ray families indicate enhanced production of multicored families together with a strong energy dependence [6]. From a 4100 m^{2}hr exposure 15% of families with $\Sigma E_\gamma > 10$ TeV were multicored compared with 50% multicore structure for events with $\Sigma E_\gamma > 100$ TeV [6].

At mountain altitudes (≥ 500 g/cm^2) emulsion chambers have shown anomalous superfamilies with unusual multicore structure [7], and without multicore structure [8]. These multicore events consist of very high energy subcores enveloped by an indeterminate number of low energy showers forming a dark background or "halo" in X-ray film. This type of event has not been reported from other experiments at higher elevations.

A few multicore events observed at aircraft altitudes resemble halo events, but detailed microscopic observation reveals that the core consists of a number of well resolved shower cores with higher average energies than those seen in a typical halo event. Multicore events observed at higher altitudes do not show the smooth background which characterizes halo events. No event observed at balloon altitudes shows this continuous background.

Another question is whether or not a cosmic limit on the energy of photons exists. It has been suggested that the highest energy cosmic rays may be photons from local point sources to explain the observed cosmic ray flux beyond the Greisen cut-off energy of $> 10^{19}$eV [4]. However, the most energetic single photon observed by emulsion chamber experiments is about 100 TeV.

There are several possible reasons why we have not observed any single photons above 100 TeV with a uniform and symmetrical core profile, even though we have no reason to expect a cutoff. The easiest explanation is distortion and deformation of the shower profile during traversal of the atmospheric overburden, with consequent cascading and shower growth before observation. Any gamma-initiated shower with this developed structure would be hard to identify as such. Moreover, the attenuation due to cascading interferes with the estimation of the initial gamma ray energy. The possibility of the onset of a new process such as an increase in the photohadronic cross-section has been suggested [4]. To investigate such an unknown reaction, a series of experiments is proposed focusing on > 10 TeV gamma rays observed at high altitudes where background interactions are minimized.

3 Characteristics of Atmospheric Shower Phenomena

The lateral development of a shower is dominated by Coulomb scattering processes. Given this predominance, the lateral spread r_m is readily evaluated:

$$r_m = \left(\frac{K}{E}\right) X \tag{1}$$

where X is the (geometrical) length corresponding to one cascade unit at a given atmospheric depth and E is either the average energy $< E >$ or the critical energy ($E_c = 81$ MeV in air). As E_c is not sensitive to external thermodynamic conditions the shower profile depends only on E_{min} for $E_{min} > E_c$. It is generally accepted that shower phenomena are well described in terms of the ratio E/E_{min}.

A typical gamma ray family shows lateral spread due to (i) transverse momenta of parent hadrons (mainly π^0) at production and (ii) Coulomb scattering of electrons in the atmosphere. The product[9] of the energy ratio E/E_{min} and the lateral spread parameter r/r_m in Moliere units can be expressed under assumptions (i) and (ii) and further assuming an isothermal, static, ideal gas atmosphere,

$$\left(\frac{E}{E_{min}}\right)\left(\frac{r}{r_m}\right) = \left(\frac{< p_{t\gamma} >}{K}\right)\left(\frac{H}{X}\right)\log\left(\frac{P}{P-p}\right) \tag{2}$$

where $< p_{t\gamma} >$, r, H, P, and p are the average transverse momenta of produced gamma rays, lateral spread of gamma ray families in geometrical length, scale height, atmospheric depth (g/cm^2) of the observation, and atmospheric overburden between event and detector, respectively.

Figure 2 illustrates the calculated energy moments (equation 2) versus p for an average gamma ray family at several observation points, ranging from a typical mountaintop altitude of 540 g/cm^2 (Mt. Chacaltaya) to the ER-2 aircraft operating ceiling of about 50 g/cm^2. When $P \geq p \approx 0$, the interaction vertex is very close to the detector and the factor $H \log(P/(P-p))$ can be approximated by $H(p/P)$ (see Fig. 1). In this case, the atmospheric overburden p(g/cm^2) and geometrical depth are directly proportional, meaning that the lateral spreads of gamma ray families at any given depth P are similar with a scale factor inversely proportional to P. Figure 2 shows this similarity for atmospheric thickness $p \leq 40$ g/cm^2 , *i.e.*, ≤ 1 c.u.

At higher altitudes ($P \leq 100$ g/cm^2) where P is comparable with the nuclear mean free path (90 g/cm^2) or the radiation length (37 g/cm^2), gamma ray families from a distant vertex $P - p \approx 0$ are observed without secondary collisions or significant atmospheric attenuation. Here even events at the top of the atmosphere can be observed without the smearing caused by similarity effects discussed above. This implies that even a very high energy binocular event ($\sim$100 TeV) resulting from a single π^0 decay, which is subject to the LPM effect, can be recorded as well separated double cores. The potential exists to study the interactions and shower development of ultra high energy gamma rays in detail.

4 Discussion

At high altitudes the primary and secondary gamma ray flux will be observed with an additional contribution from hadronically produced neutral hadrons. To reduce such contamination, we can consider only gamma ray families and eliminate events which lie in the angular region characteristic of neutron fragments from a heavy primary, or we can employ a detailed examination of 3-dimensional shower development to eliminate showers of hadronic origin. A detector size of $> .2$ m^2 and a total exposure of 1000 m^{2}hr allows the observation of entire gamma ray families. Even if only a few correlated gamma rays from each family are observed from a total exposure of 100 m^{2}hr it is still possible to examine gamma ray interactions with matter in this UHE energy range.

The purpose of the present article is to discuss the possibility of direct observation of UHE gamma rays and their properties. One of the major experimental problems at present has been how to eliminate effects of hadronic clustering (jets). In observing multicored gamma ray families (including the missing halo type events), we face a similar problem. For gamma ray families we can at least distinguish the separate cores by examining their lateral development.

Our observations on a gamma ray family observed in a balloon-borne emulsion chamber might indicate an anomalous photon interaction in which two energetic showers (1.5 to 6 TeV) originating from neutral precursors exhibit apparently hadronic shower development, with one of the two having a secondary nuclear interaction [3]. The possibility of neutron contamination via fragmentation of the hypothetical primary particle is eliminated by analysis of the lateral spread of the family. Such an analysis is possible if and only if one exposes well designed emulsion chambers to the incident cosmic ray flux at altitudes of < 100 g/cm^2. The use of an aircraft cruising at an altitude of 50 g/cm^2 could, in principle, provide improved directional and time resolution of the incident cosmic ray flux.

References

[1] V.S.Brezinsky, J.Ellis, and B.L.Ioffe, *Phys.Lett.* **172B**, 423 (1986).

[2] G.Battistoni *et al.*, *Phys.Lett.* **155B**, 465 (1985).

[3] T.A.Koss *et al.*, Contributed Paper to the 14th Int.Sympo.on Lepton and Photon Interactions, SLAC, July, 1989, Paper No.144 and Proc.XX IHEP, Madison, WI, p824, AIP, New York, 1981.

[4] F.Halzen, Proc.of Grande Workshop on the UHE Gamma Ray and Neutrino Int.(Little Rock, AR, May 1989; to be published), MAD/PH/504, July, 1989.

[5] Y.Takahashi, AIP Conf.Proc. on Cosmic Rays and Particle Physics(Bartol), 49, 166(1979).

[6] J.Iwai, *et al.*, Proc.of 14th ICRC, Kyoto, 7, 234 and 8, 1 (1979).

[7] Brazil-Japan Collaboration on Chacaltaya Emulsion Chamber Experiment, *Suppl.of Theor.Phys.* **47** 1 (1971).

[8] L.T.Baradzei, *et al.*, Proc.of Int.Sympo. on Cosmic Ray and Particle Phys., Tokyo, Mar. 1984, page 136.

[9] Y.Fujimoto, private communication with one of authors(J.I), Mar.1975.

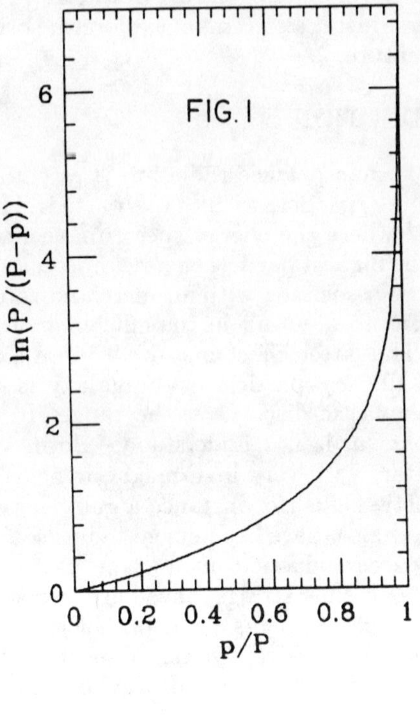

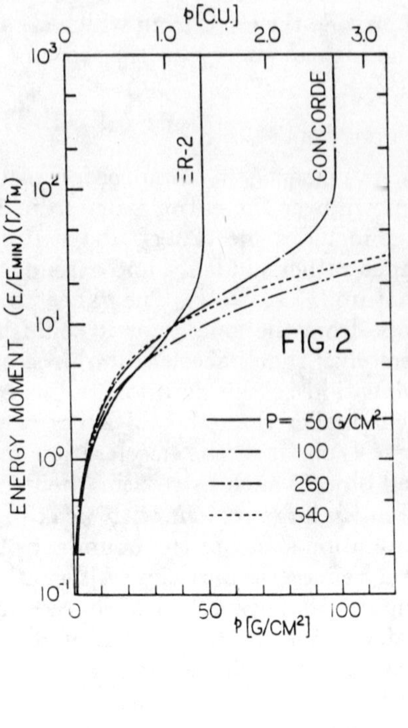

COMMENTS ON THE OBSERVATION OF COSMIC RAYS AT VERY HIGH ENERGIES

Dietrich Müller
Enrico Fermi Institute and Department of Physics
The University of Chicago, Chicago, Illinois 60637

ABSTRACT

The determination of the elemental composition and energy spectra of cosmic rays at high energies, covering the range $10^{13} - 10^{20}$ eV/particle, remains to be one of the most challenging tasks in high energy astrophysics. New experimental data will come from several advanced air shower installations presently under construction. However, direct measurements are essential to understand the particle composition. Such measurements are possible over the range $10^{13} - 10^{15}$ eV through space flight of state of the art detectors. They will provide the basis from which air shower installations must explore the origin of the highest energy particles found in nature.

1. INTRODUCTION

At present, the composition of the cosmic particle radiation is essentially unknown over the entire range from 10^{14} eV/particle to 10^{20} eV/particle. This range includes the "knee" above 10^{15} eV where the energy spectrum seems to steepen rather suddenly, and extends to the highest particle energies encountered in nature ($\sim 10^{20}$ eV). The "knee" may be associated with an increased rate of escape from the galaxy, or it could be due to a cut-off in the efficiency of the interstellar shock acceleration process. The latter effect may be a more likely explanation as the gyration radius of a 10^{15} eV particle in the galaxy is still much smaller than the thickness of the galactic disc. Over the range 10^{16} to 10^{20} eV/particle, the acceleration mechanism is not understood. Both, very small objects such as rotating neutron stars with very high magnetic fields, or extremely large regions with weak magnetic fields (for instance a galactic wind termination shock at the boundary of the halo), have been suggested as sources of very energetic particles. Alternatively, extragalactic contributions cannot be excluded, and cosmological sources such as cosmic strings, have been proposed. To determine the true origin of these particles remains to be one of the most fascinating and challenging tasks in high energy astrophysics. A sensitive tool towards accomplishing this task are reliable measurements of the composition of the high energy cosmic rays.

2. OBSERVATIONAL CONSTRAINTS

Cosmic ray observations at high energies are severely constrained by the small flux of particles and by the difficulty of measuring both, the identity (i.e. the nuclear charge Z) and the energy of each particle. The integral particle flux above 10^{15} eV/nucleus is approximately 20 for protons, 10 for He nuclei, and 5 for iron nuclei for an exposure of 1 m^2 sr year. Towards higher energies, these fluxes drop by at least a factor of 100 for each decade in energy. Therefore, the following is obvious:

- Detectors with geometry factors of a few $m^2 sr$ are required to approach energies around 10^{15} eV. Such detectors should be flown in space for the duration of about one year. They can be accommodated on state of the art spacecraft or on the planned space station.
- Detectors for heavy nuclei (such as iron) should be larger than those for protons. Even more severe are the flux limits for the less abundant species such as the primary nuclei Ne, Mg, Si or Ni and the secondary species Li, Be, B or the sub-iron elements Sc, Ti, V.
- Measurements up to energies in the 10^{16} eV region require detectors of $\sim 100\ m^2 sr$, flown for at least one year duration. This seems to be beyond the presently available space technology. However, it has been argued that a calorimeter of that size could be installed on a future lunar base. The advantage in this case would be the fact that natural lunar regolith could be used to provide the required large amount of absorber material.
- Measurements at still larger energy will remain the domain of ground based air shower installations for the foreseeable future.

3. MEASUREMENT TECHNIQUES

Direct measurements above the atmosphere permit the determination of the nuclear charge Z of the primary particle. The major challenge of such measurements is the determination of the particle energy. This is done either with a calorimeter or with a nondestructive technique such as a transition radiation detector (TRD). These two techniques are quite complementary to each other, and both are needed to advance our understanding of the composition of high energy cosmic rays:

- Calorimeters offer energy measurements over a wide range without systematic limitations, while TRD's are restricted to the range $5 \times 10^2 \leq E/mc^2 \leq 3 \times 10^4$. This corresponds to maximum energies approaching 10^{15} eV/particle. The energy calibration of a TRD can be established on the basis of accelerator measurements much more securely than that of a calorimeter (or in particular, of a calorimeter of limited thickness).
- TRD's permit good charge resolution while the charge determination with calorimeters is complicated due to backscatter from the cascade. However, the TRD technique is not practical for light particles due to large statistical fluctuations in the T.R. yield (which increases with Z^2). For protons and helium, therefore, a calorimeter is the instrument of choice.
- The most severe limitation of calorimeters is their required large mass: one interaction length of iron requires 1.3 tons per m^2 area. Calorimeters that are several interaction lengths deep can therefore have a geometric factor of at most $\sim 1 m^2 sr$ to be compatible with present U.S. launch vehicles. The non-destructive nature of a T.R. measurement permits a much more favorable area/weight ratio and therefore larger geometric factors. The mass of a TRD is just of the order of 100 kg per square meter (not counting housing and structural elements).

Somewhat special are *short* calorimeters that measure the electromagnetic cascades generated in the first interaction. While not as massive as classical calorimeters, these devices are more problematic with regard to energy calibration, energy and charge resolution, and detection efficiency.

The *indirect* technique of air shower observations makes possible detector areas of many square kilometers. Therefore, a large energy range, from $\gtrsim 10^{13}$ eV to 10^{20} eV, can be covered. However, such measurements cannot directly observe the primary particle. This limitation applies to both, the classical surface arrays, and to the air scintillation detectors (Fly's Eye) that observe the profile of the shower through the atmosphere. Any information about the primary composition must come from the analysis of secondary quantities such as height of the shower maximum in the atmosphere, ratio of electrons to muons at the surface, energetic muons and hadrons near the core, distribution of shower particles in space and time, etc. With existing arrays, it is difficult or impossible to measure a number of these parameters *simultaneously*, and no successful determination of the primary composition exists. With the installation of a number of advanced arrays presently in progress (e.g. Utah/Michigan/Chicago, Gran Sasso, Karlsruhe, Aragaz/Ani, etc.) this situation is bound to change in the coming years.

4. PROSPECTS FOR THE FUTURE

Within the next few years, significant progress is possible in the determination of the cosmic ray composition at high energies. With certainty, we can expect a substantial increase in the quality and detail of air shower data that will become available. These data will extend upwards from energies of a few 10^{13} eV/particle. Results of shower simulation calculations indicate that the parameters derived from air shower data will depend on the primary composition in a characteristic way, but they also indicate that the fluctuations are likely to be too large to permit the establishment of a well resolved "charge spectrum." Whatever distributions may be derived from the data, they will require a deconvolution using certain assumptions about the charge response of the measurement. In order to test the technique and to obtain an essential calibration of air shower measurements, a comparison with direct measurements covering a common energy region is therefore of utmost importance.

Such a comparison is possible with state of the art technology over two decades in energy, from 10^{13} to 10^{15} eV/particle. The most straightforward program would consist in the space flight of about one year duration of (1) a $1 m^2 sr$ calorimeter and (2) of the CRN transition radiation detector. The calorimeter should be optimized for the detection of light nuclei (protons and helium) up to 10^{15} eV. The continuation of the Sokol program at the Moscow State University may lead to the completion of this task. The CRN detector is an existing instrument that has been successfully flown on the Space Shuttle. It has a geometric factor of $\sim 5 m^2 sr$ and has demonstrated its capability of measuring the energies of the heavier cosmic rays ($Z > 3$) up to energies approaching 10^{15} eV.

In the more distant future, a large area calorimeter on the surface of the moon could extend the direct measurements by another order of magnitude in energy, up to $\sim 10^{16}$ eV. However, already at present, the tools are available to make substantial progress in the understanding of the composition of the cosmic ray nuclei. The direct measurements up to 10^{15} eV will have profound implications with regard to our understanding of acceleration models and of the propagation and containment of energetic particles in the galaxy. But furthermore, through their importance for the air shower work, these measurements will be essential in solving one of the major challenges in astrophysics, the origin of the highest particle energies encountered in nature.

NEUTRINO SOURCES OBSERVABLE AT ENERGIES OF 1 GeV TO 1 TeV FROM A 21ST CENTURY LUNAR BASE

Maurice M. Shapiro
University of Maryland, College Park, MD, 20472

Rein Silberberg
Code 4154, Naval Research Laboratory, Wash.,DC, 20375

ABSTRACT

High-energy neutrino (HEν) astronomy is symbiotic with gamma-ray astronomy. Both disciplines will be contributing to our knowledge of high-energy sites and processes in the universe, and to our understanding of cosmic-ray astrophysics. A lunar HEν observatory will make it possible to investigate celestial sources of neutrinos at energies of 1 GeV to 1 TeV—an interval in which detection from a terrestrial base is severely hampered by background. Candidate sources include solar flares, the interstellar medium, young neutron stars, certain binary systems, and extragalactic sources, notably active galactic nuclei (AGN). A neutrino observatory on the moon will require a massive target, and will be expensive. It may be realized only after the establishment of a substantial base for lunar exploration.

INTRODUCTION: AIMS OF NEUTRINO ASTRONOMY

At MeV energies the motivation for neutrino astronomy has sprung mainly from the important information carried by solar neutrinos[1] and by those from stellar collapse.[2, 3] At energies above 1 GeV a principal aim is to investigate sites and mechanisms of cosmic-ray production, i.e., regions where high-energy processes are dominant.[3, 4] This paper is concerned with the domain of high-energy neutrinos (HEν), and especially with those between 1 GeV and 1 TeV. The reasons for this emphasis will become evident.

Neutrinos are, in a sense, unique tracers of high-energy nuclear interactions occurring in distant celestial regions. Evidence for relativistic electrons in cosmic-ray sources reaches us in the forms of synchrotron radiation (notably radio emission), inverse-Compton radiation, and bremsstrahlung. In fact, X-ray astronomy and, to a lesser degree, gamma-ray astronomy, depend on electron interactions that give rise to hard photons. However, nearly all the cosmic-ray particles impinging on our atmosphere are nuclei ($\approx$1 per cent are electrons and positrons). It is the composition and energy spectra of these nuclei that provide most of the clues to the nature of their sources and their modes of acceleration.

COSMIC GAMMA RAYS AND COSMIC NEUTRINOS

Neutrinos are inevitable products of cosmic-ray interactions. The creation of pions and kaons in cosmic-ray collisions generates gamma rays and neutrinos. The former are mainly the daughters of π^0 mesons, while the latter are secondaries of charged π and K mesons. The gamma rays arising from the decay of neutral pions in outer space are recognized as useful tracers of cosmic rays. However, it is sometimes uncertain whether a flux of gamma rays originated in nuclear rather than in electron interactions. Moreover, some sources (e.g., AGN) immersed in dense concentrations of photons, strongly suppress the emission of gamma rays by gamma-photon interactions. Neutrinos can readily pass through regions impenetrable to hard photons. For these reasons neutrinos may be considered uniquely reliable tracers of cosmic rays. Nevertheless, the disciplines of gamma-ray astronomy, neutrino astronomy and cosmic-ray astrophysics are mutually complementary.[5]

The practical problems facing neutrino astronomy arise primarily from the low interaction cross section of neutrinos--the same attribute that makes them useful for the study of otherwise inaccessible regions (e.g., the core of the sun). At 1 GeV the cross section is $\approx 10^{-38}$ cm^2 while at 1 TeV, it is $\approx 10^{-35}$ cm^2. So large target masses are needed.

Proposed HEν observatories on earth (e.g., DUMAND,[6] Lake Baikal[7] and GRANDE[8]) exploit vast volumes of water as the target. The water also serves as a shield against ordinary cosmic-ray background, and as the medium in which neutrinos are detected via charged secondaries by means of the Čerenkov light they generate in the water, containing arrays of photo-sensors.

Among the distinctive capabilities of a neutrino observatory is the capacity to observe signals simultaneously from many sources without the need for individual pointing. This is due to the wide field of view which can approach 4π steradians. Another special feature is the uninterrupted, round-the-clock operation. In principle, this permits the measurement of temporal variations in a source over a wide range of time scales. Research in neutrino physics--e.g., on oscillations,[9] neutrino masses--can also be carried out with a neutrino observatory.

NEUTRINOS FROM GALACTIC SPACE

In terrestrial observatories the detection of HEν from some cosmic sources is degraded by the background of neutrinos produced through collisions of cosmic rays in the earth's atmosphere.[10] This is especially true of diffuse emissions at energies $<$ 1 TeV. For example, the diffuse neutrino flux expected from cosmic-ray interactions in the interstellar medium is swamped by atmospheric background up to $\approx$10 TeV. When we are dealing with discrete sources, this problem is less forbidding, but other considerations dictate that a terrestrial HEν observatory concentrate on detection of neutrinos above 1 TeV. At these higher energies the decay of atmospherically generated mesons that yields neutrinos is suppressed. Owing to the time dilation of the progenitors, these tend to be captured before they decay. Another reason for choosing the 1 TeV threshold is the need for good angular resolution ($\lesssim 1^\circ$).

However, the fluxes of the parent charged particles, and hence of the neutrinos, are more copious at lower energies,[11] and the prospect of detecting enough of them is enhanced despite the lower cross section.

Thus there are incentives for setting up an HEν observatory at a site free of atmosphere--e.g., at a lunar base.[11, 12.] This was already realized by F. Reines in 1965.[13]

HIGH-ENERGY NEUTRINOS "SEEN" FROM A LUNAR BASE

Some time in the 21st century, lunar exploration will have progressed sufficiently so that it will become feasible to exploit the tunnels believed to exist beneath the lunar surface. A neutrino observatory could then be assembled in a shielded cavern. We may ask, would not the background neutrinos generated in the lunar regolith interfere with observations? It turns out that, because of the density of the lunar soil, the background would be very low. In the competition between capture and decay of pions, the former process would dominate, even in the short life span of a low-energy pion. The interaction mean free path would be shorter (<1 m) than the decay path.[14]

We mentioned above that the diffuse neutrino emission from the Galactic disk at energies below 10 TeV is overwhelmed by atmospheric background at the earth.[3,10] On the moon, a 10^6 ton target array would yield (according to our estimate) some 10 neutrino events per year above 10^8 eV. As larger detectors become possible, the data base would grow proportionately.

What are some other potential neutrino sources that might be profitably investigated from the moon at energies precluded for a terrestrial observatory--i.e., at energies between 1 GeV and 1 TeV?

SOLAR FLARES

An attractive possibility is the detection of HEν from solar flares.[12a,15] This includes the tantalizing prospect of observing flares on the far side of the sun. The most intense solar flares--those especially hazardous to astronauts--occur about once per decade. However, less violent and more frequent solar outbursts can also deliver dangerous doses of energetic charged particles. It would be useful to detect these promptly in order to give timely warning to space travellers. The observation of "hidden" flares on the far side of the sun by means of neutrinos (which readily penetrate a solar diameter) should actually be favored over that of flares on the near side. This is because flare protons moving toward the denser layers of solar atmosphere on the far side would generate more pions (and secondary muons and neutrinos) going toward the observer than those on the near side.

DISCRETE CANDIDATE SOURCES OF COSMIC HEν

Among the likely HEν sources within the Galaxy or nearby, are young pulsars, compact binary systems (e.g., objects like Her X-1, Cyg X-3, SS433, Vela X-1, LMC X-4), and the Galactic center. The prospect of detecting some of these with an expanded DUMAND appears promising. The minimum detectable flux is $5 \times 10^{-10} cm^{-2} s^{-1}$ at energies above 1 TeV. Assuming that at least one muon-neutrino should arrive at earth for each arriving high-energy gamma ray, then the various models used to calculate the neutrino event rates give values of 1 - 10^3 events per year, with a background rate lower than one event per year.[16]

As mentioned above, the study of cosmic sources of HEν below 1 TeV from an earth-based observatory is unprofitable, mainly because of the rapidly rising atmospheric background. On the moon, the principal advantage would be the detectability of neutrinos with $<$ 1 TeV. The event rates per ton of target at these lower HEν energies would be appreciably higher than those above the 1 TeV threshold adopted for the DUMAND poject.

Among the types of candidate sources cited above are young pulsars. HEν generated by the bombardment of expanding supernova shells by cosmic rays from central neutron stars have been analyzed.[17] The process is akin to the production of extensive air showers by ultra-high-energy cosmic-ray nuclei striking the earth's atmosphere. HEν are a component of these showers, and they may be emitted with significant intensities for months up to a few years until the expanding shell of debris becomes too thin.

Potential extragalactic sources of HEν have also been studied. Among these are AGN, notably the highly luminous and relatively nearby Seyferts NGC 4151 and NGC 1068. These galaxies are believed to be powered by central, supermassive black holes, and the physics of these objects can be investigated by means of the gamma rays and neutrinos they emit.[18] Cen A[19] and 3C 273 have also been considered as candidate sources.

A LUNAR HE OBSERVATORY

Practical considerations in the design of an HEν observatory on the moon have been discussed by Cherry[12b] and by Cherry and Lande.[12c] The mass of target material required is so great as to preclude transport from earth. It would, instead, consist of lunar mantle material. A considerable mass of detector apparatus would still have to be brought from earth. Caverns created by volcanic activity seem to occur on the moon at moderate sub-surface depths of a few meters. The overlying rock and soil would shield astronauts and instruments from radiation impinging on the surface. An observatory for cosmic neutrinos on the moon would clearly involve a costly and ambitious program that may be feasible only after a very substantial lunar base will have been established--probably not before several decades hence. Nevertheless, it will probably materialize eventually--unless methods less expensive, now unforeseen, will be devised for exploring the important domain of neutrino energies between 1 GeV and 1 TeV in which most of the cosmic rays are found.

REFERENCES

1. R. Davis, Jr. et al., Phys. Rev. Ltrs. 20, 1205 (1968); R. Davis, Jr., et al., Ann. Rev. Nucl. Part. Sci. 39 (1989).
2. H. Y. Chiu, Ann. Phys. 16, 321 (1961); W. A. Fowler and F. Hoyle, Ap. J. Suppl. 9, 201 (1964); Y. Oyama et al., Phys. Rev. Ltrs., 59, 2604 (1987); V. Trimble, Rev. Mod. Phys. 60, 859 (1988); W. D. Arnett et al., Ann. Rev. Astron. and Astroph. 29 (1989); T. K. Gaisser and. T. Stanev, Phys. Rev. Ltrs. 58, 1695 (1987).

3. M. M. Shapiro and R. Silberberg, in Relativity, Quanta and Cosmology, M. Pantaleo and F. de Finis, eds., Johnson Reprint Corp., New York (1979), p. 475, ff.
4. S. H. Margolis, D. N. Schramm, and R. Silberberg, Ap. J. 221, 990 (1978); V. J. Stenger, Ap. J. 284, 810 (1984); Observational Neutrino Astronomy, World Scientific, Singapore (1989), ed. D. Cline.
5. M. M. Shapiro and R. Silberberg, Space Sci. Rev. 36, 51 (1983).
6. V. Z. Peterson, in Composition and Origin of Cosmic Rays, ed. M. M. Shapiro, Reidel Publ. Co. Dordrecht (1982), p. 251 ff.
7. L. B. Bezrukov et al., Proc. 20th Internat. Cosmic Ray Conf., 6, 292 (1987); G. V. Domogatsky, in Conf. Proc. Neutrino 86 (1986).
8. T. Haines and GRANDE Collaboration, in Cosmic Gamma Rays, Neutrinos and Related Astrophysics, eds. M. M. Shapiro and J. P. Wefel, Reidel Publ. Co., Dordrecht (1989), p. 427, ff.
9. R. Silberberg and M. M. Shapiro, Proc. 1980 DUMAND Workshop, Univ. of Hawaii, ed. V. Stenger, 2, 66 (1981); J. G. Learned and V. Z. Peterson, Hawaii DUMAND Center Publication HDC-5-89 (1989).
10. R. Silberberg, M. M. Shapiro and F. W. Stecker, Proc. DUMAND 1978 Workshop , La Jolla, ed. A. Roberts, 2, 183 (1979).
11. J. H. Adams, Jr. and M. M. Shapiro, in Lunar Bases and Space Activities of the 21st Century,Nat. Acad. Sci. (1985), ed. W. W. Mendell, p. 315, ff.
12. (a) M. M. Shapiro and R. Silberberg, ibid., p. 329, ff.
(b) M. L. Cherry and K. Lande, ibid., 335 ff.
(c) M. L. Cherry, in this volume.
13. F. Reines, as reported by M. M. Shapiro, "Galactic Cosmic Rays," in NASA 1965 Summer Conference on Lunar Exploration and Science, NASA SP-88, p. 317, ff.
14. See Fig. 3 of Ref. 12(b).
15. I. N. Efrofeeva et al., Proc. 18th Internat. Cosmic Ray Conf., 7, 104 (1983); see slso Ref. 12 (a).
16. V. Z. Peterson et al., DUMAND presentation, HEPAP Sub-panel on Non-Accelerator Phys., U.S. Dept. of Energy, March 20, 1989.
17. V. S. Berezinsky and O. F. Prilutsky, Proc. Internat. Conf. (Neutrino 76), Aachen (1976); M. M. Shapiro and R. Silberberg, Adv. Space Res. 1, 141 (1981).
18. R. Silberberg and M. M. Shapiro, in Proc. 1978 DUMAND Workshop, ed. A. Roberts, p. 237 ff.; see also Ref. 5; V. S. Berezinsky and V. L. Ginzburg, Mon. Not. Roy. Astron. Soc., 194, 3 (1981).
19. R. Silberberg and M. M. Shapiro, Adv. Space Res., 1, 63 (1981)

SOURCE COMPOSITION OF COSMIC RAYS

R. Silberberg and C.H. Tsao
Code 4154, E.O. Hulburt Center for Space Research
Naval Research Laboratory, Washington, DC 20375-5000

M.M. Shapiro
University of Maryland, College Park, MD 20472

ABSTRACT

A theory is developed that yields great improvement in deriving the cosmic-ray source abundances for energies below 10^{12} eV/u. In addition, based on the acceleration theory of Völk and Biermann and on nucleosynthesis processes in pre-supernova stars, a theory is presented for the source composition at 10^{12}-10^{15} eV/u. The strong shock wave of young supernova remnant accelerates the wind particles of the pre-supernova red, blue supergiant stars and Wolf-Rayet (WR) stars to energies up to 10^{15} eV/u. They contain the nucleosynthesis products of the CNO cycle and of He-burning. They accelerate the flare particles in interstellar space. The composition below 10^{12} eV/u differs from that of the general stellar photosphere by: (1) Suppression of elements with a large FIP (> 10 eV) by a factor of 4; (2) The depletion of light nuclei ($Z \lesssim 10$); (3) A large contribution of WC stars to ^{12}C, ^{16}O and ^{22}Ne, with a renormalization of the initial $(Z>2)/(Z\lesssim2)$ abundances of Prantzos et al., based on general elemental abundances.

INTRODUCTION

At energies of 10^8 to 10^{11} eV/u the cosmic-ray abundances, when corrected back to the sources, resemble the general abundances but also show some significant differences. The elements H, He and N are underabundant in cosmic rays by factors of 20 to 30; O, S, and Ar by 4; and ^{20}Ne by 6. The elements C and Zn and the nuclide ^{22}Ne are underabundant by 2. At energies 10^{12} to 10^{13} eV/u, the source abundances are not yet well known. The data of Burnett et al.[1], Jones et al.[2], and Swordy[3], imply differences from the composition at lower energies. The He/H ratio is larger by a factor of about 2, the (C,O)/H by 5±3, and Ar/Fe by 2±1.

ACCELERATION MODEL

Acceleration of cosmic rays at supernova remnants has long been investigated, e.g. by Ginzburg and Syrovatskii[4]. The acceleration process by shock waves has been proposed and explored by Axford et al.[5]. We consider the high-energy (10^{13} -10^{15}eV/u) particles to be produced from particles in the stellar wind cavities around the star that becomes the supernova, using the shock-wave acceleration model of Völk and Biermann[6] in the close vicinity of the supernova. These cosmic rays would reflect the surface composition of the supernova precursors, the red and blue supergiants and the WR stars which display the effects of the CNO cycle and He-burning nucleosynthesis on the surface. As the shock waves pass beyond the wind cavity, they become weaker and accelerate mainly the particles that end up in the energy interval 10^8 to 10^{12} eV/u. The composition at these energies resembles that of the solar corona and the flare particles. Shapiro[7] shows that the flare stars (generating particles near energies of 1 MeV/u) are plausible sources of the injected material. A minor contribution from WR stars has been proposed by Meyer[8] to explain the enhancement of ^{22}Ne, ^{25}Mg and ^{26}Mg in cosmic rays. The winds of the WR stars are energetic ($\sim 10^5$ eV/u), with $v/c \sim 0.01$.

SOURCE COMPOSITION, $10^8 \lesssim E/(\mathrm{eV/u}) \lesssim 10^{12}$

The cosmic-ray source composition below 10^{12} eV/u resembles that of normal stars like the sun, with 3 modifications. Elements with a high first ionization potential (FIP) $\gtrsim$ 10.5 eV are suppressed by a factor of 4. This implies pre-injection origin in photospheres of normal stars, with temperatures near 10^4 °K, and an easier escape of the charged particles into the chromosphere and corona.

A second difference is the underabundance of H and He in cosmic rays, and the underabundance of N, reported by Webber and Gupta[9]. The suppression of light ions, proposed by Silberberg and Tsao[10] would result in an underabundance of C and O, which is compensated by contributions from WC stars. The suppression of light nuclei could be due to rigidity cutoff or deceleration at boundaries between stellospheres and interstellar space, due to the effective charge dependence of the rigidity, i.e. (momentum/nucleon)$\times(A/Z_{eff})$. Near 1 MeV multiple electron pickup, with $(Z_{eff}/Z)<1$, becomes appreciable in the interval C to Mg. From the table of Northcliffe and Schilling[11], and their equation on page 236, the ratio of mass number to Z_{eff} (near 1 MeV/u) A/Z_{eff} = 2 for He and 2.95 for Mg. Thus, if there is an effective rigidity cutoff R_o near the boundary of the stellosphere in the neighborhood of energies of 1 MeV/u, the momentum/nucleon at the cutoff is $P_o = R_o \times Z_{eff}/A = 0.5R_o$ for He and 0.339 R_o for Mg; i.e. the P_o at the cutoff is numerically less than the cutoff rigidity. If the exponent of the integral momentum spectrum of stellar flare particles is 5, i.e. $J(P>P_o) \propto P_o^{-5}$, then He is suppressed relative to Mg by a factor of $(2.95/2)^5 = 7$. The determination of the suppression factor from first principles would require detailed knowledge of the energy spectrum of stellar flares, and of processes at the stellosphere boundary shocks. Meanwhile, we adopt an empirical approach which suggests a suppression factor S(Z), where $0.12 \lesssim [(S(Z) = 0.15Z-0.93] \lesssim 1$; i.e. S(Z) = 0.12 for $Z \lesssim 7$ and 1 for $Z \gtrsim 13$. This factor explains the values of H, He, N and ^{20}Ne.

A third difference between the cosmic-ray source composition and that of general abundances is the enhancement of ^{22}Ne, ^{25}Mg and ^{26}Mg, explained by Meyer[8] in terms of a 2% contribution of WR stars to the cosmic-ray source material. However, the calculations of Prantzos et al.[12] require modification. The mass fraction for ^{22}Ne of Prantzos et al.[12] is 1.6 times higher than that of Maeder[13] and Maeder and Meynet[14]. Prantzos et al.[12] assume the mass fraction $(Z>2)/(Z\lesssim2)$ to be 0.03, while the general solar system abundances of Grevesse and Anders[15] yield 0.019. The initial CNO abundances of Prantzos et al.[12] hence should be reduced by 0.019/0.03=0.63. The abundance of N for the WN phase calculations of Prantzos et al.[12] should be reduced by 0.63, and also of ^{22}Ne, ^{25}Mg and ^{26}Mg during the WC phase. The C and O abundances during the WC phase are not affected.

The cosmic-ray source abundances are presented in the review of Lund[16], in his Table 1. For ^{22}Ne, ^{25}Mg, ^{26}Mg, they are supplemented by the data reviewed by Mewaldt[17] and for N, by Webber and Gupta[9]. These are to be compared with the calculated cosmic-ray source abundances $<CR>=10^{-4}<G>(I)(S) + 2.1\times10^{4}\ A^{-1}[1.0 \text{ or } 0.63](WC + a\ WN)$. $<G>$ is the general abundance of Grevesse and Anders[15] normalized to 10^6 for Si. The factor 10^{-4} changes it to Lund's normalization. (I) and (S) are FIP and light-ion suppression factors, WC and WN the compositions of ejecta from the C-rich and N-rich WR stars, shown in Table 2 of Prantzos et al.[12]. "a" is the fraction of mass loss during the WN phase. Integrating over the initial mass function by Prantzos et al.[12], a = 0.5. A is for the mass fraction to number conversion. The factor 0.63 is for N and products of N-burning in He. For nuclei not produced via N-burning (e.g. ^{12}C, ^{16}O and ^{4}He), the factor 1.0 is used. The factor 2.1×10^4 is for normalization. A comparison of the observationally deduced cosmic-ray source abundances and those calculated is given in Table 1. They agree in general to within 20%.

Table 1: Comparison of Calculated and Observationally Deduced Cosmic-Ray Source Abundances (Normalized to Si=100)

Element or isotope	Source Obs.	Source Calc.	Element or isotope	Source Obs.	Source Calc.
H	8.9×10^4	9.0×10^4	^{24}Mg	79	73
He	1.2×10^4	1.3×10^4	^{25,26}Mg	27	23
C	431	490	Al	10.2	8.5
N	15	22	Si	100	100
O	511	469	S	12.6	13.9
^{17}O	<1	0.3	Ar	3	2.7
^{20}Ne	46	47	Ca	6	6.1
^{22}Ne	18	21.3	Fe	93	90
Na	5.6	4.8	Ni	5.1	4.9

SOURCE COMPOSITION, 10^{12} TO 10^{15} eV/u

The composition at these energies, based on the acceleration model of Völk and Biermann[6] is taken to be that of the winds of presupernova stars: the red supergiants of mass about 9-23 $M_\odot$; blue supergiants, 23-35 $M_\odot$; and WN and WC stars of 35-100 $M_\odot$. Their contributions are weighted by (a) the initial mass function from Humphreys and McElroy[18], $\phi(M)\ dM \propto M^{-2.5}\ dM$, where $\phi(M)$ is per $\log(M/M_\odot)$ interval, and (b) the stellar mass loss function evaluated from table 2 of Maeder and Meynet[19]. The stellar wind composition (assuming the general abundances of Grevesse and Anders[15]) is modified as follows: (1) for red supergiants, FIP applies and CN cycle has converted most C into N, which has convected to stellar surface; (2) for blue supergiants, FIP is displaced due to higher temperature from 10 to 20 eV, and CNO cycle has converted most C and O into N, with a composition like Fig. 4 of Maeder[13]; (3) In WC stars, the He-burning products ^{12}C, ^{16}O and ^{22}Ne are on surface, with a composition like Prantzos et al.[12], but adjusting the metallicity mass fraction Z = 0.03 to 0.019 on basis of Grevesse and Anders[15].

CONCLUSIONS AND CONSEQUENCES NEAR 10^{13} eV/u

Since H is depleted during the WC and WN stages and reduced to ~ 1/2 in blue supergiants, the He/H ratio is increased by a factor of 2 in agreement with Burnett et al.[1]. C and O are enhanced by factor of 4 relative to H, due to He-burning into C and O during the WC stage in agreement with Burnett et al.[1], and Swordy[3]. Ar/Fe ratio is enhanced by a factor of 2 due to absence of FIP effect in blue supergiants and WN and WC stars in agreement with Jones et al.[2]. N/O would increase from the low-energy value of 0.03 to 0.06 or 0.07 due to large N abundance in blue supergiants and WN stars. S/Fe should increase by factor of 3 due to absence of FIP effect in blue supergiants and WR stars. Most Ne should be ^{22}Ne, as a result of WC contribution, $^{22}Ne/H$ is enhanced by a factor of 4. Na/Si should increase from 0.06 at 10^9 eV/u to about 0.26 at 10^{13} eV/u, due to nucleosynthesis in blue supergiants and WN stars; see Fig. 8b of Prantzos et al.[12].

Acknowledgement

The authors are indebted to Prof. P. Biermann for a discussion of the acceleration model.

REFERENCES

1. T.H. Burnett et al., Cosmic Abundances of Matter, denoted below as CAM, (AIP Conf. Proc. 183, N.Y. 1989), p. 396.
2. M.D. Jones et al., 19th Internat. Cosmic Ray Conf. (ICRC), La Jolla, 2, 28 (1985).
3. S. Swordy, Cosmic Gamma Rays, Neutrinos and Related Astrophysics (Kluwer Academic Publishers, Dordrecht, (1989), p. 481.
4. V.L. Ginzburg and S.I. Syrovatskii, The Origin of Cosmic Rays (MacMillan, N.Y. 1964).
5. W.I. Axford, E. Leer and G. Skadron, 15th ICRC, 11, 132 (1977).
6. H.J. Völk and P.L. Biermann, Ap. J. Letters 333, L65 (1988).
7. M.M. Shapiro, 21st ICRC, Adelaide, (1990), to be published.
8. J.P. Meyer, 17th ICRC, Paris 2, 265 (1981).
9. W.R. Webber and M. Gupta, 20th ICRC, Moscow 2, 129 (1987).
10. R. Silberberg and C.H. Tsao, Ap. J. (1990), to be published.
11. L.C. Northcliffe and R.F. Scilling, Nucl. Data Tables A7, 233 (1970).
12. N. Prantzos et al., Ap. J. 304, 695 (1986).
13. A. Maeder, Astron. Ap. 173, 247 (1987).
14. A. Maeder, and G. Meynet, Astron. Ap. 182, 243 (1987).
15. N. Grevesse and E. Anders, (CAM) p.1.
16. N. Lund, (CAM) p.111.
17. R. Mewaldt, (CAM) p.124.
18. R.M. Humphreys and B. McElroy, Ap. J. 284, 565 (1984).

CURRENT PROSPECTS FOR MEASUREMENTS OF COSMIC RAYS WITH ENERGIES ABOVE 10^{19} EV

D.B. Kieda
Department of Physics, University of Utah
Salt Lake City, Utah 84112 USA

ABSTRACT

We review the possibilities for using the details of the cosmic ray primary energy spectrum above 10^{19} eV to provide information about the sources of these particles. A measurement of the anisotropy and primary composition of the particles provides important clues which are useful for distinguishing between various origin theories. The poor event statistics at these energies point out the need for a new detector with a substantial improvement in detection area. We describe the capabilities and status of the High Resolution Fly's Eye; a proposed large collection area instrument which can provide an order of magnitude increase in detection area as well as improvements in directional and composition sensitivity.

SOURCES OF COSMIC RAYS WITH $E > 10^{19}$ EV.

Several theories have been proposed to explain the origin of the highest energy cosmic rays. Fichtel and Linsley[1] have proposed that the majority of cosmic rays above 10^{16} eV are due to discrete sources (probably compact objects) located in the galaxy. This naturally leads to an anisotropy increase with increasing energy as the Larmor radius of the cosmic ray approaches the Galactic distance scale. Accelerated nuclei are expected to be photodissociated in the X-ray field of the source; consequently these cosmic rays are expected to consist of mainly protons. It is difficult for these type of sources to generate particles with energies greater than $10^{18} - 10^{19}$ eV.

Jokipii and Morfill[2] have proposed that the highest energy cosmic rays are accelerated by a termination shock wave of a hypothetical Galactic wind at the Galactic halo, analogous to lower energy solar wind particle acceleration observed in the solar system. This model predicts a 70-100 EeV cutoff in the energy spectrum with a small (~5%) large scale anisotropy. The acceleration cutoff is rigidity dependent, thus the highest energy cosmic rays are expected to be iron nuclei.

The existence of primordial superconducting cosmic ray strings[3,4] may also result in a substantial flux of cosmic rays with energies up to 10^{24} eV. The cosmic rays are produced in this scenario by quark production from heavy fermion decay, thus no heavy element production is expected. The extragalactic nature of this source predicts an isotropic distribution of the cosmic rays. The predicted energy spectrum may be modified by the Greisen-Zatsepin cutoff[5,6], a suppression above 6×10^{19} eV due to interactions of the cosmic rays with the intergalactic 2.7° microwave background. Hill and Schramm [7] have proposed that the cutoff will be preceded by a slight flattening of the spectrum due to a pileup of cosmic

Model	Energy Spectrum	Anisotropy	Composition
Compact Object	$10^{18} - 10^{19}$ eV Energy Cutoff	Galactic Plane Enhancement	Protons
Galactic Wind	7-10 $\times 10^{19}$ eV Energy Cutoff	Slight ($\sim$ 5%) Enhancement	Heavy (Iron)
Extragalactic Cosmic Strings	6×10^{19} eV Cutoff, Bump Before Cutoff	No Enhancement (Isotropic)	Protons, 1×10^{17} eV ν's

Table 1: Summary of Various High Energy Cosmic Ray Source Model Predictions.

rays scattered to lower energies from above the cutoff. Extragalactic cosmic ray interactions with the microwave background will also create residual 10-100 TeV gamma rays[8] and 10^{17} eV neutrinos[9,10,11] which in principle can be measured.

Table 1 summarizes the various theories and the predictions for the energy spectrum, anisotropy, and composition. It is obvious that the various source models described here can lead to near identical predictions for the shape of the cosmic ray spectrum. Consequently, the primary composition and large scale anisotropy of the cosmic rays must also be measured to distinguish among competing theories. We note that previous and present experimental apparatuses are event rate limited and cannot provide a meaningful constraint on the anisotropy. With the exception of the Fly's Eye detector measurements, there also is little primary composition information in this energy range.

PRESENT STATUS OF EXPERIMENTAL KNOWLEDGE

Table 2 summarizes the present cosmic ray statistics[12] with $E > 10^{19}$ eV. Summed over all past and present experiments, there are approximately 800 events with $E > 10^{19}$ eV and only 14 events with $E > 10^{20}$ eV. It is impossible to perform any meaningful anisotropy or composition measurement for these events with such poor statistics. The primary spectrum measurement itself is also severely compromised by the poor statistics as well as by the systematic energy uncertainties inherent in the conversion from ground level electron/muon density to primary energy. The Fly's Eye experiment substantially reduces these biases by calculating the primary energy from the electron air shower profile as it develops in the atmosphere. We note that the three operational experiments are currently collecting about 75 events/yr with $E > 10^{19}$ eV, and *0 events/yr with* $E > 10^{20}$ *eV*. There is an obvious need for an experiment which can provide an order of magnitude increase in statistics with good energy and primary composition resolution.

THE HIGH RESOLUTION FLY'S EYE

The construction of the proposed High Resolution (Hi Res) Fly's Eye will provide a substantial increase in detection capability for the highest energy cosmic

Name	Volcano Ranch†	Haverah Park†	SUGAR†	Yakutsk	Fly's Eye	Akeno
No. of EAS $> 10^{19}$ eV	44	108	423	147	63	19
No. of EAS $> 10^{20}$ eV	1	5	8	0	0	0

Table 2: World Statistics of Cosmic Ray Events with $E > 10^{19}$ eV.
† Array no longer operational.

rays. The Hi Res Eye consists of 3 Fly's Eye installations with a 15 km separation between each installation. Each installation consists of 58 detection units which detect extensive air showers using the nitrogen fluorescence technique employed with the present Fly's Eye. The depth of air shower maximum X_{max} is measured for each event; this quantity is sensitive to the primary composition as a proton and an iron primary with the same energy will have an 90 gm/cm^2 difference in $< X_{max} >$ between the two species. Stereo event reconstruction between adjacent installations is employed to eliminate reconstruction biases. Phototube apertures are reduced from the present 5° down to 1° in the Hi Res Eye in order to gain sufficient track length for measuring the most distant events. The Hi Res mirrors are also larger to provide an improved signal-to-noise ratio. The total system requires about 45,000 electronics channels.

Figure 1 compares the present Fly's Eye detection aperture with the Hi Res Eye aperture[13]. The new instrument has almost two orders of magnitude more aperture for events with $E > 10^{18}$ eV and X_{max} resolution $\delta X_{max} < 30$ gm/cm^2. With this aperture, the Hi Res Eye will measure 270 events/yr compared to 1 event/yr with the present Fly's Eye. This is a factor of 4 increase over the present world collection rate. The Hi Res Eye will detect 5 events/yr with $E > 10^{20}$ eV and 30 gm/cm^2 X_{max} resolution‡ , resulting in 50% increase in the worldwide total after less than two years of operation. If the X_{max} resolution requirement is relaxed to 90 gm/cm^2, the Hi Res Eye will provide 600 events/yr with $E > 10^{19}$ eV and 15 events/yr with $E > 10^{20}$ eV. It is important to note that 90% of the events in the worldwide total have no X_{max} measurement.

A prototype 4 mirror installation of the Hi Res Eye is currently under construction by a collaborative effort between the University of Utah, Columbia University, and the University of Illinois. The four mirrors will be installed in fixed garages located on top of Little Granite Mountain at the Dugway Army base, next to the site of the present Fly's Eye 1 detector. The mirrors will be oriented to measure extensive air shower tracks that strike the region surrounding the present Fly's Eye 2/Chicago Air Shower Array/Michigan Muon Array area, 3.4 km to the northeast. Coincident events between the Hi Res Eye prototype and the other

‡ assuming an integral spectral index of -3.0 above 10^{19} eV

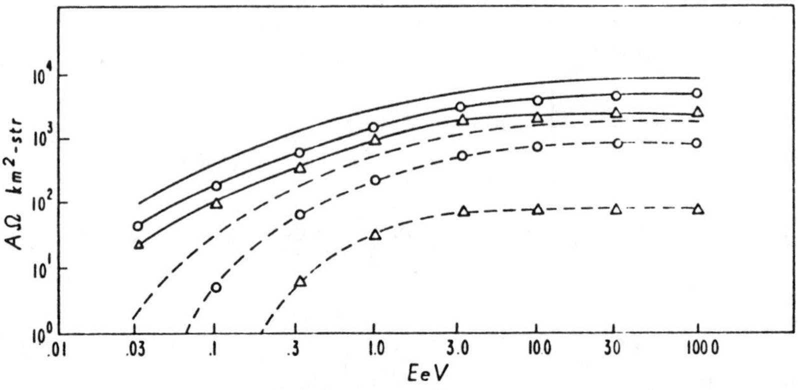

Figure 1: Aperture calculation for Hi Res Eye (—) and present Fly's Eye (- - -). — : Minimum trigger (< 10° track length), ○ : $\delta X_{max} < 90$ gm/cm², and △: $\delta X_{max} < 30$ gm/cm². 1 EeV = 10^{18} eV.

detectors will provide the measurements necessary to evaluate the performance of the Hi Res device. Preliminary calculations indicate that several hundred coincident events per year will be measured. Currently two phototube clusters are fully constructed and undergoing test with the remaining two clusters are slated for completion by 2/90. Several mirrors have been fabricated and are undergoing spot-size tests. The mirror housing sites have been selected and design of the mirror housings is nearing completion; construction is slated to begin in 1/90. The data acquisition electronics has already been through the first prototype testing stage; construction of the full system is expected to be complete by 4/90. We plan to have the working prototype detector in the field by mid-1990.

REFERENCES

1. C. E. Fichtel and J. Linsley, *Ap. J.* **300**, 474 (1986).
2. J. R. Jokipii and G. E. Morfill, *Proc.* 19th *ICRC*, (La Jolla) **3**, 132 (1985).
3. J. P. Ostriker et al., *Phys. Lett.* **B180**, 231 (1986).
4. C. T. Hill et al., *Phys. Rev.* **D36**, 1007 (1987).
5. K. Greisen, *Phys. Rev. Lett.* **16**, 748 (1966).
6. G. T. Zatsepin and V. A Kuz'min, *JETP Lett.* **4**, 114 (1966).
7. C. T. Hill and D. Schramm, *Phys. Rev.* **D31**, 564 (1984).
8. F. Halzen et al., Univ. of Wisc.-Mad. preprint MAD/PH/470 BA-89-12 (1989).
9. F. W. Stecker, *Phys. Rev. Lett.*, **21**, 1016 (1963).
10. C. T. Hill and D. N. Schramm, *Phys. Lett.* **131B** 247 (1983).
11. G. T. Zatsepin, *Sov. J. Nucl. Phys.*, **11**, 111 (1970).
12. J. Wdodczyk and A. W. Wolfendale, preprint, to appear in *Ann. Rev. Nucl. Part. Sci.* (1989).
13. R. M. Baltrusaitis et al., Nucl. Inst. and Meth. **A264**, 87 (1988).

HEAP - An Instrument to Measure the Elemental Abundances above 10^{15}eV at a Lunar Base

Simon P. Swordy

Enrico Fermi Institute, University of Chicago.

LASR, 933 E56th St, Chicago, Il 60637

At $\approx 10^{15}eV$ the slope of the energy spectrum of cosmic rays becomes significantly steeper than at lower energies. The measurement of relative elemental abundances at these energies is expected to provide a means to resolve the origin of this feature and greatly contribute to the understanding of the sources of cosmic rays at high energies. We describe a moon based detector, HEAP, for making well resolved elemental measurements at these energies using hadronic calorimetry. This detector is particularly well suited for a site on the lunar surface because there is no overlying layer of atmosphere and the large mass required can be provided by the lunar regolith.

Introduction

The differential flux of cosmic ray particles as a function of energy is shown in Figure 1. (The flux measurements have been multiplied by $(\text{Energy})^{2.5}$ to compress the vertical scale). The measurements show that the total particle flux falls as a power law of slope $\approx E^{-2.7}$ below 10^{15}eV, but at a steeper rate of $\approx E^{-3.2}$ at higher energy. The observation of this 'break' in the energy spectrum is tantilizing, since it is at an energy above which the mechanism of shock acceleration is predicted to become less effective. It is also possible that the 'break' is due to a decrease in the efficiency of particle trapping in the galaxy, since the gyroradii of these particles in the galactic magnetic field is becoming significant compared to the thickness of the galactic disk. To further investigate these questions, we would like to know in detail how the abundances of various elements in the cosmic rays vary in this energy range. At present this is not known, with estimates ranging from pure protons to pure iron nuclei. The variations in elemental composition with energy may indicate a radical change, which implies the sources of the highest energy cosimic rays are indeed different from those at lower energy. The main obstacles to performing these measurements are the low particle fluxes and the presence of the Earth's atmosphere. Because of interactions in the $\approx$1000 g/cm^2 of overlying atmosphere the elemental character

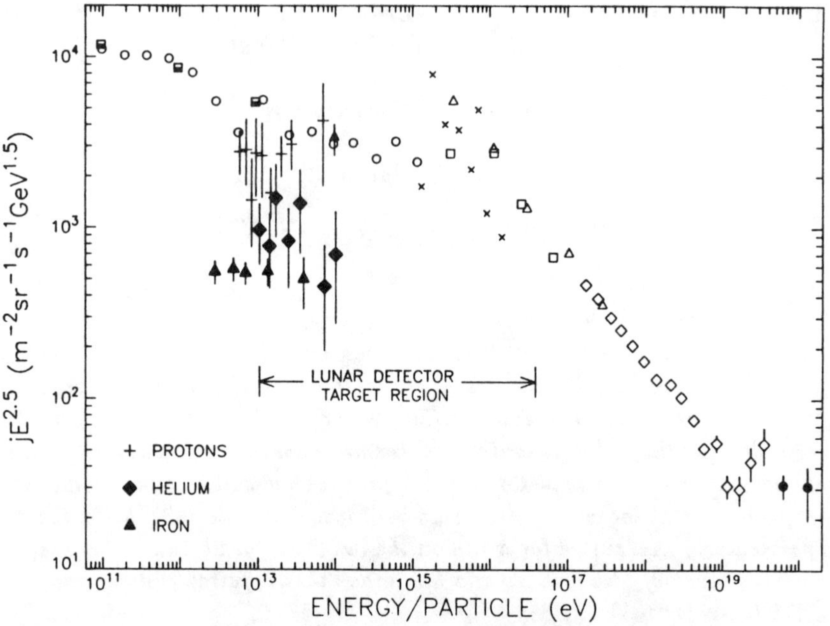

Figure 1. The differential flux of all particles in the cosmic rays as a function of energy, (note flux is multiplied by $(\text{Energy})^{2.5}$), a compilation from Linsley 1983[3]. Also shown in larger symbols are measured elemental components H and He from Burnett et. al. 1983[4], and Fe from Grunsfeld et. al. 1988[5].

---

of a primary cosmic ray is lost and only a shower of secondary particles survives at ground level. These air showers can be analysed to make some estimate of the nuclear composition of the incident particles, but detailed studies are only possible by direct measurements.

Direct observation of the primary nuclei is limited to instruments exposed in space or on high altitude balloons near the top of the atmosphere. Examples of the highest energy direct measurements and the contribution to the overall flux for protons, helium and iron nuclei are shown in Figure 1. The steeply falling spectrum of particles means a detector must increase in collecting power by a factor of 50 or more for each decade increase in energy, to achieve the same statistical significance in a flux measurement. Since detector systems on spacecraft cannot easily be made large, the direct elemental composition has only been determined up to energies of $\approx 10^{14}$eV. Here we propose an instrument for a lunar base - the High Energy Abundance Project, HEAP, which would make direct measurements with a collecting power far larger than previous or planned experiments. The target region indicated for these measurements would explore new territory where the detailed elemental abundances are essentially unknown.

Why a Lunar Site ?

The only practical device to measure the energy of the particles at such large energies over a large area appears to be a sampled hadronic calorimeter, such as has been proposed for the US Space Station[1]. In such a detector a hadronic shower develops in much the same way as in the Earth's atmosphere. The calorimeter measurement is preceded by an ionization loss detector which measures the incoming particle charge before the first interaction in the calorimeter. For this type of experiment a lunar site is ideal. Firstly, there is no overlying atmosphere so direct observations of cosmic rays is possible with large, surface mounted detectors. Secondly the lunar regolith can be used as the bulk of the detector, eliminating the need to lift large masses of material into Earth orbit. The following sections discuss the detailed design and expected performance of HEAP, a calorimeter built from lunar material.

A Detector for the Lunar Surface

The schematic design of HEAP is shown in Figure 2. All signals are provided by thin, ($\approx$1cm thick), plastic scintillation counter layers viewed by photomultiplier tubes. The uppermost layers of scintillator are used to determine the charge of the incident cosmic ray before interactions occur. These provide an output of light per unit pathlength which increases as Z^2 for relativistic nuclei of low atomic number, Z. For $Z>10$ there is some saturation of the yield, but the resolution is adequate even for iron nuclei, ($Z=26$). The charge resolution of plastic scintillators has been demonstrated[2] to be better than 0.4 charge units for iron nuclei of energy $\approx 3\text{x}10^{13}$eV. The hadronic calorimeter is formed from layers of compressed lunar material separated by scintillator strips which provide the hadron shower samples. The hadronic interaction length, Λ, in the lunar material is $\approx$80 g/cm^2. Since the regolith is readily compressed to a density of 2.5 g/cm^3, each of the 10 layers in the detector is approximately an interaction length. The expected depth of hadronic shower maximum at these energies can be interpolated from accelerator measurements at lower energies and estimates from the Fly's Eye air shower experiment at $\approx 10^{18}$eV. At 10^{15}eV the shower maximum should occur at a depth of $\approx 5\Lambda$. A significant part of the shower will be contained within the detector depth of $\approx 10\Lambda$ for a vertically incident particle. The

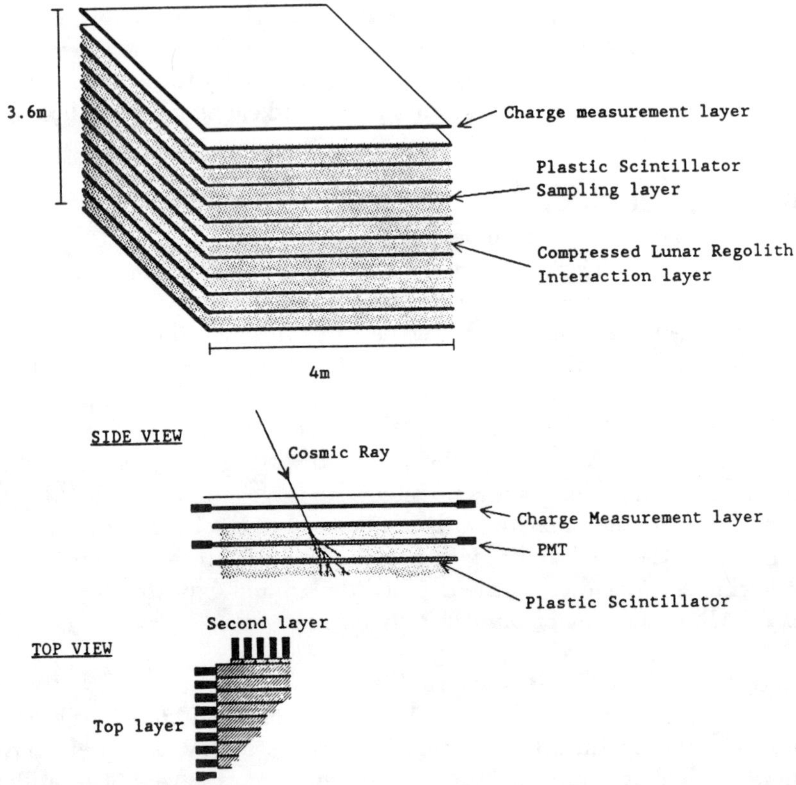

Figure 2. Schematic of a high energy cosmic ray detector for the lunar surface.

expected energy resolution at this energy is $\Delta E/E \approx 15\%$. The profile of shower development measured in HEAP will be extremely valuable for interpretation of air shower measurements made on Earth. To normalize the signals from the detector elements some determination of the cosmic ray trajectory is required. This is provided by segmentation of the scintillator layers into strips, as shown in the top view of the detector. Each strip is viewed by two photomultipliers. The orientation of these strips changes in each successive layer to provide views of the trajectory on orthogonal planes, these can be used to find the pathlength of the event in each detector element. The accuracy of the trajectory measurement with HEAP is expected to be $<2^{o}$ which will allow an investigation of the anisotropy in arrival direction of the particles in addition to the composition measurements. The triggering scheme for the detector can be extremely simple since all that is required is an event which crosses the charge determining scintillators and deposits a certain minimum energy in the calorimeter.

The aperture of HEAP, as described here, is estimated to be $\approx$24 m^{2}str. If positioned near a lunar base it could easily collect data for 3 years or more since it only requires electrical power for operation. Table 1 lists the expected number of events for protons, helium, and iron nuclei predicted in the detector over this time period, (these calculations assume a simple extrapolation of fluxes measured at lower energies).

It is estimated that the trigger rate, for a threshold energy in the calorimeter of 10^{13}eV, will be $\approx$10 per minute. The event information can be readily transferred to Earth for processing through a modest telemetry link which operates through the lunar base systems.

The execution of HEAP at a lunar base would be relatively straightforward. The scintillation counters would be preassembled and transferred to the lunar base complete with electronics. The counters would then be stacked to form the detector at a lunar site by interleaving the scintillator layers with trays of compressed lunar material, forming the "HEAP". Physical and electronic characteristics of the detector are given in Table 2.

Conclusions

The idea of measuring the elemental abundance of cosmic rays above 10^{15}eV with a hadronic calorimeter in space is not new. However, the lunar base provides a unique opportunity for such an experiment because the extremely massive component of the detector is already available in the lunar regolith. It seems the relatively simple HEAP detector can be designed and built with presently available technology. This instrument would finally resolve the long standing scientific question of the composition of particles at and above the 'break' in the cosmic ray spectrum at 10^{15}eV.

The particular design given here has an aperture of $\approx$24 m^{2}str, although it could be made significantly larger than this if the resources were available. Some tradeoff could also be made of area versus number of instrumented layers - for example the same scintillator could instrument twice the area if samples were only taken every two interaction lengths. It is also conceivable that a number of HEAP modules could be built, leading to detailed direct measurements above 10^{16}eV.

References

1. J. F. Ormes and R. E. Streitmatter, Proceedings of Workshop on cosmic ray and high energy gamma ray experiments for the Space Station era, Louisiana State Univ., 1985, pg 340.

2. J. L'Heureux, J. M. Grunsfeld, P. Meyer, D. Müller, and S. P. Swordy, Proc. 20th Int. Cosmic Ray Conf., vol 2, 1987, pg 366.

3. J. Linsley, Proc. 18th Int. Cosmic Ray Conf., vol 12, 1983, pg 135.

4. T. H. Burnett et. al. (JACEE collab.), Phys. Rev. Lett., vol 51, 1983, 11, pg 1010.

5. J. M. Grunsfeld, J L'Heureux, P. Meyer, D. Müller and S. P. Swordy, Ap. J. Lett., vol 327, 1988, L31

Table 1. Expected Number of Events in 3 Years

PARTICLE	$> 10^{13}$	$> 10^{15}$	$> 10^{16}$eV
Protons	$3x10^6$	10^3	25
Helium	$1.5x10^6$	600	12
Iron	$8x10^5$	300	6

Table 2. Detector Characteristics

MASSES	(kg)
Scintillator	1,700
PMTs/Electronics	1,000
Mechanical Support	1,000
Lunar Regoltih	160,000
POWER	1,000 W
TELEMETRY	1 kbit/s

AIP Conference Proceedings

		L.C. Number	ISBN
No. 171	Vacuum Design of Advanced and Compact Synchrotron Light Sources (Upton, NY, 1988)	88-71824	0-88318-371-4
No. 172	Advances in Laser Science–III: Proceedings of the International Laser Science Conference (Atlantic City, NJ, 1987)	88-71879	0-88318-372-2
No. 173	Cooperative Networks in Physics Education (Oaxtepec, Mexico 1987)	88-72091	0-88318-373-0
No. 174	Radio Wave Scattering in the Interstellar Medium (San Diego, CA 1988)	88-72092	0-88318-374-9
No. 175	Non-neutral Plasma Physics (Washington, DC 1988)	88-72275	0-88318-375-7
No. 176	Intersections Between Particle and Nuclear Physics (Third International Conference) (Rockport, ME 1988)	88-62535	0-88318-376-5
No. 177	Linear Accelerator and Beam Optics Codes (La Jolla, CA 1988)	88-46074	0-88318-377-3
No. 178	Nuclear Arms Technologies in the 1990s (Washington, DC 1988)	88-83262	0-88318-378-1
No. 179	The Michelson Era in American Science: 1870–1930 (Cleveland, OH 1987)	88-83369	0-88318-379-X
No. 180	Frontiers in Science: International Symposium (Urbana, IL 1987)	88-83526	0-88318-380-3
No. 181	Muon-Catalyzed Fusion (Sanibel Island, FL 1988)	88-83636	0-88318-381-1
No. 182	High T_c Superconducting Thin Films, Devices, and Application (Atlanta, GA 1988)	88-03947	0-88318-382-X
No. 183	Cosmic Abundances of Matter (Minneapolis, MN 1988)	89-80147	0-88318-383-8
No. 184	Physics of Particle Accelerators (Ithaca, NY 1988)	87-07208	0-88318-384-6
No. 185	Glueballs, Hybrids, and Exotic Hadrons (Upton, NY 1988)	89-83513	0-88318-385-4
No. 186	High-Energy Radiation Background in Space (Sanibel Island, FL 1987)	89-083833	0-88318-386-2

No. 187	High-Energy Spin Physics (Minneapolis, MN 1988)	89-083948	0-88318-387-0
No. 188	International Symposium on Electron Beam Ion Sources and their Applications (Upton, NY 1988)	89-084343	0-88318-388-9
No. 189	Relativistic, Quantum Electrodynamic, and Weak Interaction Effects in Atoms (Santa Barbara, CA 1988)	89-084431	0-88318-389-7
No. 190	Radio-frequency Power in Plasmas (Irvine, CA 1989)	89-045805	0-88318-397-8
No. 191	Advances in Laser Science–IV (Atlanta, GA 1988)	89-085595	0-88318-391-9
No. 192	Vacuum Mechatronics (First International Workshop) (Santa Barbara, CA 1989)	89-045905	0-88318-394-3
No. 193	Advanced Accelerator Concepts (Lake Arrowhead, CA 1989)	89-045914	0-88318-393-5
No. 194	Quantum Fluids and Solids—1989 (Gainesville, FL, 1989)	89-81079	0-88318-395-1
No. 195	Dense *Z*-Pinches (Laguna Beach, CA, 1989)	89-46212	0-88318-396-X
No. 196	Heavy Quark Physics (Ithaca, NY, 1989)	89-81583	0-88318-644-6
No. 197	Drops and Bubbles (Monterey, CA, 1988)	89-46360	0-88318-392-7
No. 198	Astrophysics in Antarctica (Newark, DE, 1989)	89-46421	0-88318-398-6
No. 199	Surface Conditioning of Vacuum Systems (Los Angeles, CA, 1989)	89-82542	0-88318-756-6
No. 200	High T_c Superconducting Thin Films: Processing, Characterization, and Applications (Boston, MA, 1989)	90-80006	0-88318-759-0
No. 201	QED Stucture Functions (Ann Arbor, MI, 1989)	90-80229	0-88318-671-3
No. 202	NASA Workshop on Physics From a Lunar Base (Stanford, CA, 1989)	90-55073	0-88318-646-2